Springer Proceedings

The series Springer Proceedings in Energy covers a broad range of multidisciplinary subjects in those research fields closely related to present and future forms of energy as a resource for human societies. Typically based on material presented at conferences, workshops and similar scientific meetings, volumes published in this series will constitute comprehensive state-of-the-art references on energy-related science and technology studies. The subjects of these conferences will fall typically within these broad categories:

- Energy Efficiency
- Fossil Fuels
- Nuclear Energy
- Policy, Economics, Management & Transport
- Renewable and Green Energy
- Systems, Storage and Harvesting
- Materials for Energy

eBook Volumes in the Springer Proceedings in Energy will be available online in the world's most extensive eBook collection, as part of the Springer Energy eBook Collection. Please send your proposals/inquiry to Dr. Loyola DSilva, Senior Publishing Editor, Springer (loyola.dsilva@springer.com)

More information about this series at http://www.springer.com/series/13370

Abdallah Khellaf
Editor

Advances in Renewable Hydrogen and Other Sustainable Energy Carriers

 Springer

Editor
Abdallah Khellaf
Centre de Développement des Energies
Renouvelables
Algiers, Algeria

ISSN 2352-2534   ISSN 2352-2542 (electronic)
Springer Proceedings in Energy
ISBN 978-981-15-6597-7   ISBN 978-981-15-6595-3 (eBook)
https://doi.org/10.1007/978-981-15-6595-3

© The Editor(s) (if applicable) and The Author(s), under exclusive license to Springer Nature Singapore Pte Ltd. 2021
This work is subject to copyright. All rights are solely and exclusively licensed by the Publisher, whether the whole or part of the material is concerned, specifically the rights of translation, reprinting, reuse of illustrations, recitation, broadcasting, reproduction on microfilms or in any other physical way, and transmission or information storage and retrieval, electronic adaptation, computer software, or by similar or dissimilar methodology now known or hereafter developed.
The use of general descriptive names, registered names, trademarks, service marks, etc. in this publication does not imply, even in the absence of a specific statement, that such names are exempt from the relevant protective laws and regulations and therefore free for general use.
The publisher, the authors and the editors are safe to assume that the advice and information in this book are believed to be true and accurate at the date of publication. Neither the publisher nor the authors or the editors give a warranty, expressed or implied, with respect to the material contained herein or for any errors or omissions that may have been made. The publisher remains neutral with regard to jurisdictional claims in published maps and institutional affiliations.

This Springer imprint is published by the registered company Springer Nature Singapore Pte Ltd.
The registered company address is: 152 Beach Road, #21-01/04 Gateway East, Singapore 189721, Singapore

Preface

With the advent of the energy transition, hydrogen field use has widen. Indeed initially hydrogen, as an energy vector and an alternative fuel, has been considered in conjunction with renewable energy in order to address the issue of conventional energy sources depletion and the problem of pollution resulting from the use of these conventional energy sources. With the emergence of energy transition and the massive introduction of renewables in the energy mix, the hydrogen role has expanded. System energy stability and mismatch of energy supply and demand has become acute. Recourse to hydrogen and its technologies and processes such power-to-gas offer suitable solutions to address these issues. This situation has led to an increase need for hydrogen, giving an impetus to its technology development and offering divers applications to the other sectors.

It is in light of these developments that the third International Symposium on Sustainable Hydrogen (ISSH2'2019) has been organized November 27–28, 2019, in Algiers, Algeria.

The main topic of the symposium is hydrogen as an energy vector, an alternative fuel, a storage medium and a buffer product. Hydrogen is considered as one of the keys to the successful transition to sustainable and clean energy and one of the undertakings where development is completely compatible with environment protection.

This year's special topic is "promising alternative fuels." It has been selected to draw attention to the new concepts and new resources and technologies considered to overcome the practical challenges in achieving low-carbon energy society. With their technologies improving at fast paces, these fuels are likely to play a central role in socio-economic development, energy access and climate change mitigation.

The various thematic sessions have dealt with the scientific advances, the experience gained, the provided opportunities, the encountered hurdles and the actions to undertake to ensure sustainable development in the fields of alternative fuels and energy.

A panel discussion addressing the topic of "Hydrogen: Energy storage and Energy transition" has been organized on the first day of the symposium with the active participation of all the attendees. A debate has been carried out on advances in hydrogen as a storage medium and as a buffer product and its role in the success of the energy transition.

We are grateful to our sponsors (IAHE, TIKA, NAFTAL, Sonelgaz, BDL, CASH, Lamaraz, Envilab) who have assisted us in organizing this event.

We also kindly thank the Algerian National Library (Bibliothèque Nationale d'Algérie) for graciously providing their conference facilities for the organization of the symposium.

Algiers, Algeria
March 2020

The Steering Committee
Abdallah Khellaf
Rafika Boudries
Fethia Amrouche
Nourdine Kabouche
Abdelhamid Mraoui
Fatiha Lassouane
Ilyes Nouicer
Sabah Menia
Mounia Belacel
Yasmina Bakouri

Contents

1 **Hydrogen Storage: Different Technologies, Challenges and Stakes. Focus on TiFe Hydrides** 1
David Chapelle, Anne Maynadier, Ludovic Bebon, and Frédéric Thiébaud

2 **Preparation of ZrO_2–Fe_2O_3 Nanoparticles and Their Application as Photocatalyst for Water Depollution and Hydrogen Production** 11
M. Benamira, N. Doufar, and H. Lahmar

3 **Preparation of Anode Supported Solid Oxide Fuel Cells (SOFCs) Based on BIT07 and $Pr_2NiO_{4+\delta}$: Influence of the Presence of GDC Layer** 19
M. Benamira, M. T. Caldes, O. Joubert, and A. Le Gal La Salle

4 **Preparation of the Spinel $CuCo_2O_4$ at Low Temperature. Application to Hydrogen Photoelectrochemical Production** 25
R. Bagtache, K. Boudjedien, I. Sebai, D. Meziani, and Mohamed Trari

5 **Synthesis and Electrochemical Characterization of Fe-Doped $NiAl_2O_4$ Oxides** 33
Warda Tibermacine and Mahmoud Omari

6 **Review on the Effect of Compensation Ions on Zeolite's Hydrogen Adsorption** 41
Redouane Melouki and Youcef Boucheffa

7 **Catalytic Reforming of Methane Over Ni–La_2O_3 and Ni–CeO_2 Catalysts Prepared by Sol-Gel Method** 47
Nora Yahi, Kahina Kouachi, Hanane Akram, and Inmaculada Rodríguez-Ramos

8	**Hydrogen Effect on Soot Formation in Ethylene-Syngas Mixture Opposed Jet Diffusion Flame in Non-conventional Combustion Regime** .... Amar Hadef, Selsabil Boussetla, Abdelbaki Mameri, and Z. Aouachria	55
9	**A Two-Dimensional Simulation of Opposed Jet Turbulent Diffusion Flame of the Mixture Biogas-Syngas** .... Abdelbaki Mameri, Selsabil Boussetla, and Amar Hadef	61
10	**Numerical Evaluation of NO Production Routes in the MILD Combustion of the Biogas-Syngas Mixture** .... Selsabil Boussetla, Abdelbaki Mameri, and Amar Hadef	69
11	**Effect of H_2/CO Ratio and Air N_2 Substitution by CO_2 on CH_4/Syngas Flameless Combustion** .... Amar Hadef, Abdelbaki Mameri, and Z. Aouachria	77
12	**Chaotic Bacterial Foraging Optimization Algorithm with Multi-cross Learning Mechanism for Energy Management of a Standalone PV/Wind with Fuel Cell** .... Issam Abadlia, Mohamed Adjabi, and Hamza Bouzeria	85
13	**Mechanical Properties of the Tetragonal $CH_3NH_3PbI_3$ Structure** .... Kamel Benyelloul, Smain Bekhechi, Abdelkader Djellouli, Youcef Bouhadda, Khadidja Khodja, and Hafid Aourag	93
14	**Hydrogen Production by the *Enterobacter cloacae* Strain** .... Azri Yamina Mounia, Tou Insaf, and Sadi Meriem	99
15	**Visible Light Hydrogen Production on the Novel Ferrite $CuFe_2O_4$** .... S. Attia, N. Helaïli, Y. Bessekhouad, and Mohamed Trari	105
16	**Prediction of New Hydrogen Storage Materials: Structural Stability of $SrAlH_3$ from First Principle Calculation** .... Youcef Bouhadda, Kamel Benyelloul, N. Fenineche, and M. Bououdina	113
17	**CFD Analysis of the Metal Foam Insertion Effects on SMR Reaction Over Ni/Al_2O_3 Catalyst** .... Ali Cherif, Rachid Nebbali, and Lyes Nasseri	121
18	**Photocatalytic Evolution of Hydrogen on $CuFe_2O_4$** .... H. Lahmar, M. Benamira, L. Messaadia, K. Telmani, A. Bouhala, and Mohamed Trari	129
19	**CFD Study of ATR Reaction Over Dual Pt–Ni Catalytic Bed** ..... Ali Cherif, Rachid Nebbali, and Lyes Nasseri	137

20	Optimization of the Ni/Al$_2$O$_3$ and Pt/Al$_2$O$_3$ Catalysts Load in Autothermal Steam Methane Reforming..................	145
	Ali Cherif, Rachid Nebbali, and Lyes Nasseri	
21	Proton Exchange Membrane Fuel Cell Modules for Ship Applications	151
	S. Tamalouzt, N. Benyahia, and A. Bousbaine	
22	Analysis of off Grid Fuel Cell Cogeneration for a Residential Community	161
	A. Mraoui, B. Abada, and M. Kherrat	
23	Tri-generation Using Fuel Cells for Residential Application	171
	A. Mraoui, B. Abada, and M. Kherrat	
24	Response Surface Methodology Based Optimization of Transesterification of Waste Cooking Oil	177
	R. Alloune, M. Y. Abdat, A. Saad, F. Danane, R. Bessah, S. Abada, and M. A. Aziza	
25	Numerical Investigation on Concentrating Solar Power Plant Based on the Organic Rankine Cycle for Hydrogen Production in Ghardaïa...	185
	Halima Derbal-Mokrane, Fethia Amrouche, Mohamed Nazim Omari, Ismael Yahmi, and Ahmed Benzaoui	
26	Optimization Study of the Produced Electric Power by PCFCs	195
	Youcef Sahli, Abdallah Mohammedi, Monsaf Tamerabet, and Hocine Ben-Moussa	
27	Accurate PEM Fuel Cell Parameters Identification Using Whale Optimization Algorithm	203
	Mohammed Bilal Danoune, Ahmed Djafour, and Abdelmoumen Gougui	
28	Hydrogen Versus Alternative Fuels in an HCCI Engine: A Thermodynamic Study	211
	Mohamed Djermouni and Ahmed Ouadha	
29	Thermodynamic Study of a Turbocharged Diesel-Hydrogen Dual Fuel Marine Engine	221
	Fouad Selmane, Mohamed Djermouni, and Ahmed Ouadha	
30	Effect of Bluff-Body Shape on Stability of Hydrogen-Air Flame in Narrow Channel	231
	Mounir Alliche, Redha Rebhi, and Fatma Zohra Khelladi	

31 Experimental Validation of Fuel Cell, Battery
and Supercapacitor Energy Conversion System for Electric
Vehicle Applications 239
R. Moualek, N. Benyahia, A. Bousbaine, and N. Benamrouche

32 Compromise Between Power Density and Durability of a PEM
Fuel Cell Operating Under Flood Conditions 247
H. Abdi, N. Ait Messaoudene, and M. W. Naceur

33 Optimal Design of Energy Storage System Using Different
Battery Technologies for FCEV Applications 255
B. Bendjedia, N. Rizoug, and M. Boukhnifer

34 Simultaneous Removal of Organic Load and Hydrogen Gas
Production Using Electrodeposits Cathodes in MEC........... 263
Amit Kumar Chaurasia and Prasenjit Mondal

35 An Improved Model for Fault Tolerant Control of a Flooding
and Drying Phenomena in the Proton Exchange Membrane
Fuel Cell 271
A. A. Smadi, F. Khoucha, A. Benrabah, and M. Benbouzid

36 Design of a Microbial Fuel Cell Used as a Biosensor of Pollution
Emitted by Oxidized Organic Matter 279
Amina Benayyad, Mostefa Kameche, Hakima Kebaili,
and Christophe Innocent

37 Bioelectricity Production from *Arundo Donax*-MFC
and *Chlorophytum Comosum*-MFC 285
L. Benhabylès, Y. M. Azri, I. Tou, and M. Sadi

38 Implementation of Fuel Cell and Photovoltaic Panels Based DC
Micro Grid Prototype for Electric Vehicles Charging Station..... 291
N. Benyahia, S. Tamalouzt, H. Denoun, A. Badji, A. Bousbaine,
R. Moualek, and N. Benamrouche

39 Application of Hydrotalcite for the Dry Reforming Reaction
of Methane and Reduction of Greenhouse Gases 299
Nadia Aider, Fouzia Touahra, Baya Djebarri, Ferroudja Bali,
Zoulikha Abdelsadek, and Djamila Halliche

40 Processing of CO_x Molecules in CO_2/O_2 Gas Mixture
by Dielectric Barrier Discharge: Understanding the Effect
of Internal Parameters of the Discharge 307
L. Saidia, A. Belasri, and S. Baadj

41 Performance Comparison of a Wankel SI Engine Fuelled
with Gasoline and Ethanol Blended Hydrogen 315
Fethia Amrouche, P. A. Erickson, J. W. Park, and S. Varnhagen

42	Parametric Study of SO_3 Conversion to SO_2 in Tubular Reactor for Hydrogen Production *via* Sulfuric Cycle .... F. Lassouane	323
43	Preparation and Physical/Electrochemical Characterization of the Hetero-System 10% $NiO/\gamma-Al_2O_3$ .... I. Sebai, R. Bagtache, A. Boulahaouache, N. Salhi, and Mohamed Trari	331
44	Synthesis and Characterization of the Double Perovskite La_2NiO_4-Application for Hydrogen Production................ S. Boumaza, R. Brahimi, L. Boudjellal, Akila Belhadi, and Mohamed Trari	339
45	Optimal Design and Comparison Between Renewable Energy System, with Battery Storage and Hydrogen Storage: Case of Djelfa, Algeria .... Ilhem Nadia Rabehi	347
46	New Neural Network Single Sensor Variable Step Size MPPT for PEM Fuel Cell Power System .... Abdelghani Harrag	355
47	Modified P&O-Fuzzy Type-2 Variable Step Size MPPT for PEM Fuel Cell Power System .... Abdelghani Harrag	363
48	A GIS-MOPSO Integrated Method for Optimal Design of Grid-Connected HRES for Educational Buildings .... Charafeddine Mokhtara, Belkhir Negrou, Noureddine Settou, Abdessalem Bouferrouk, Yufeng Yao, and Djilali Messaoudi	371
49	A Comparison Between Two Hydrogen Injection Modes in a Metal Hydride Reactor .... Bachir Dadda, Allal Babbou, Rida Zarrit, Youcef Bouhadda, and Saïd Abboudi	379
50	Effect of the Complexing Agent in the Pechini Method on the Structural and Electrical Properties of an Ionic Conductor of Formula $La_{1-x}Sr_xAlO_{3-\delta}$ (x = 0, 0.05, 0.1, 0.15) .... F. Hadji, F. Bouremmad, S. Shawuti, and M. A. Gulgun	387
51	Production of Bio-Oil for Chemical Valorization by Flash Pyrolysis of Lignocellulosic Biomass in an Entrained Bed Reactor .... Imane Ouarzki, Aissa Ould Dris, and Mourad Hazi	395

52	**Suitable Sites for Wind Hydrogen Production Based on GIS-MCDM Method in Algeria** Djilali Messaoudi, Noureddine Settou, Belkhir Negrou, Belkhir Settou, Charafeddine Mokhtara, and Chetouane Mohammed Amine	405
53	**Liquefaction of Hydrogen: Comparison Between Conventional and Magnetic Refrigeration Systems** Mustapha Belkadi and Arezki Smaili	413
54	**Numerical Study of Heat and Mass Transfer During Absorption of H_2 in a $LaNi_5$ Annular Disc Reactor Crossed by a Tubular Heat Exchanger** Abdelaziz Bammoune, Samir Laouedj, and Bachir Dadda	421
55	**Photocatalytic Hydrogen Production on 5% CuO/ZnO Hetero-Junction** Meriem Haddad, Akila Belhadi, and Mohamed Trari	429
56	**Field Enhancement Factor Around Hydrogen-Negative Index Metamaterial Waveguide** Houria Hamouche and Mohammed M. Shabat	435
57	**Effect of Silicates and Carbonates on the Structure of Nickel-Containing Hydrotalcite-Derived Materials** Baya Djebarri, Nadia Aider, Fouzia Touahra, Ferroudja Bali, Juan Paul Holgado, and Djamila Halliche	443
58	**On the Effect of the Inlet Hydrogen Amount on Hydrocarbons Distribution Produced via Fischer-Tropsch Synthesis** Abdelmalek Bellal and Lemnouer Chibane	451
59	**Predictive Current Control in Grid-Connected Fuel Cell–Photovoltaic Based Hybrid System for Power Quality Improvement** M. R. Bengourina, L. Hassaine, and M. Bendjebbar	459
60	**Hydrogen Production by Photo Fermentation via *Rhodobacter sp.*** Sabah Menia, Ilyes Nouicer, and Hammou Tebibel	467
61	**Estimation of Hydrogen Production in Three Cities in the North of Algeria** Ilyes Nouicer, M. R. Yaiche, Sabah Menia, Fares Meziane, and Nourdine Kabouche	473

Contents

62 Comparative Study of Different PV Systems Configurations Combined with Alkaline and PEM Water Electrolyzers for Hydrogen Production 481
Nourdine Kabouche, Fares Meziane, Ilyes Nouicer, and Rafika Boudries

63 Wind Power System for Large-Scale Energy and Hydrogen Production in Hassi R'mel 491
Fares Meziane, F. Chellali, K. Mohammedi, Nourdine Kabouche, and Ilyes Nouicer

64 Analysis and Design of PEM Fuel Cell/Photovoltaic System to Supply a Traffic Light Signals in Ouargla City Based on Field Experience 499
Abdelmoumen Gougui, Ahmed Djafour, Taha Hamidatou, S. Eddine Khennour, and Mohammed Bilal Danoune

65 Assessment of Hydrogen Production from Geothermal Thermoelectric Generator 507
M. M. Hadjiat, S. Ouali, K. Salhi, A. Ait Ouali, and E. H. Kuzgunkaya

Author Index 515

Chapter 1
Hydrogen Storage: Different Technologies, Challenges and Stakes. Focus on TiFe Hydrides

David Chapelle, Anne Maynadier, Ludovic Bebon, and Frédéric Thiébaud

Abstract The share of renewable energies in the energy mix is gradually increasing. This transition brings many challenges in the management of electricity grids, especially because of the fluctuating and intermittent nature of renewable energies. Therefore, hydrogen represents one of the keystones for the sustainable exploitation of our energy resources. Hydrogen allows storing in the long term not consumed but available electricity, and hydrogen is a 'fuel' for mobile, nomadic and remote site applications. Once produced and awaiting consumption, the hydrogen must be stored in optimal conditions of safety and efficiency with regard to the application and its location. The most mature solution to date is the storage under the compressed form, which consists in keeping the hydrogen gas in a container at increasing pressures in order to increase the energy density; cryogenic storage is now well controlled but generally reserved for very specific applications for reasons inherent to the technology and because of significant costs; and finally the so-called 'solid storage', to which the scientific community has been showing a marked interest for several decades in the hope of identifying a lasting solution likely to replace advantageously other solutions. In this paper, these storage media are introduced by evoking their technological characteristics and their fields of application often justified by inherent limitations of the technology. We will also discuss the challenges still posed by these storage solutions today by linking them with the research work carried out in the Department of Applied Mechanics in FEMTO ST Institute.

Keywords Hydrogen storage · LH2 · CGH2 · Solid storage · Metal hydride

---

D. Chapelle (✉) · A. Maynadier · L. Bebon · F. Thiébaud
Department of Applied Mechanics, FEMTO ST Institute, Bourgogne Franche-Comté University, 24 Rue de L'Epitaphe, 25000 Besançon, France
e-mail: david.chapelle@univ-fcomte.fr

Federation for Fuel Cell Research FCLAB (FR CNRS 3539), Belfort, France

© The Editor(s) (if applicable) and The Author(s), under exclusive license to Springer Nature Singapore Pte Ltd. 2021
A. Khellaf (ed.), *Advances in Renewable Hydrogen and Other Sustainable Energy Carriers*, Springer Proceedings in Energy,
https://doi.org/10.1007/978-981-15-6595-3_1

1.1 Introduction

For decades, one can read the emergency to act against the climate change, to reduce drastically our greenhouse gas emission and at the same time to develop a sustainable management of our energy resources. Undoubtedly, even if it has taken too much time, the energy transition is occurring in many countries and the part of renewable energy in the energy mix is getting greater and greater month after month. One say, it cannot happen overnight, all the more this new energetic balance comes with new challenges, especially regarding the management of electricity grids. By nature, renewable energies have fluctuating and intermittent features but are also desynchronized between production and consumption. Without any consideration on technological or cost issues, the previous situation being established, hydrogen carrier comes as a really promising candidate so as to create synergies between renewable energies and hydrogen production and use. Hydrogen allows storing in the long term not consumed but available electricity, and hydrogen is a 'fuel' for mobile, nomadic and remote site applications. In the first case, hydrogen can regulate and stabilize our electricity distribution networks at different scales; in the second case, it should progressively replace petroleum products. The recent report [1] of IRENA (International Renewable Energy Agency) from September 2019, prepared for the 2nd Hydrogen Energy Ministerial Meeting in Tokyo, gives a clear, synthetic but complete overview of the topic and establishes the analysis of potential pathways to a hydrogen-enabled clean energy future. It emphasis "Clean hydrogen is enjoying unprecedented political and business momentum, with the number of policies and projects around the world expanding rapidly". As an illustration (Fig. 1.1), "the cost

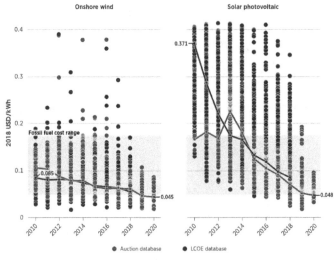

Fig. 1.1 Global cost trends for onshore wind and solar PV [1]

of renewable power generation has fallen dramatically in recent years".

Among all challenges, the hydrogen must be stored safely and efficiently with regard to the application and its location. We list three major media to store hydrogen: the most mature solution to date is the storage under the compressed form, which consists in keeping the hydrogen gas in a container at increasing pressures in order to increase the energy density; cryogenic storage is now well controlled but generally reserved for very specific applications for reasons inherent to the technology and because of significant costs; and finally we identify a third media, the so-called 'solid storage', to which the scientific community has been showing a marked interest for several decades in the hope of identifying a lasting solution likely to replace other solutions, if not advantageously, at least by greatly reducing constraints for the end user.

Here we discuss these storage media by evoking their technological characteristics and their fields of application often justified by inherent limitations of the technology. We will also discuss the challenges still posed by these storage solutions by linking them with the research activities with a specific focus on hydrogen solid storage.

1.2 Hydrogen Storage Media

Among the various ways to store hydrogen, we decide to pay attention on the three major solutions. Nowadays, they are the more mature from both a technological and cost point of view. They are used for decades in various domains of activities.

Because of its high volumetric storage density, the cryogenics or liquid hydrogen (LH2) has been considered even for automotive implementation till years 2010. Due to the low required operating temperature, around 20 K, the tank has to be designed in order to diminish all the more heat exchanges. As illustrated in Fig. 1.2, even if well

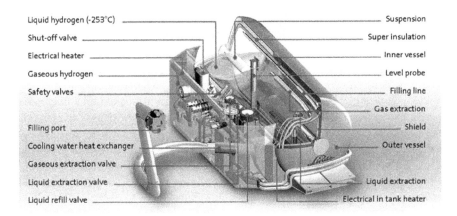

Fig. 1.2 Schematic architecture of a liquid hydrogen (LH2) tank including all devices to ensure its control and monitoring [2]

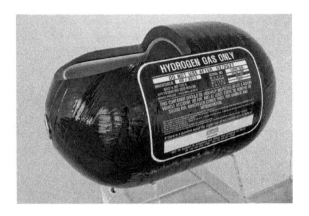

Fig. 1.3 CGH2 tank produced by Korean manufacturer ILJIN [3]

established the architecture of a LH2 tank is undoubtedly the more complex among all hydrogen storage solutions. It includes some heat exchanger limiting such as an efficient multi-layer vacuum super insulator (40 layers of metal foil); safety devices such as safety valve to vent hydrogen gas and prevent gas pressure increase. Due to unavoidable heat loss, an amount of hydrogen gas is rejected to environment, this is called boil-off. To complete previous drawbacks, energy balance for LH2 tank is clearly unfavorable because 30% of the stored chemical energy is required to liquefy hydrogen.

In comparison, pressurized hydrogen gas consumes less of this stored chemical energy to be produced: one say 15% for a 700 bar hydrogen compressed gas (CGH2) and 12% for a 350 bar CGH2. This matter of fact is enough to justify the CGH2 technology is the most suitable for many applications involving hydrogen energy. Moreover the manufacturing process that is the filament winding process is perfectly mastered and machine abilities are fully developed to design any geometry with high accuracy of fiber positioning and the production time also decreases. Figure 1.3 shows a commercial CGH2 tank manufactured by the Korean ILJIN. These products are implemented in fuel cell vehicles.

As mentioned above, the mechanical reinforcement of a CGH2 tank is obtained by the filament winding process. The technological state of art leads to manufacture Type IV (type I is a full metal tank) CGH2 tanks because this technology has proven to be the unique allowing to store for long term hydrogen and to withstand multiple emptying and refilling cycles. The reinforcement is here deposited on a polymeric liner including metallic heads so as to mount fittings. Gravimetric density of 5–6 wt% may be reached with this technology, considering tank and fittings. Some tricky challenges remain considering the use of such tanks. Among them, we have to note the evolution of temperature during filling, the collapse of the liner during emptying of hydrogen all the more kinetics is high etc. Some issues are coming from field experience: one expect refilling time to be short but it leads to a potentially critical increase of temperature, involving to ensure a refreshing of gas before filling; due to decrease of pressure, we observe a peeling off between the liner and the composite part, obviously detrimental for safety reason. Undoubtedly, the CGH2 tank is the most mature solution considering technology, cost, emptying and refilling kinetics, nevertheless works are now concentrating on the global efficiency, ease of use and safety.

Solid hydrogen storage in metals allows reaching the higher volumetric density because mean distance between hydrogen atoms is then the lowest (see Fig. 1.4). Expectedly, due to storage medium the gravimetric density is also the lowest if the three different media are compared. However, solid storage has undoubtedly some strong advantages among which it is worth notice the opportunity to choose the working pressure and temperature according the application (see Fig. 1.5). That means low pressure, between 1 and 10 bar, and temperature, between 10 and 80 °C, are commonly attainable what is particularly attractive for safety reason and public acceptance. Moreover, the chemical reaction between hydrogen and metals

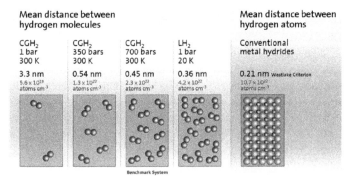

Fig. 1.4 Hydrogen density according the storage medium [4, 5]

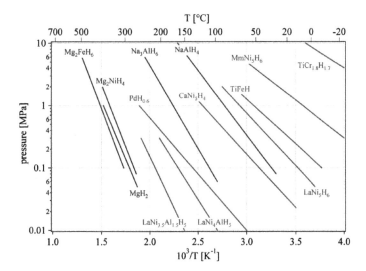

Fig. 1.5 Van't Hoff plots of some hydrides [6]. Slope relates to the enthalpy of formation

is exothermic or endothermic during absorption or desorption respectively. The amount of heat is sufficiently high to be managed at all times and necessarily during desorption if a sufficient hydrogen flow (for fueling a fuel cell, for example) is required.

As a synthetic consequence of above comments, one say the hydrogen storage medium is application dependent. High technological application may use LH2 solution, when hydrogen for mobility, more precisely for automotive, require pressurized gas, and solid hydrogen is more suitable for stationary applications, eventually for nomad application.

1.3 TiFe and Its Hydrides

Here we pay attention on solid storage solution, more precisely on Titanium-Iron alloys which are very good candidate to replace well-known Lanthanum-Nickel system. The TiFe system has the main advantages to offer a solid solution at lower cost and more easily available and recyclable in Europe. Figure 1.6 represents the phase stability domains of the binary alloy Ti–Fe as a function of temperature and Ti composition. Two intermetallic compounds exist: FeTi which is of cubic structure similar to that of CsCl and Fe_2Ti of structure C14 similar to the structure of $MgZn_2$. Fe_2Ti does not absorb hydrogen, but absorption becomes possible for compounds that are richer in Ti than $TiFe_2$ according Reilly et al. [7].

These authors studied the Fe–Ti–H system and in particular the reaction between the TiFe compound and the hydrogen resulting in the formation of two hydrides $TiFeH_1$ and $TiFeH_2$. Isothermal curves (Pressure-Composition-Isotherm, PCI) give the hydrogen equilibrium pressure as a function of the amount of hydrogen absorbed

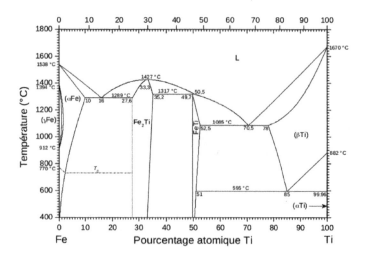

Fig. 1.6 Phase diagram of TiFe. Phase stability domains according Ti atomic percentage [8]

by the material at a defined temperature. Figure 1.3a shows PCI curve at 40 °C of TiFe compound from [7] and Fig. 1.3b represents PCI curves obtained on TiFeMn0.1 at 8, 22 and 45 °C with our own equipment.

As previously mentioned, when designing a vessel to welcome hydrides, one has to consider heat exchange to manage the inlet or outlet flow, but also kinetics of reaction [9, 10]. Another challenging issue regarding hydride is the decrepitation phenomenon that occurs after repeated absorption/desorption cycles. The particle size of the pristine powder bed decreases while the number of cycles increases. This is of first interest for the designer because distribution inside the vessel consequently evolves leading to high level of stress on the tank wall and a potential failure [11, 12] (Fig. 1.7).

To investigate the phenomenon, we develop specific observation bench test with hydrogen reactor and CCD camera to visualize the way particle sizes diminish during absorption. Figure 1.8 illustrates the decrepitation of a 2 mm diameter particle of LaNi$_5$ when submitted to 30 bar hydrogen pressure. The LaNi$_5$ is here chosen because the phenomenon has clearly a higher dynamics with more visible effect on a shorter time.

In parallel, we carry out Xray tomography analysis of TiFeMn particles at successive time to evidence the effect of hydrogen on the morphology (see Fig. 1.9). This reveals a fracture network appears due to the hydride phase transformation. The integrity of the TiFeMn particle is maintained when the fracture network is widely spread what is directly correlated to mechanical properties of the alloy, more precisely

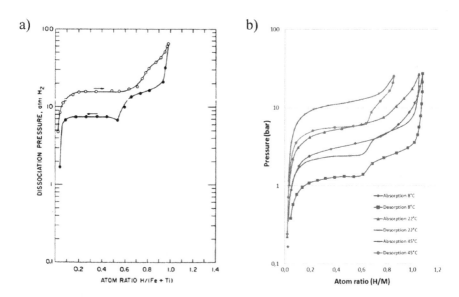

Fig. 1.7 Pressure-composition-isotherm curves of TiFe alloys; **a** Hysteresis in the TiFe–H system at 40 °C according [7], **b** TiFeMn alloy characterized at 8, 22 and 45 °C on our bench test

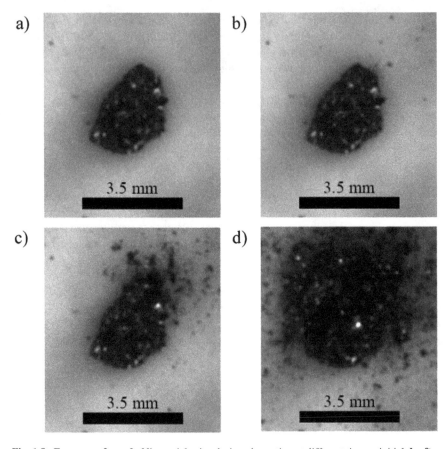

Fig. 1.8 Focus on a 2 mm LaNi$_5$ particle size during absorption at different time: **a** initial, **b after 2 h**, **c 4 h** and **d** 6 h

Fig. 1.9 X-ray tomography of a TiFeMn particle after a 20 h exposure to hydrogen

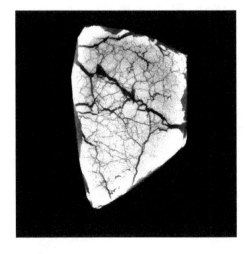

resilience. This respond seems really different of the respond of LaNi$_5$ alloy (see Fig. 1.8) for which energetic expulsion of very small particles is observed.

These observations, briefly introduced here, feed our reflections and our modeling to predict the behavior of the single particle but also the particle in the powder bed. The Different Element Modeling is peculiarly convenient for such mechanical issue, and we presently develop routines for YADE [13] to simulate the particle response depending on mechanical parameters.

1.4 Conclusion

In a world where renewable energies have to become the rule, hydrogen benefits a high potential in order to store electricity under gas form for long term and thus answers main drawbacks of these energies, that are their intermittence and fluctuating nature. We introduced the three main hydrogen storage media: liquid hydrogen LH2, pressurized hydrogen CGH2 and solid storage solutions. Considering the state of art, these solutions are rather mature, even if at different stages. The disadvantages but also the domains of application were recalled. As an illustration, a focus was made on the research activities at the Department of Applied Mechanics in Besançon that is the decrepitation phenomenon of metal hydride occurring while the material is submitted to repeated hydrogen cycles. We aim at increasing knowledge and at modeling the decrepitation in order to develop methodologies and tools to be used when designing solid storage tank.

References

1. IRENA, Hydrogen: a renewable energy perspective 2019. https://www.irena.org/-/media/Files/IRENA/Agency/Publication/2019/Sep/IRENA_Hydrogen_2019.pdf
2. The Linde Group, Wiesbaden
3. ILJIN Group, Seoul
4. U. Eberle, G. Arnold, R. von Helmolt, Hydrogen storage in metal–hydrogen systems and their derivatives. J. Power Sources **154**(2), 456–460 (2006)
5. U. Eberle, G. Arnold, R. von Helmolt, Fuel cell vehicles: status 2007. J. Power Sources **165**(2), 833–843 (2007)
6. Zuttel, A.: Materials for hydrogen storage. materialstoday **6**(9), 24–33 (2003)
7. J.J. Reilly, R.H. Wiswall, Jr., Formation and properties of iron titanium hydride, Inorg. Chem. **13**, 218–222 (1974)
8. H. Okamoto, *Phase Diagrams for Binary Alloys*, 2nd edn. (ASM International, Ohio, 2000)
9. A. Zeaiter, D. Chapelle, F. Cuevas, A. Maynadier, M. Latroche, Milling effect on the microstructural and hydrogenation properties of TiFe$_{0.9}$Mn$_{0.1}$ alloy. Powder Technol. **339**, 903–910 (2018)
10. C. Lexcellent, G. Gay, D. Chapelle, Thermomechanics of a metal hydride-based hydrogen tank. Continuum Mech. Thermodyn. **27**, 379–397 (2015)
11. K. Nasako, Y. Ito, N. Hiro, M. Osumi, Stress on a reaction vessel by the swelling of a hydrogen absorbing alloy. J. Alloy. Compd. **264**, 271–276 (1998)

12. Qin, F.: Pulverization, expansion of $La_{0.6}Y_{0.4}Ni_{4.8}Mn_{0.2}$ during hydrogen absorption–desorption cycles and their influences in thin-wall reactors. Int. J. Hydrogen Energy **33**, 709–717 (2008)
13. V. Šmilauer et al., *Yade Documentation*, 2nd edn. The Yade Project (2015). https://doi.org/10.5281/zenodo.34073. http://yade-dem.org/doc/

Chapter 2
Preparation of ZrO_2–Fe_2O_3 Nanoparticles and Their Application as Photocatalyst for Water Depollution and Hydrogen Production

M. Benamira, N. Doufar, and H. Lahmar

Abstract This study is deducted to the treatment of wastewater from the organic pollutants and hydrogen production using a photocatalyst composed of Fe_2O_3 and ZrO_2. The structural and photophysical properties of the catalysts have been characterized by X-ray diffraction (XRD), scanning electron microscopy (SEM), Brunauer–Emmett–Teller (BET) and UV–Vis diffuse reflectance spectrometry. The photocatalyst acquired p type conductivity, due to oxygen insertion in the layered lattice with an activation energy of 0.15 eV. The flat band potential (E_{fb}, -0.44 V_{SCE}), close to the photocurrent onset potential (-0.2 V_{SCE}). A total degradation of Methyl orange is achieved within 90 min under sunlight irradiation. The total degradation of phenol is also achieved in less than 120 min. Based on the energy band positions, the detailed reaction mechanism has been discussed. The position of the conduction band in the energy diagram shows the possibility of H_2 production under visible light.

Keywords Photocatalytic activity · ZrO_2–Fe_2O_3 · DRX

2.1 Introduction

Zirconia, ZrO_2, is a ceramic material with a property-structure dependence. For that, by playing on the structure with judicious additions or by using the appropriate preparation method, it is possible to modify its properties of use to make it a functional material used in various applications. Zirconia has a wide variety of application due to its flexible structural characteristics [1–4]. This flexibility allows new areas of functional applications to emerge.

M. Benamira (✉) · N. Doufar
Laboratory of Materials Interaction and Environment (LIME), University of Mohamed Seddik Benyahia, BP 98 Ouled Aïssa, Jijel, Algeria
e-mail: benamira18@gmail.com; m_benamira@univ-jijel.dz

H. Lahmar
University of Mohamed Seddik Benyahia, BP 98 Ouled Aïssa, Jijel, Algeria

© The Editor(s) (if applicable) and The Author(s), under exclusive license to Springer Nature Singapore Pte Ltd. 2021
A. Khellaf (ed.), *Advances in Renewable Hydrogen and Other Sustainable Energy Carriers*, Springer Proceedings in Energy,
https://doi.org/10.1007/978-981-15-6595-3_2

Our main objective in this work is to study a photocatalytic process for water treatment as an alternative to existing processes using the sun. This work consists in synthesizing new material based on zirconia that can be used in water depollution. Our choice is focused on the ZrO_2–Fe_2O_3 system. The addition of Fe increases the catalytic activity of zirconia for hydrocarbon isomerization reactions. However, the addition of ZrO_2 increases the catalytic activity of Fe_2O_3 by increasing the stability of the Fe^{3+} cation. Catalysts based on the Fe_2O_3–ZrO_2 system are also used in ammonium synthesis, Fischer-Tropic synthesis and carbon monoxide hydrogenation. Iron as a dopant was able to stabilize cubic zirconia, c-ZrO_2, but due to its smaller ionic radius compared to the rare earth cations, this stabilization is not effective. In this work, we were interested in the synthesis of Fe-doped ZrO_2 material by co-precipitation for the degradation of organic pollutants and also for the production of H_2 under visible light.

2.2 Experimental

Fe-doped ZrO_2 (3 g) was synthesized by co-precipitation using $ZrOCl_2 \cdot 8H_2O$ (99.5%; Sigma–Aldrich), $FeCl_3 \cdot 6H_2O$ (97%; Sigma–Aldrich) and NH_4OH as starting materials. The starting reagents were mixed according to the following equation:

$$(1-x)ZrOCl_2 \cdot 8H_2O + xFeCl_3 \cdot 6H_2O$$
$$+ (2+x)NH_4OH \rightarrow Zr_{(1-x)}Fe_xO_{(2-x/2)} + (2+x)NH_4Cl$$
$$+ 18\,H_2O\,(x=2/3) \qquad (2.1)$$

3.255 g of $ZrOCl_2 \cdot 8H_2O$ was dissolved in distilled water (100 mL) and stirred vigorously for 40 min. After that, 5.46 g of $FeCl_3 \cdot 6H_2O$ were added to the first solution. The aqueous suspensions were mixed and stirred slowly for 60 min at 65 °C. NH_4OH (3.5 M) was added dropwise under continuous stirring for 60 min to adjust the pH solution at 10, and to form a solid precipitate with a brown color. The resulting mixture was centrifuged for 15 min; the obtained precipitate was filtered and washed several times with deionized water. After that, the resulting precipitate was dried in an oven at 100 °C for 24 h and calcined at 600 °C for 3 h. TiO_2 powder used in the hetero-junction was synthesized as reported in our previous work [5]. The hetero-junction $ZrFe_2O_5/TiO_2$ was obtained by mixing both catalysts in the mass ratio (1/1).

X-ray diffraction (XRD) was used to characterize the structure of the as prepared powders. The patterns were obtained using a Brüker "D8 Advance" powder diffractometer with a Cu Kα radiation ($\lambda = 1.54056$ Å) and 2θ varying from 20 to 80° by steps increments of 0.0146° and 0.2 s counting time per step. The Scanning Electron Microscopy (SEM) was performed on a FEG-SEM JEOL 7600 in order to evaluate the size distribution as well as the morphology of the nanoparticles. The FT-IR spectra were recorded on a VERTEX 70/70v spectrophotometer (BRUKER) in transmission mode between 400 and 4000 cm^{-1}. The UV–vis diffuse reflectance spectra were obtained with a UV-visible spectrophotometer (Specord 200 Plus). The BET surface areas of the powders were carried out on a Micromeritics 3Flex analyzer determined by N_2 adsorption–desorption isotherm measurement at -196 °C. The electrochemical characterization was performed in 0.1 M Na_2SO_4 electrolyte using a standard three-electrode cell using Solartron 1287 potentiostat and frequency response analyzer Solartron 1260; the sweep rate was set at 5 mV s^{-1}. The working electrode is prepared as 10 mm diameter pellet using uniaxial pressure of 3.5 MPa and sintered at 600 °C for 2 h. It was introduced into resin epoxy, polished with SIC paper (1200) and washed with water. The electrical contact was achieved by the painted silver lacquer. The photocatalytic tests were conducted in May and June between 9 am and 5 pm. The sunlight irradiation was evaluated as 700 W/m^2, while the temperature averaged 28 °C. The photocatalytic tests were carried out in an open Pyrex glass reactor with a double wall. Before irradiation, the suspensions composed of 50 mL of organic pollutant, 50 mg of catalyst and H_2O_2 (10^{-5} M) were kept in the dark for 60 min under magnetic stirring to ensure good adsorption equilibrium on the catalyst surface. Then, the reactor was exposed to solar irradiation. The absorbance was measured with a UV-visible spectrophotometer (Shimadzu UV-2400). After irradiation, the solutions were centrifuged and filtered through a membrane filter in order to measure the decomposition rate. The production of H_2 by photocatalysis was carried out at 50 °C [6]. The photocatalyst powder (100 mg) was introduced in a Pyrex reactor equipped with a cooling system and dispersed in 200 mL of $Na_2S_2O_3$ solution. Before each test, O_2 was purged by passing N_2 for 20 min. Three tungsten lamps with a total intensity of 30 mW cm^{-2} were used as the light source. Hydrogen gas was identified by gas chromatography using Clarus® 680 GC PerkinElmer Gas Chromatograph and the volume of H_2 was quantified in a water manometer.

2.3 Results and Discussion

2.3.1 Characterization of Material

Figure 2.1a shows the XRD pattern of $ZrFe_2O_5$ synthesized by co-precipitation. The substitution of Zr by Fe confirms the formation of the tetragonal solid solution $Zr_{1-x}Fe_xO_{2-x/2}$ in agreement with JCPDS Card N° 17-0923. The incorporation of high Fe content reveals the presence of new peaks of the hematite (α-Fe_2O_3, JCPDS N° 01-1053) and confirms the results obtained by Davison et al. [7] and Botta et al. [8]. The substitution of Zr^{4+} by Fe^{3+} ions in the ZrO_2 lattice was reported using different preparation methods. These studies confirm the formation of tetragonal phase at low Fe^{3+} concentrations (<10 at.%). However, for high Fe^{3+} content (>20 at.%), the XRD patterns indicate the presence of α-Fe_2O_3 phase in total agreement with our results.

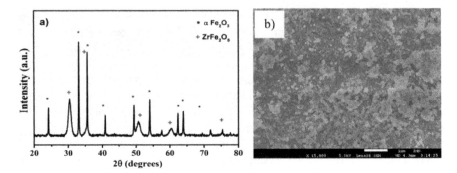

Fig. 2.1 a Powder X-ray diffraction results of ZrFe$_2$O$_5$ powder. **b** Scanning electron microscopy of ZrFe$_2$O$_5$ powder

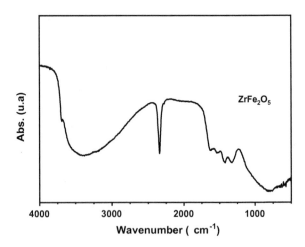

Fig. 2.2 FT-IR spectra of ZrFe$_2$O$_5$ prepared by the co-precipitation method

The infrared Fourier transform spectrum is illustrated in Fig. 2.2; the wide band at 3400 cm^{-1} is attributed to the folding mode of OH groups due to adsorbed water on the powder surface. The peaks observed below 750 cm^{-1} are attributed to the Fe–O and Zr–O stretching vibrations while the peaks at 1470 and 1560 cm^{-1} are assigned to adsorbed water molecules. A weak peak at 2315 cm^{-1} is assigned to adsorbed carbon dioxide when the sample was handled in air [9, 10].

The converted UV-visible absorption data are used to evaluate the optical band gap (E_g) by extrapolating the linear portion of the plots $(\alpha h\nu)^2$ versus $h\nu$ to $(\alpha h\nu)^2 = 0$. The optical band gap of the as-prepared powder is 1.74 eV.

2.3.2 Photocatalytic Activity

The ZrFe$_2$O photocatalyst was used for the degradation of the azo dye methyl orange and the organic compound phenol under solar light irradiation.

The standard redox potential of •OH radical (+2.8 V) classifies it as a powerful oxidizing agent, able to oxidize the majority of organic compounds into mineral end products. An initial concentration of 30 mg/L was used for the two products tested (phenol and methyl orange); Fig. 2.3 summarizes the kinetic results of the degradation in the presence of H$_2$O$_2$ under solar light irradiation. The two pollutants are totally degraded after 90 min for methyl orange (30 mg/L) and 120 min for phenol (30 mg/L). The results obtained in this study are promising compared to that those reported in the literature. Murcia et al. [11] studied the photocatalytic degradation of methyl orange and phenol over TiO$_2$ and Pt–TiO$_2$. Total elimination of phenol and methyl orange was obtained after 2 h under UV light; while degradation of only 60% of phenol within 2 h under sunlight irradiation was obtained by Al-Hamdi et al. [12].

The relevant reactions of photocatalysis mechanism occur at the surface of the photocatalyst causing the degradation of these two organic compounds (OC) can be expressed as follows:

$$ZrFe_2O_5 + h\nu \rightarrow ZrFe_2O_5\left(e_{CB}^- + h_{VB}^+\right) \qquad (2.2)$$

$$ZrFe_2O_5 + \left(h_{VB}^+\right) + H_2O \rightarrow ZrFe_2O_5 + H^+ + OH^- \qquad (2.3)$$

$$ZrFe_2O_5 + \left(h_{VB}^+\right) + OH^- \rightarrow ZrFe_2O_5 + {}^\bullet OH \qquad (2.4)$$

$$ZrFe_2O_5 + \left(e_{CB}^-\right) + O_2 \rightarrow ZrFe_2O_5 + O_2^{\bullet -} \qquad (2.5)$$

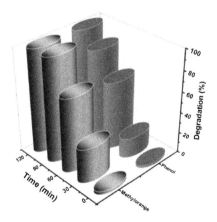

Fig. 2.3 Degradation of methyl orange and phenol by ZrFe$_2$O$_5$ catalyst in the presence of H$_2$O$_2$ ~ 0.2 mL/L, $T = 25$ °C, pH = 7, catalyst dose = 1 mg/mL) under solar light irradiation

$$O_2^{\bullet -} + H^+ \rightarrow HO_2^{\bullet} \qquad (2.6)$$

$$2HO_2^{\bullet} \rightarrow H_2O_2 + O_2^{\bullet -} \qquad (2.7)$$

$$H_2O_2 + (e_{CB}^-) \rightarrow {}^{\bullet}OH + OH^- \qquad (2.8)$$

$$OC + OH^{\bullet}/O_2^{\bullet -} \rightarrow \text{(Degradation products) } CO_2 + H_2O \qquad (2.9)$$

The use of the hetero-junction system is one of the effective ways to improve the photocatalytic activity under visible light irradiation. The new hetero-junction $ZrFe_2O_5/TiO_2$ with a mass ratio (1/1) was used for the H_2 production. Figure 2.4a presents the current density-potential J(V) characteristics of $ZrFe_2O_5$ at pH ~ 7. The cathodic reduction potential of water (production of H_2) appears at -0.16 V_{SCE} and

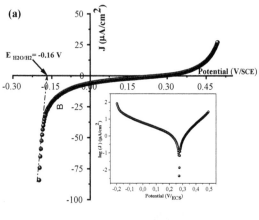

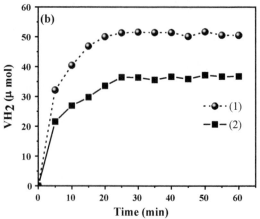

Fig. 2.4 **a** The intensity potential J(V) characteristic of $ZrFe_2O_5$ (Inset-Semi-logarithmic plot over $ZrFe_2O_5$ electrode). **b** Volume of evolved H_2 versus illumination time with $ZrFe_2O_5$: (1) with Methyl orange, (2) without Methyl orange

corresponds to the intercept point of current density curve with the potential-axis. The anodic oxidation potential of water appears at 0.3 V_{SCE} (Inset Fig. 2.4a). The volume of H_2 obtained with Methyl orange increased compared to $ZrFe_2O_5$ alone by 60% (Fig. 2.4b).

2.4 Conclusion

The degradation of organic pollutants by $ZrFe_2O_5$ under solar light irradiation confirms the high photocatalytic efficiency. The use of p–n hetero-junction improves the photocatalytic activity of H_2 production by enhancing the separation of electron-hole pairs and reduces the recombination process. The photocatalytic mechanism of the degradation was proposed. A total photodegradation of methyl orange and phenol was obtained in less than 120 min.

References

1. M.H. Khajezadeh, M. Mohammadi, M. Ghatee, Hot corrosion performance and electrochemical study of CoNiCrAlY/YSZ/YSZ-La_2O_3 multilayer thermal barrier coatings in the presence of molten salt. Mater. Chem. Phys. **220**, 23–34 (2018)
2. M. Benamira, A. Ringuede, M. Cassir, D. Horwat, J.F. Pierson, P. Lenormand, J. Fullenwarth, Comparison between ultrathin films of YSZ deposited at the solid oxide fuel cell cathode/electrolyte interface by atomic layer deposition, dip-coating or sputtering. Open Fuels Energy Sci. J. **2**(1), 87–99 (2009)
3. M. Benamira, A. Ringuedé, M. Cassir, D. Horwat, P. Lenormand, F. Ansart, J.P. Viricelle, Enhancing oxygen reduction reaction of YSZ/$La_2NiO_{4+\delta}$ using an ultrathin $La_2NiO_{4+\delta}$ interfacial layer. J. Alloy. Compd. **746**, 413–420 (2018)
4. E. Lay, M. Benamira, C. Pirovano, G. Gauthier, L. Dessemond, Effect of Ce-doping on the electrical and electrocatalytical behavior of La/Sr chromo-manganite perovskite as new SOFC anode. Fuel Cells **12**, 265–274 (2012)
5. N. Doufar, M. Benamira, H. Lahmar, M. Trari, I. Avramova, M.T. Caldes, Structural and photochemical properties of Fe-doped ZrO_2 and their application as photocatalysts with TiO_2 for chromate reduction. J. Photochem. Photobiol. A **386**, 112105 (2020)
6. H. Lahmar, M. Trari, Photocatalytic generation of hydrogen under visible light on La_2CuO_4. Bull. Mater. Sci. **38**(4), 1043–1048 (2015)
7. S. Davison, R. Kershaw, K. Dwight, A. Wold, Preparation and characterization of cubic ZrO_2 stabilized by Fe (III) and Fe (II). J. Solid State Chem. **73**, 47–51 (1988)
8. S.G. Botta, J.A. Navío, M.C. Hidalgo, G.M. Restrepo, M.I. Litter, Photocatalytic properties of ZrO_2 and Fe/ZrO_2 semiconductors prepared by a sol–gel technique. J. Photochem. Photobiol. A Chem. **129**, 89–99 (1999)
9. S. Douafer, H. Lahmar, M. Benamira, G. Rekhila, M. Trari, Physical and photoelectrochemical properties of the spinel $LiMn_2O_4$ and its application in photocatalysis. J. Phys. Chem. Solids **118**, 62–67 (2018)
10. S. Douafer, H. Lahmar, M. Benamira, L. Messaadia, D. Mazouzi, M. Trari, Chromate reduction on the novel hetero-system $LiMn_2O_4$/SnO_2 catalyst under solar light irradiation. Surf. Interfaces **17**, 100372 (2019)

11. J.J. Murcia, M.C. Hidalgo, J.A. Navío, J. Araña, J.M. Doña-Rodríguez, Correlation study between photo-degradation and surface adsorption properties of phenol and methyl orange on TiO_2 vs platinum-supported TiO_2. Appl. Catal. B **150**, 107–115 (2014)
12. A.M. Al-Hamdi, M. Sillanpää, J. Dutta, Photocatalytic degradation of phenol by iodine doped tin oxide nanoparticles under UV and sunlight irradiation. J. Alloy. Compd. **618**, 366–371 (2015)

Chapter 3
Preparation of Anode Supported Solid Oxide Fuel Cells (SOFCs) Based on BIT07 and $Pr_2NiO_{4+\delta}$: Influence of the Presence of GDC Layer

M. Benamira, M. T. Caldes, O. Joubert, and A. Le Gal La Salle

Abstract In this work, we study the electrochemical performance of anode supported solid oxide fuel cells (SOFCs) based on perovskite-type materials: $BaIn_{0.3}Ti_{0.7}O_{2.85}$ (BIT07) as an electrolyte, BIT07-Ni as a cermet anode and $Pr_2NiO_{4+\delta}$ as a cathode. Anode/electrolyte assemblies have been prepared by tape casting and co-firing and the cathode has been deposited by screen-printing. The performance of BIT07-Ni/BIT07/$Pr_2NiO_{4+\delta}$ cells has been determined between 600 and 750 °C under humidified (3% H_2O) hydrogen as fuel and air as oxidant. The presence of an interfacial layer of gadolinia doped ceria (GDC) is also tested. Impedance Spectroscopy (EIS) measurements have also been carried out and allowed to differentiate between the series and polarization resistances. The power density obtained from the cell with GDC was 119.21 mW/cm^2 at 700 °C, compared with 28.1 mW/cm^2 for the cell without GDC thin layer. These results confirmed that the presence of a dense thin layer of GDC at the interface electrolyte/cathode is a very promising method for intermediate temperature SOFCs (IT-SOFCs) to increase the performance.

Keywords Tape casting · SOFC · EIS · $Pr_2NiO_{4+\delta}$

3.1 Introduction

Nowadays, solid oxide fuel cells (SOFCs) operate at 800–1000 °C. Reducing the operating temperature of solid oxide fuel cells requires the development of new electrolyte materials with good performance at intermediate temperatures (500–700 °C). $BaIn_{0.3}Ti_{0.7}O_{2.85}$ (BIT07) has the required transport properties and redox stability to be considered as a promising electrolyte material with an ionic conductivity of 10^{-2}

M. Benamira (✉)
Laboratory of Materials Interaction and Environment (LIME), University of Mohamed Seddik Benyahia, BP 98 Ouled Aïssa, Jijel, Algeria
e-mail: m_benamira@univ-jijel.dz; benamira18@gmail.com

M. Benamira · M. T. Caldes · O. Joubert · A. Le Gal La Salle
Institut Des Matériaux Jean Rouxel (IMN), Université de Nantes, CNRS, 2, rue de la Houssinière, BP 32229, 44322 Nantes Cedex 3, France

© The Editor(s) (if applicable) and The Author(s), under exclusive license to Springer Nature Singapore Pte Ltd. 2021
A. Khellaf (ed.), *Advances in Renewable Hydrogen and Other Sustainable Energy Carriers*, Springer Proceedings in Energy,
https://doi.org/10.1007/978-981-15-6595-3_3

S cm^{-1} at 700 °C and chemical stability under a wide range of partial pressures of oxygen [1–4]. The praseodymium nickelate $Pr_2NiO_{4+\delta}$ (so-called PNO) is a mixed ionic electronic conducting (MIEC) oxide widely studied as oxygen electrode for solid oxide fuel cells and solid oxide electrolysis cells (SOEC) as well. MIEC oxide attracted particular attention since the oxygen reduction reaction (ORR) has been extended to the whole volume of the electrode, compared to a pure electronically conducting cathode where the ORR is restrained to the triple phase boundary (TPB) points located at the cathode/electrolyte interface [5–9].

This paper aims to study the influence of the presence of the material used as an interlayer between the PNO electrode and the BIT07 electrolyte on the electrochemical performance of the SOFC. $Ce_{0.8}Gd_{0.2}O_{1.9}$ (namely GDC) has been selected as interlayer material. It has also been demonstrated that BIT07 can be used with well-known cathode materials, such as $La_{0.58}Sr_{0.4}Co_{0.2}Fe_{0.8}O_{3-\delta}$ (LSCF) or $Nd_2NiO_{4+\delta}$, and that best results regarding electrochemical performance have been obtained with LSCF.

Our main objective in this work is to prepare a complete Ni-BIT07/BIT07/PNO and Ni-BIT07/BIT07/GDC//PNO cells by tape casting and screen printing. They have been tested under air on the cathode side and wet H_2 (3% H_2O) on the anode side. The performance has been discussed in terms of total resistance determined from current density/voltage curve (J/U) and Electrochemical Impedance Spectroscopy (EIS) measurements at the Open Circuit Voltage (OCV) to understand the origin of this resistance, and to determine the relevant parameters that can be modified in order to minimize it. The stability of the cell presenting the highest performance has been studied at the operating temperature of 700 °C.

3.2 Experimental

$BaIn_{0.3}Ti_{0.7}O_{2.85}$ (BIT07) was synthesised by solid state reaction by using stoichiometric amounts of $BaCO_3$ (Alfa Aesar), In_2O_3 (Alfa Aesar) and TiO_2 (Merck). Reactants were thoroughly mixed using acetone in an agate mortar and calcined at 1200 °C for 24 h. The obtained powder was again ground well, mixed, pressed into disks and sintered at 1350 °C for 24 h, ground and passed through mesh 100. $Pr_2NiO_{4+\delta}$ powder was provided by Marion Technologie and NiO powder (grain size 0.5–1 µm) was provided by Pharmacie Centrale de France. X-ray diffraction (XRD) was used to characterize the structure of the as-prepared powders. The patterns were obtained using a Brüker "D8 Advance" powder diffractometer with a Cu Kα radiation ($\lambda = 1.54056$ Å) and 2θ varying from 20 to 80° by steps increments of 0.0146° and 0.2 s counting time per step. The Scanning Electron Microscopy (SEM) was performed on a FEG-SEM JEOL 7600. By Brunauer–Emmett–Teller (BET), the surface areas of the powders were measured and carried out on a Micromeritics 3Flex analyzer determined by N_2 adsorption–desorption isotherm measurement at -196°C.

Before cell fabrication and in order to decrease the grain sizes, the as-prepared powders of BIT07 were ball-milled at 500 rpm: 60 h. The anode composition has been

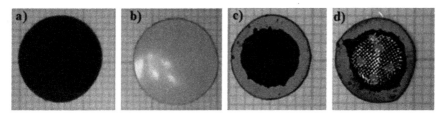

Fig. 3.1 Preparation of Ni-BIT07/BIT07/PNO anode supported cell. **a** Anode of BIT07-NiO, **b** electrolyte BIT07, **c** deposition of PNO cathode, **d** current collector of gold grid

optimized to reach the best electrochemical performance, obtained with a BIT07-Ni cermet (BIT07:NiO 50:50 wt%), realized with a BIT07 powder ball-milled and 5 wt% CB as pore forming agent. The button cell (Ø 10 mm) is composed of an anode supported electrolyte/anode half-cell, prepared by tape casting and co-fired. Then a PNO cathode is deposited on the electrolyte by screen-printing (Fig. 3.1).

Current collectors, made of discs (5 mm in diameter) of gold grid (Goodfellow, wire diameter: 60 μm) are attached on both the electrodes using gold ink. The cell is placed at 120 °C for overnight to evaporate the solvents and to obtain good electrical contacts. The current–voltage characteristic is measured with the use of a laboratory-made testing system. The cell was sealed on the top of an alumina sample tube. Thus separating the atmospheres between the inside and outside of the tube. The fuel-gas supply tube is situated inside the sample tube. The system was kept vertically in a tubular furnace. The measurements of current density (J) and voltage (U) were done by digital multimeters Keithley 197 and Protek 506, respectively. Current drawn in the circuit was varied using a rheostat. Prior to the measurements, nickel oxide is reduced in situ at 700 °C for 2 h under wet H_2 and measurements were made at 700 °C, under wet (3% H_2O) H_2 on the anode side, and air on the cathode side.

Electrochemical performances of single cells were studied by electrochemical impedance spectroscopy (EIS). The measurements were realized using a frequency response analyzer Solartron 1260. The impedance spectra were recorded over a frequency range 2 MHz to 0.01 Hz with a signal amplitude of 10 mV and with 10 points per decade under open circuit conditions from 500 to 700 °C. The EIS diagrams were normalized to the electrode area and fitted using equivalent circuits with the ZView® software (Scribner Associates). Post mortem (after measurements) analyses of the cells have been carried out by SEM, using a JEOL 7600.

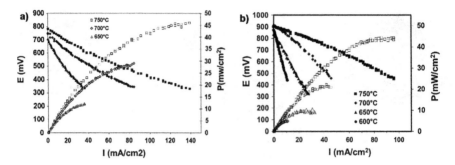

Fig. 3.2 Performance of a single cell consisting in: **a** Ni-BIT07/BIT07 and $Pr_2NiO_{4+\delta}$ sintered at 1050 °C, **b** Ni-BIT07/BIT07 and $Pr_2NiO_{4+\delta}$ sintered at 1100 °C as the cathode; I-V curves and resulting power densities

3.3 Results and Discussion

3.3.1 Characterization of Material

Figure 3.2 presents the voltage (U) and power density (P) versus current density (J) characteristics obtained for two single cells at different temperatures 1050 and 1100 °C, respectively, Fig. 3.2a, b. In both cases, the polarisation curves (U/J) are clearly linear in the studied current density range. In both cases, the OCV is lower than the theoretical value expected, suggesting a small leakage between gas chambers. Table 3.1 gathers the cells characteristics obtained from the voltage and power density versus current density characteristics and also from the Nyquist diagrams of the single cells.

The extrapolated maximum power densities are of about 28.1 and 22.06 mW cm^{-2} for the two cells, respectively. The ASR value obtained from the EIS measurements, 10.41 for cells A at 700 °C, is in agreement with the ones obtained from the U/J

Table 3.1 Characteristics of the cell Ni-BIT07/BIT07 and $Pr_2NiO_{4+\delta}$ sintered at 1050 °C recorded at OCV, ASR(U/J) calculated from U/J curves, ASR(EIS), Rs and Rp obtained from the EIS measurements

T (°C)	OCV (V)	P_{max} (mW cm^{-2})	ASR$_{(U-J)}$ (Ω cm^2)	ASR$_{(Nyquist)}$ (Ω cm^2)	Re.S (Ω cm^2)	Rp.S (Ω cm^2)
600	0.875	4.83	41.34	43.39	3.46 (7.4%)	39.93 (92.6%)
650	0.843	9.85	18.81	20.27	2.20 (9.8%)	18.07 (90.2%)
700	0.914	22.06	9.75	10.41	1.01 (8.8%)	9.40 (91.2%)
750	0.900	44.13	4.64	5.63	1.14 (16.8%)	4.49 (83.2%)

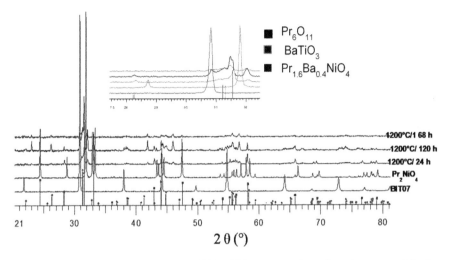

Fig. 3.3 X-ray diffraction patterns of BIT07/Pr$_2$NiO$_{4+\delta}$ composite pellets that were calcined at 1100 °C for different times

curves, 9.95. Those values are too high with a major contribution of polarization resistance (91.2%) and need more improvement. Thus indicates that the polarization resistance played an important role in cell performance.

In order to understand the origin of the low performances of the two cells, the reactivity test between BIT07 and Pr$_2$NiO$_{4+\delta}$ powders for more than 168 h at 1100 °C was done as shown in Fig. 3.3. Indeed, the DRX analysis reveals the significant presence of a new phase (Pr$_{1.6}$Ba$_{0.4}$NiO$_4$) due to the reaction between the electrolyte and cathode. Considering the reactivity of the Pr$_2$NiO$_{4+\delta}$ with BIT07, the higher power densities would expect to be achieved by using a thin interfacial layer of GDC.

A thin interfacial layer (5 μm) of GDC is deposited to the electrolyte/anode half-cell by screen-printing and co-fired at 1350 °C for 8 h. Then a PNO cathode is deposited on the electrolyte by screen-printing. To demonstrate the benefit of incorporating a thin layer at the interface between the electrode and electrolyte, the I–V–P curves of the fuel cell are shown in Fig. 3.4. The power density obtained from the cell with GDC thin film was about 119.21 mW/cm^2 at 700 °C, compared with about 28.1 mW/cm^2 for the cell without thin film.

3.4 Conclusion

In this work, anode supported cells based on BIT07 as an electrolyte, BIT07-Ni as anode and Pr$_2$NiO$_{4+\delta}$ as cathode have been successfully realized by tape casting and screen-printing. The reactivity test between BIT07 and Pr$_2$NiO$_{4+\delta}$ powders reveals the presence of a new phase (Pr$_{1.6}$Ba$_{0.4}$NiO$_4$) which explain the low performances of the single cell. The power density of a single cell with GDC thin film at the

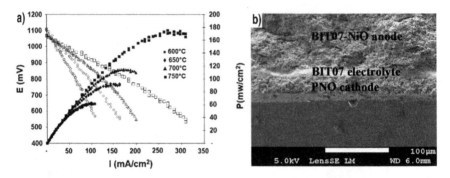

Fig. 3.4 a Performance of a single cell consisting in Ni-BIT07/BIT07/GDC/$Pr_2NiO_{4+\delta}$; I-V curves and resulting power densities, **b** scanning electron microscope image (cross-section) of single cell

interface electrolyte/cathode was significantly higher than that of a single cell without interfacial layer. This is attributed to increased interfacial area. The factors such as the better electrocatalytic activity of ceria and the intrinsic properties of nanostructured thin films.

References

1. F. Moser, M.T. Caldes, M. Benamira, J.M. Greneche, P. Leone, O. Joubert: Development of new anodes compatible with the solid oxide fuel cell electrolyte $BaIn_{0.3}Ti_{0.7}O_{2.85}$. J. Power Sour **201**, 103–111 (2012)
2. M. Benamira, M. Letilly, E. Quarez, O. Joubert, A.L.G. La Salle, Optimization of SOFC anode/electrolyte assembly based on $BaIn_{0.3}Ti_{0.7}O_{2.85}$ (BIT07)/Ni-BIT07 using an interfacial anodic layer. J. Power Sour. **251**, 66–74 (2014)
3. M. Benamira, L. Thommy, F. Moser, O. Joubert, M.T. Caldes, New anode materials for IT-SOFC derived from the electrolyte BaIn0.3Ti0.7O2.85 by lanthanum and manganese doping. Solid State Ionics **265**, 38–45 (2014)
4. M. Letilly, A. Le Gal La Salle, M. Caldes, M. Marrony, O. Joubert, Validation of $BaIn_{0.3}Ti_{0.7}O_{2.85}$ as SOFC electrolyte with Nd2NiO4, LSM and LSCF as cathodes. Fuel Cells **9**, 622–629 (2009)
5. E. Lay, M. Benamira, C. Pirovano, G. Gauthier, L. Dessemond, Effect of Ce-doping on the electrical and electrocatalytical behavior of La/Sr chromo-manganite perovskite as new SOFC anode. Fuel Cells **12**, 265–274 (2012)
6. M. Benamira, A. Ringuedé, M. Cassir, D. Horwat, P. Lenormand, F. Ansart, J.P. Viricelle, Enhancing oxygen reduction reaction of $YSZ/La_2NiO_{4+\delta}$ using an ultrathin $La_2NiO_{4+\delta}$ interfacial layer. J. Alloy. Compd. **746**, 413–420 (2018)
7. M.T. Caldes, K.V. Kravchyk, M. Benamira, N. Besnard, O. Joubert, O. Bohnke, N. Dupré, Metallic nanoparticles and proton conductivity: improving proton conductivity of $BaCe_{0.9}Y_{0.1}O_{3-\delta}$ and $La_{0.75}Sr_{0.25}Cr_{0.5}Mn_{0.5}O_{3-\delta}$ by Ni-doping. ECS Trans **45**, 143–154 (2012)
8. J.M. Bassat, M. Burriel, O. Wahyudi, R. Castaing, M. Ceretti, P. Veber, J.A. Kilner, Anisotropic oxygen diffusion properties in $Pr_2NiO_{4+\delta}$ and Nd2NiO4 + δ single crystals. J. Phys. Chem. C **117**, 26466–26472 (2013)
9. C. Allançon, A. Gonthier-Vassal, J.M. Bascat, J.P. Loup, P. Odier, Influence of oxygen on structural transitions in $Pr_2NiO_{4+\delta}$. Solid State Ionics **74**, 239–248 (1994)

Chapter 4
Preparation of the Spinel CuCo$_2$O$_4$ at Low Temperature. Application to Hydrogen Photoelectrochemical Production

R. Bagtache, K. Boudjedien, I. Sebai, D. Meziani, and Mohamed Trari

Abstract We reported for the first time a facile method for the preparation of the spinel CuCo$_2$O$_4$ at 140 °C by using sulfates of copper and cobalt in KOH. The compound crystallizes in a cubic system (SG: F d $\bar{3}$ m) with a lattice constant of 8.0590 Å. The material was characterized by physical and photo-electrochemical techniques. The X-ray diffraction (XRD) pattern confirms the single phase with a high purity. The UV–Visible spectroscopy of the black product revealed a direct optical transition of 1.38 eV. The capacitance-potential (C^{-2}-E) displays a negative slope and a flat band potential (+0.28 V$_{SCE}$) characteristic of p-type semiconductor behavior. Further, the Nyquist plot shows a semicircle followed by straight line at low frequencies, indicating a diffusion process. To show effectively the photocatalytic performance, we tested our product in the hydrogen production upon visible light irradiation. H$_2$ evolution rate of 7 μmol h^{-1} mg^{-1} and a quantum efficiency of 1.1% were obtained upon visible light illumination.

Keywords Spinel CuCo$_2$O$_4$ · Photo-electrochemical characterization · Hydrogen · Visible light

4.1 Introduction

Nowadays, many efforts have been devoted in the field of environment which consists to develop new processes for the purification of air and treatment of water [1]. In this

---

R. Bagtache (✉) · K. Boudjedien · D. Meziani
Faculty of Chemistry, Laboratory of Electrochemistry-Corrosion, Metallurgy and Inorganic Chemistry, USTHB, BP 32, 16111 Algiers, Algeria
e-mail: bagtacheradia@yahoo.fr

I. Sebai
Faculty of Chemistry, USTHB, LCGN, BP 32, Algiers 16111, Algeria

M. Trari
Laboratory of Storage and Valorisation of Renewable Energies, Faculty of Chemistry, (USTHB), BP 32, 16111 Algiers, Algeria

© The Editor(s) (if applicable) and The Author(s), under exclusive license to Springer Nature Singapore Pte Ltd. 2021
A. Khellaf (ed.), *Advances in Renewable Hydrogen and Other Sustainable Energy Carriers*, Springer Proceedings in Energy,
https://doi.org/10.1007/978-981-15-6595-3_4

regard, the photo-catalysis has attracted much attention for the degradation of organic pollutants into small molecules and/or total mineralization [2]. In recent years, the photo-catalysis has become a challenge to face the modern world as it provides an alternative simple using a source of energy (artificial: lamp or natural: sunlight) to achieve the chemical transformation [3]. Currently, compounds with a spinel structure AB_2O_4 such as $ZnCo_2O_4$, $NiCo_2O_4$, $CuAl_2O_4$, $CuCr_2O_4$, $ZnCr_2O_4$, $MnCo_2O_4$ and $CuCo_2O_4$ received special attention because of their numerous applications in many fields such as: electronics, magnetism, catalysis, electro-catalysis, photo-catalysis, etc.

As examples these oxides, $CuCo_2O_4$ is a promising oxide in the field of storage energy; this is due to its properties: high chemical stability, electrochemical stability, ecofriendly, low cost, refractor, super-capacitor, moisture sensor and negative electrode in Li-ion battery materials [4–8]. We report our contribution by synthesizing for the first time $CuCo_2O_4$, by co-precipitation at low temperature from sulfates reagents and the physical and photoelectrochemical characterizations are reported. Moreover, we successfully tested the as-prepared material in the hydrogen production.

4.2 Experimental

The co-precipitation was used to prepare $CuCo_2O_4$ in presence of sulphate by mixture of 3.2394 g of $CoSO_4$ and 5.0872 g of $CuSO_4$; $5H_2O$ (Merck 99%) in 250 mL of distilled water; the pH was adjusted with KOH at 12.25. Then, product is treated in air oven at 140 °C for 3 days. The recovered product was filtered, washed and dried overnight at 60 °C.

The X-ray diffraction (XRD) pattern was recorded with a Panalytical with Cu Kα anticathode ($\lambda = 0.154128$ nm) in the range [2θ: 5–90°] to identify the synthesized phase. The optical absorption spectrum of $CuCo_2O_4$ was performed with a Jasco 650 UV–Vis spectrophotometer, $BaSO_4$ was used as reference.

The photo-electrochemical properties were studied in a standard cell with Pt auxiliary electrode; all potentials were reported against a saturated calomel electrode (SCE); the latter was separated from the working solution by a fritted bridge filled with the same solution and the potential was uncorrected for the junction potential. The electrode potential was piloted by a computer aided Versa STAT3 potentiostat and the interfacial capacitance was measured at 10 kHz.

The photocatalytic experiments were carried out in a Pyrex reactor (200 mL capacity) connected to a thermo-stated bath (50 °C); no temperature increase of the solution was observed during irradiation. 100 mg of $CuCo_2O_4$ are suspended in 200 mL of aqueous solution (Na_2SO_3 or $Na_2S_2O_3$, 10^{-3} M) under magnetic agitation (210 rpm); the pH is adjusted by NaOH (pH ~ 12).

4.3 Results and Discussion

The powder XRD pattern of $CuCo_2O_4$ (Fig. 4.1a) shows the peaks observed at $2\theta \sim 19°$, $31.1°$, $36°$, $38°$, $45°$, $59°$, $65°$ are typical of spinel cubic structure with space group of F d $\bar{3}$ m and the result perfectly agree with the standard values of the JCPDS N° 00-001-1155. No other impurity peaks were observed, which indicates that high purity of $CuCo_2O_4$. The crystallite size of $CuCo_2O_4$ (D = 32 nm) is calculated from the full width at half maximum (FWHM, $\beta = 0.26°$) of the strongest XRD peak (311):

$$D = 0.9\ \lambda\ (\beta \cos \theta) \tag{4.1}$$

To probe the optical properties, we have determined the forbidden band (E_g) of $CuCo_2O_4$, by using the Pankov formula:

$$(\alpha h\nu)^{2/m} = \text{Const} \times (h\nu - E_g + E_{ph}) \tag{4.2}$$

where α (cm^{-1}) is the optical absorption coefficient and $h\nu$ (eV) the energy of incident photon; the exponent m = 4 and 0.5 respectively for direct or indirect transitions; E_{ph} is the phonon energy for indirect transitions when the bottom of CB (Conduction Band) does not coincide with the top of VB (Valence Band).

The extrapolation of the straight line $(\alpha h\nu)^2$ to the abscissa axis ($h\nu = 0$) provides a direct transition at 1.38 eV (Fig. 4.1b), due to the degeneracy lift of d-orbital of Co^{3+} in octahedral site into an upper t_{2g} level and higher e_g level obeying the selection rules. The existence of the gap is in conformity with a semiconducting behavior. The other important factor in photo-catalysis is the semi conductivity which produces a space charge region near the interface where most if not all pairs the electron/hole (e^-/h^+) are separated.

The electrical conductivity (σ) against 1000/T, determined on sintered disc, obeys to the law $\sigma = \sigma_o \exp \{E_a/RT\}$ with a room temperature value ($\sigma_{300\ K}$) of 10^{-4} (Ω

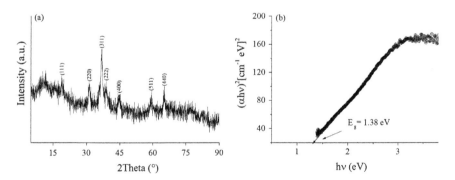

Fig. 4.1 **a** X-ray diffraction pattern and **b** direct optical transition of $CuCo_2O_4$ prepared by co-precipitation route from sulfate precursors

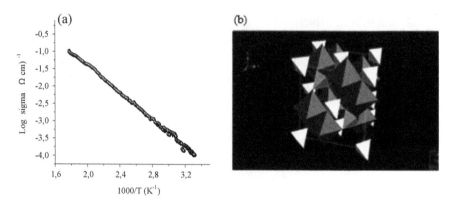

Fig. 4.2 **a** The logarithm of the electrical conductivity (σ) versus reciprocal absolute temperature and **b** projection of structure of $CuCo_2O_4$

cm)$^{-1}$ (Fig. 4.2a). The activation energy of 0.38 eV, deduced from the slope d logσ/dT (Fig. 4.2a) confirms the semi conductivity and where the acceptors are no longer ionized. The holes move in a narrow with deriving from Co^{3+}: 3d a mobility of 2.5×10^{-6} cm^2 V^{-1} s^{-1}, obtained from the conductivity of $CuCo_2O_4$ ($\sigma = e \mu h N_A$). Such low mobility comes from to the obstruction of O^{2-} ions to the electron's movement across octahedra/tetrahedra sharing respectively common edges/corners in the spinel structure (Fig. 4.2b). The conduction mechanism is governed by low polaron jump with phonon assisted conduction through mixed valences $Co^{3+/2+}$.

The intensity-potential J(E) profile in the dark is asymmetrical, indicating an irreversible and sluggish system (Fig. 4.3a); the peaks at ~−0.28 V along the negative potential, correspond to the irreversible reduction of dissolved oxygen ($O_2 + 2H_2O + 2e^- \rightarrow H_2O_2 + 2OH^-$) to peroxide H_2O_2, such potential is close to that cited in the literature (−0.38 V).

The drastic decrease of the dark current J_d below −0.6 V is due to H_2 evolution. On the other, the augmentation beyond 1 V is due to oxygen liberation ($H_2O + 2h^+$

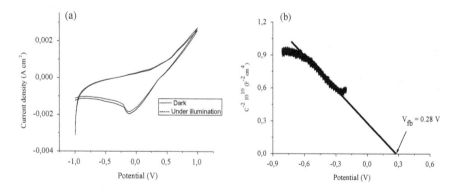

Fig. 4.3 **a** The current J versus (E) characteristic and **b** the Mott-Schottky of $CuCo_2O_4$ at pH ~ 12

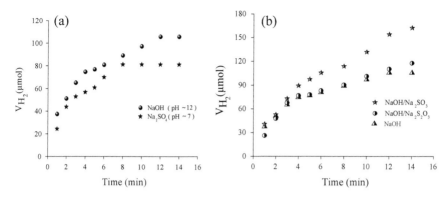

Fig. 4.4 a Volume of evolved oxygen versus illumination time at different pH ~ 7 and 12 and **b** by various electrons scavengers $S_2O_3^{2-}$ and SO_3^{2-} of $CuCo_2O_4$

$\rightarrow 0.5O_2 + 2H^+$) and this clearly indicates that the photo holes do not oxidize water because of the narrow band gap E_g of the spinel $CuCo_2O_4$ (1.38 eV).

The flat band potential (E_{fb}) and the carriers density ($N_{A/D}$) is obtained by measuring the capacitance (C) versus applied potential (E) at the interface $CuCo_2O_4$/NaOH:

$$C^{-2} = \pm(0.5 \times e\varepsilon\varepsilon_o N_A)^{-1}\{E - E_{fb}\} \tag{4.3}$$

The sign ± correspond respectively to p- or n-type behavior, e the electric charge, ε and ε_o the permittivity of material and vacuum respectively.

The negative slope is characteristic of p-type semi-conductivity; the potential E_{fb} (0.28 V) and the holes density N_A (8.7 × 1018 cm^{-3}) are simply deduced from the intercept with the abscissa axis and the slope of the straight line (C^{-2} – E) respectively (Fig. 4.3b).

The E_{fb} value comes from the small ionization energy of 3d metal and provides the position of the conduction band (-0.72 V = $E_{fb} + E_a/e - E_g/e$) [1], such value is more cathodic than the potential of the H_2 evolution (~-0.8 V), the latter was simply deduced from the J(E) profile. Accordingly, the potential valence band is equal to 0.66 V (=$E_{CB} + E_g$). The holes can react with sulfite SO_3^{2-} through a valence band process; such reaction favors the charges separation and improves the photo-activity. The spinel alone gives an appreciable H_2 volume because of the difference between CB (-072 V) and H_2 level which moderate the electrons transfer, (Fig. 4.4a) is found with an evolution rate of 340 μmol min^{-1} (g catalyst)$^{-1}$. The pH influences the photocatalysis; indeed, the bands of $CuCo_2O_4$ formed from 3d orbital are pH-independent while that of the H_2O/H_2 follows the nerstian variation of 0.06 V/pH. This property has been judiciously exploited to get an optimal band bending at the junction $CuCo_2O_4$/electrolyte equal to the difference $\{E_{H_2O/H_2} - E_{CB}\}$. The lower

---

[1] The activation energies of $CuCo_2O_4$ (0.38 eV) was measured from the electrical conductivity on sintered pellets.

the over potential, the rougher the active area and the co-precipitation yield a porous structure with an improved surface area. The crystallite size (32 nm) should give a minimal surface area of 37 m² g⁻² {=6 (ρD)⁻¹}, ρ is the experimental density (5.05 g cm⁻³). The photo-voltage for the H_2O splitting is smaller than the theoretical value ($\Delta G°_R/nF = 1.23$ V) of the total reaction and the splitting cannot occur because of O_2 and H_2 over-voltage. The H_2 evolution occurs concomitantly with the SO_3^{2-} oxidation; SO_3^{2-} was selected in the goal to contribute to the charges separation and to improve the photo-activity (Fig. 4.4b). The H_2 liberation requires 2 photons by H_2O molecule and the quantum efficiency (η) is given by:

$$\eta = 2 \times V \times N/(\Phi \times V_m \times t) \tag{4.4}$$

V and V_m are respectively the volume of evolved H_2 formed and the molar volume, N the Avogadro number and Φ the photons flux absorbed by $CuCo_2O_4$.

A light flux of 2.09×10^{19} photons s⁻¹ was measured, leading to η value of 1.1%. Initially, the hydrogen production occurs with a high evolution rate and slows down over time; the saturation is observed beyond 14 min; indicating the occupation of photoelectrochemical sites by H_2 molecules.

4.4 Conclusion

The spinel $CuCo_2O_4$ was prepared by co-precipitation in KOH medium at low temperature, not exceeding 140 °C using sulfates of copper and cobalt as precursors. The compound crystallizes in a cubic system with a lattice constant of 8.0590 Å. The X-ray diffraction (XRD) pattern confirms the single phase and the diffuse reflectance of the black product indicated a direct transition of 1.38 eV coming from the crystal field splitting. $CuCo_2O_4$ was characterized photo-electrochemically; the measurement done in alkaline electrolyte showed an electrochemical stability with a cathodic peak assigned to the oxygen reduction to peroxide. The capacitance-potential indicated p-type semi-conductivity. The spinel was successfully tested for the hydrogen formation under visible light illumination using SO_3^{2-} as sacrificial agent; a quantum yield of 1.1% was obtained.

References

1. R. Bagtache, K. Abdmeziem, K. Dib, M. Trari, Synthesis and photoelectrochemical characterization of KZn 2 (HPO 4) PO 4: application to rhodamine B photodegradation under solar light. Int. J. Environ. Sci. Technol. **16**(7), 3819–3828 (2019)
2. R. Bagtache, K. Abdmeziem, G. Rekhila, M. Trari, Synthesis and semiconducting properties of Na2MnPO4F. Application to degradation of Rhodamine B under UV-light. Mater. Sci. Semicond. Process. **51**, 1–7 (2016)

3. C. Belabed, B. Bellal, A. Tab, K. Dib, M. Trari, Optical and dielectric properties for the determination of gap states of the polymer semiconductor: application to photodegradation of organic pollutants. Optik **160**, 218–226 (2018)
4. M.H. Habibi, Z. Rezvani, Photocatalytic degradation of an azo textile dye (CI Reactive Red 195 (3BF)) in aqueous solution over copper cobaltite nanocomposite coated on glass by Doctor Blade method. Spectrochim. Acta Part A Mol. Biomol. Spectrosc. **147**, 173–177 (2015)
5. M.J. Iqbal, B. Ismail, C. Rentenberger, H. Ipser, Modification of the physical properties of semiconducting MgAl2O4 by doping with a binary mixture of Co and Zn ions. Mater. Res. Bull. **46**(12), 2271–2277 (2011)
6. R. Rahmatolahzadeh, M. Mousavi-Kamazani, S.A. Shobeiri, Facile co-precipitation-calcination synthesis of $CuCo_2O_4$ nanostructures using novel precursors for degradation of azo dyes. J. Inorg. Organomet. Polym Mater. **27**(1), 313–322 (2017)
7. S. Sun, Z. Wen, J. Jin, Y. Cui, Y. Lu, Synthesis of ordered mesoporous $CuCo_2O_4$ with different textures as anode material for lithium ion battery. Microporous Mesoporous Mater. **169**, 242–247 (2013)
8. S. Vijayakumar, S.-H. Lee, K.-S. Ryu, Hierarchical $CuCo_2O_4$ nanobelts as a supercapacitor electrode with high areal and specific capacitance. Electrochim. Acta **182**, 979–986 (2015)

Chapter 5
Synthesis and Electrochemical Characterization of Fe-Doped NiAl$_2$O$_4$ Oxides

Warda Tibermacine and Mahmoud Omari

Abstract A new spinel solid solution system of Ni$_{1-x}$Fe$_x$Al$_2$O$_4$ (0.0 ≤ x ≤ 0.5) was synthesized through sol-gel method. The effect of Fe doping on the electrocatalytic properties of nickel aluminate was investigated. The synthesized powders were characterized by means of X-ray diffraction, scanning electron microscopy and electrochemical measurements. From the preceding analysis, it can be shown that compounds show a single spinel phase in the temperature range of 650–1000 °C and the solubility of iron in the NiAl$_2$O$_4$ structure was limited to x ≤ 0.5. The electrochemical measurements indicate that the catalytic activity is strongly influenced by iron doping. The highest electrode performance is achieved with Ni$_{0.7}$Fe$_{0.3}$Al$_2$O$_4$ (i = 86.84 mA/cm^2) which is ~27 times greater than that of NiAl$_2$O$_4$ (i = 3.22 mA/cm^2) at E = +0.8 V. After one hundred cycles, the stability of the doped electrode with 30% of iron is much better than that of the undoped electrode. These results indicate clearly that Ni$_{0.7}$Fe$_{0.3}$Al$_2$O$_4$ electrode has promising potential for cost-effective potential generation.

Keywords Spinel oxide · Sol-gel · Oxygen evolution reaction

5.1 Introduction

Nanocrystalline metal aluminates possess important applications in various fields such as heterogeneous catalysis, pigments, sensors and ceramics [1–6]. Aluminate spinels have been used as catalysts in the decomposition of methane, steam reforming dehydration of saturated alcohols to olefins, dehydrogenation of alcohols, etc. These oxides have also been reported as good photocatalysts, e.g. for the degradation of methyl orange [7–9].

Nickel aluminate has been used in various catalytic applications and high temperature fuel cells, due to its high melting point, high activity and resistance to corrosion

---

W. Tibermacine (✉) · M. Omari
Laboratory of Molecular Chemistry and Environment, University of Biskra, 07000 Biskra, Algeria
e-mail: m2omari@yahoo.fr

© The Editor(s) (if applicable) and The Author(s), under exclusive license to Springer Nature Singapore Pte Ltd. 2021
A. Khellaf (ed.), *Advances in Renewable Hydrogen and Other Sustainable Energy Carriers*, Springer Proceedings in Energy,
https://doi.org/10.1007/978-981-15-6595-3_5

[10]. It has been proposed as a promising candidate for an anode in aluminum production and as an anode in an internal reforming solid oxide fuel cell (IR-SOFC) [11], in addition it has been used as good electrocatalysts for the oxidation of organic compounds and nitrous oxide [10], and also as inert anodes in aluminum electrolysis [12]. On the other hand, the spinel $NiAl_2O_4$ is photosensitive to visible light [13], and it presents an attractive property in photocatalysis [14].

This oxide can be properly modified by the partial substitution of atoms at A and/ or B sites which may affect strongly its physical property. Nickel aluminate ($NiAl_2O_4$) oxide doped on the A site with various metal ions such as Cu [15], Cd [16], Mg [17], Ce [18], were previously studied. A few years ago, it has been reported that the oxygen evolution reaction (OER) indicates that substitution of Ni by Fe in $Ni_{0.9}Fe_{0.1}Co_2O_4$ spinel increases the electrocatalytic activity of the resulting material significantly [19]. On the other hand, another work on mixed Fe-Ni oxide catalyst showed much higher activity toward oxygen evolution and methanol oxidation than either of the pure oxides with a peak in activity occurring near 10 mol% Fe [20]. In alkaline solution some substituted ferrites which the foreign element was added to the B site such as $CoFe_{1.7}Ni_{0.3}O_4$, $CoFe_{1.6}Mn_{0.4}O_4$ [21] and $NiFe_{2-x}Cr_xO_4$ ($0 \leq x \leq 1$) [22], manifest a reduced oxygen over-potential. Despite all these precedent works on spinel oxides based on iron and nickel, there is no report on the solid solution and activity of $Ni_{1-x}Fe_xAl_2O_4$ materials for the oxygen evolution reaction.

In the present study, we examine the effect of partial substitution of nickel by iron on structural, grain morphology, surface area, electrocatalytic activity and stability of $Ni_{1-x}Fe_xAl_2O_4$ ($0 \leq x \leq 0.5$) electrodes prepared by the sol-gel method.

5.2 Experimental Procedure

Different nickel aluminate powders were prepared according to the formula $Ni_{1-x}Fe_xAl_2O_4$ ($0 \leq x \leq 0.6$) by a sol-gel process. $Fe(NO_3)_3 \cdot 9H_2O$ (BIOCHEM), $Ni(NO_3)_2 \cdot 6H_2O$ (BIOCHEM), $Al(NO_3)_3 \cdot 9H_2O$ (FLUKA) and citric acid (JANSSEN CHIMICA) were used as salt precursors. The calculated amount of $Fe(NO_3)_3 \cdot 9H_2O$, $Ni(NO_3)_2 \cdot 6H_2O$, $Al(NO_3)_3 \cdot 9H_2O$ was dissolved in C_2H_5OH 99%. Then, the proper amount of citric acid dissolved in ethanol was added where the mole ratio of total metal ions and citric acid is 1:2:3. The resulting solution was slowly stirred, heated and concentrated by evaporating the ethanol at 80 °C until a gel was obtained. This last was then dried in an oven slowly upon increasing the temperature to 110 °C for 12 h in order to produce a solid amorphous citrate precursor. The resulting precursor was calcined in air for 6 h in the temperature range 400–1000 °C with a heating rate of 5 °C min^{-1}.

X-ray diffraction (XRD) was performed with a D8 Advance Brucker using a Cu K_α line at 0.1540 nm in 2Θ range of 10°–90° in steps of 0.010°. Linear sweep and cyclic voltammetry experiments for O_2 evolution were performed in potassium hydroxide 1 M using a Parstat 4000 potentiostat-galvanostat with oxide powders, Pt plate and Hg/HgO as working, auxiliary and reference electrodes, respectively.

Fig. 5.1 XRD patterns of the $Ni_{1-x}Fe_xAl_2O_4$ ($0 \leq x \leq 0.6$) samples calcined at (650–1000 °C) (s): spinel; (*): $FeAl_2O_4$

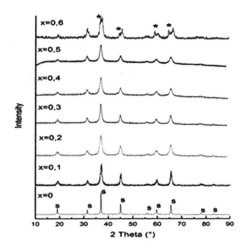

5.3 Results and Discussion

5.3.1 XRD Study

The XRD patterns of the $Ni_{1-x}Fe_xAl_2O_4$ (x = 0, 0.1, 0.2, 0.3, 0.4, 0.5 and 0.6) calcined at 650–1000 °C for 6 h in air are shown in Fig. 5.1. The results confirm that all samples with $0 \leq x \leq 0.5$ are consistent with the standard data for $NiAl_2O_4$ spinel phase (JCPDS card No. 10-0339), indicating the formation of a single phase with space group Fd3m and with no detectable secondary phase. These peaks can be indexed as (111), (220), (311), (400), (422), (511) and (440) planes, respectively. These planes are associated with the nickel aluminate spinel with a cubic structure. For x = 0.6, the main phase was also cubic spinel with another phase $FeAl_2O_4$ (JCPDS card 00-007-0068) indicating a Fe solubility limit of ~0.5 in the $Ni_{1-x}Fe_xAl_2O_4$.

5.3.2 SEM Analysis

The crystallite size (Dhkl) of the samples $0 \leq x \leq 0.5$ was calculated using Scherrer's equation (Eq. 1) [23].

$$D = k\lambda/\beta\cos\theta \quad (1)$$

where D is the average size of crystallites (nm), k Scherrer constant (≈ 0.9), λ wavelength of the incident radiation (nm), θ half of the angular position of the peak concerned and β full width at half maximum. The crystallite size was calculated for different samples in the range of 13.4–43 nm indicating that the spinel powders prepared by sol-gel method are composed of nanometric particles. The crystallite

Fig. 5.2 SEM micrographs of $Ni_{1-x}Fe_xAl_2O_4$. **a** x = 0.2; **b** x = 0.3 calcined at 800 °C

size decreases with increasing iron content. A similar result was also found for $Ni_{0.9}Fe_{0.1}Co_2O_4$ [19]. This is probably due to the incorporation of iron into the $NiAl_2O_4$ lattice, which leads to the formation of either cation or oxygen vacancies reducing the crystallite size.

The SEM micrographs of $Ni_{1-x}Fe_xAl_2O_4$ samples are shown in Fig. 5.2a,b. Particles have different shapes, sizes and the powders are agglomerated. The particle size is appreciated between 0.6 and 2.3 μm. The formation of agglomerate is probably due to the nature of the solvent used in the preparation of samples [24].

5.3.3 Electrochemical Properties

Polarization studies under potentiostatic conditions for $Ni_{1-x}Fe_xAl_2O_4$ ($0 \leq x \leq 0.5$) catalysts were carried out (Fig. 5.3). The highest electrode performance is achieved, for anodic current density with $Ni_{0.7}Fe_{0.3}Al_2O_4$. Oxygen evolution reaction shows an important jump for ($0.0 \leq x \leq 0.3$), where the current density of $Ni_{0.7}Fe_{0.3}Al_2O_4$ (i = 86.84 mA/cm^2) is ~27 times greater than that of $NiAl_2O_4$ (i = 3.22 mA/cm^2) at E = +0.8 V. As Fe content is increased beyond 30%, catalytic activity starts to decrease. The improvement of catalytic activity with incorporation of iron ($x \leq 0.3$) is probably due to the amelioration of the conductivity and the crystallinity of the doped material [25]. Burke et al. have also reported the role of iron in activating OER catalysts [26]. It has been shown that oxidized nickel (oxy) hydroxide is conductive and thus electrically connects the dispersed Fe sites to the conductive electrode. On the other hand, the electronic interaction between Ni and Fe likely further activates the Fe site for the OER. For higher iron content ($x > 0.3$), the trend reverses and the catalytic activity becomes lower. This can be probably due to that these catalysts have not the optimal M–O bond strength that is, this bond which constitutes an

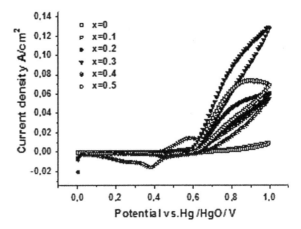

Fig. 5.3 I–E polarization curves of oxygen evolution reactions of $Ni_{1-x}Fe_xAl_2O_4$ electrodes in 1 M KOH

intermediate specie of the OER mechanism is too strong or too weak. On the other side, Friebel et al. [27] have reported that for Fe content of 25–50%, the presence of the phase-segregated FeOOH that lowered the overall geometric activity. This is due to that FeOOH is electrically insulating and thus OER is less active [26].

The chemical stability of $Ni_{1-x}Fe_xAl_2O_4$ oxides under oxygen evolution reaction conditions was tested. Figure 5.4 shows the cyclic voltammograms of the 1st and 100th cycle for $NiAl_2O_4$ and $Ni_{0.7}Fe_{0.3}Al_2O_4$ electrodes towards oxygen evolution reaction. In the two cases, after one hundred cycles, the curves show almost similar peaks with a slight decrease in current density for the undoped sample while it becomes higher for $Ni_{0.7}Fe_{0.3}Al_2O_4$ at E > 0.7 V.

Indeed, during 100 cycles, the current density decreases from 3.22 to 2.11 mA/cm² (~34%) for $NiFeAl_2O_4$ and increases from 86.84 to 91.75 mA/cm² for $Ni_{0.7}Fe_{0.3}Al_2O_4$ (~5.65%) at E = 0.8 V. This result indicates clearly that the stability of the electrode doped with 30% of iron is much better than the undoped

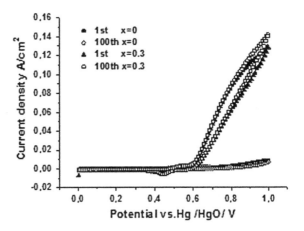

Fig. 5.4 Cyclic voltammograms of $NiAl_2O_4$ and $Ni_{0.7}Fe_{0.3}Al_2O_4$ for the 1st and 100th cycles in 1 M KOH media

electrode. This is probably due to the improved crystallinity of the doped sample compared to the undoped one. Furthermore, the current density of $Ni_{0.7}Fe_{0.3}Al_2O_4$ electrode after 100 cycles is 5.65% higher than that of the first cycle indicating that this electrode is more activated which explains its better catalytic activity.

5.4 Conclusion

The catalytic properties of $Ni_{1-x}Fe_xAl_2O_4$ ($0 \leq x \leq 0.5$) spinel oxides for the oxygen evolution reaction were explored. XRD analysis reveals that all samples crystallize with cubic structure at the temperature range (650–1000 °C). A single-phase spinel was observed for $Ni_{1-x}Fe_xAl_2O_4$ in the composition range $0 \leq x \leq 0.5$. The electrochemical behavior of these samples reveals that the doped electrode with 30% of iron exhibits a higher electroactivity. This indicates that $Ni_{0.7}Fe_{0.3}Al_2O_4$ oxide is among the investigated series the best electrocatalyst for oxygen evolution reaction. These results show clearly that there is a close relationship between the activity of the catalyst and the iron content. After one hundred cycles, the stability of the doped electrode with 30% of iron is much better than the undoped one. The catalytic activity is improved with ~5.65% for $Ni_{0.7}Fe_{0.3}Al_2O_4$ while it is reduced with ~34% for $NiAl_2O_4$. These results indicate clearly that the $Ni_{0.7}Fe_{0.3}Al_2O_4$ electrode can play an important role in water splitting process.

References

1. D. Dhak, P. Pramanik, Particle size comparison of soft-chemically prepared transition metal (Co, Ni, Cu, Zn) aluminate spinels. J. Am. Ceram. Soc. **89**(3), 1014–1021 (2006)
2. C. Feldmann, Preparation of Nanoscale Pigment Particles. Adv. Mater. **13**, 1301–1303 (2001)
3. C.R. Michel, CO and CO_2 gas sensing properties of mesoporous $CoAl_2O_4$. Sens. Actuat. B Chem. **147**(2), 635–641 (2010)
4. C.R. Michel, J. Rivera, A.H. Martinez, M. Santana-Aranda, Effect of the preparation method on the gas sensing properties of nanostructured $CoAl_2O_4$. J. Electrochem. Soc. **155**(10), 263–269 (2008)
5. P.M.T. Cavalcante, M. Dondi, G. Guarini, M. Raimondo, G. Baldi, Colour performance of ceramic nano-pigments. Dye. Pigment **80**(2), 226–232 (2009)
6. V. Lorenzelli, V.S. Escribano, R. Guidetti, Study of the surface properties of the spinels $NiAl_2O_4$ and $CoAl_2O_4$ in relation to those of transitional aluminas. J. Catal. **131**(1), 167–177 (1991)
7. L. Dussault, J.-C. Dupin, C. Guimon, M. Monthioux, N. Latorre, T. Ubieto, E. Romeo, C. Royo, A. Monzon, Development of Ni–Cu–Mg–Al catalysts for the synthesis of carbon nanofibers by catalytic decomposition of methane. J. Catal. **251**(1), 223–232 (2007)
8. N. Salhi, A. Boulahouache, C. Petit, A. Kiennemann, C. Rabia, Steam reforming of methane to syngas over $NiAl_2O_4$ spinel catalysts. Int. J. Hydrogen Energy **36**(17), 11433–11439 (2011)
9. L. Nakka, J.E. Molinari, I.E. Wachs, Surface and bulk aspects of mixed oxide catalytic nanoparticles: oxidation and dehydration of CH_3OH by polyoxometallates. J. Am. Chem. Soc. **131**(42), 15544–15554 (2009)

10. M.K. Nazemi, S. Sheibani, F. Rashchi, V.M. Gonzalez-DelaCruz, A. Caballero, Preparation of nanostructured nickel aluminate spinel powder from spent NiO/Al_2O_3 catalyst by mechanochemical synthesis. Adv. Powder Technol. **23**(6), 833–838 (2012)
11. R.D. Peelamedu, R. Roy, D.K. Agrawal, Microwave-induced reaction sintering of $NiAl_2O_4$. Mater. Lett. **55**(4), 234–240 (2002)
12. T.E. Jentoftsen, O. Lorentsen, E.W. Dewing, J.M. Haarbeg, J. Thonstad, Solubility of some transition metal oxides in cryolite-alumina melts: Part I. Solubility of FeO, $FeAl_2O_4$, NiO, and $NiAl_2O_4$. Metall. Mater. Trans. B **33**(6), 901–908 (2002)
13. I. Sebai, N. Salhi, G. Rekhila, M. Trari, Visible light induced H_2 evolution on the spinel $NiAl_2O_4$ prepared by nitrate route. Int. J. Hydrogen Energy **42**(43), 26652–26658 (2017)
14. Y. Yang, Y. Sun, Y. Jiang, Structure and photocatalytic property of perovskite and perovskite-related compounds. Mater. Chem. Phys. **96**(2–3), 234–239 (2006)
15. C.O. Arean, J.S.D. Vinuela, Structural study of copper-nickel aluminate ($Cu_xNi_{1-x}Al_2O_4$) spinels. J. Solid State Chem. **60**(1), 1–5 (1985)
16. C.O. Arean, M.L.R. Martinez, A.M. Arjona, Structural study of $Cd_xNi_{1-x}Al_2O_4$ spinels. Mater. Chem. Phys. **8**(5), 443–450 (1983)
17. F. Aupretre, C. Descorme, D. Duprez, D. Casanave, D. Uzio, Ethanol steam reforming over $Mg_xNi_{1-x}Al_2O_3$ spinel oxide-supported Rh catalysts. J. Catal. **233**(2), 464–477 (2005)
18. C.O. Augustin, K. Hema, L.J. Berchmans, R. Kalai Selvan, R. Saraswathi, Effect of Ce^{4+} substitution on the structural, electrical and dielectric properties of $NiAl_2O_4$ spinel. Phys. Status Solidi **202**(6), 1017–1024 (2005)
19. D. Chanda, J. Hnát, M. Paidar, K. Bouzek, Evolution of physicochemical and electrocatalytic properties of $NiCo_2O_4$ (AB_2O_4) spinel oxide with the effect of Fe substitution at the A site leading to efficient anodic O_2 evolution in an alkaline environment. Int. J. Hydrogen Energy **39**(11), 5713–5722 (2014)
20. J. Landon, E. Demeter, N. Inoglu, C. Keturakis, I.E. Wachs, R. Vasic, Spectroscopic characterization of mixed Fe-Ni oxide electrocatalysts for the oxygen evolution reaction in alkaline electrolytes. ACS Catal. **2**(8), 1793–1801 (2012)
21. M.I. Godinho, M.A. Catarino, M.I. Silva Pereira, M.H. Mendonça, F.M. Costa, Effect of the partial replacement of Fe by Ni and/or Mn on the electrocatalytic activity for oxygen evolution of the $CoFe_2O_4$ spinel oxide electrode. Electrochim. Acta **47**(27), 4307–4314 (2002)
22. R.N. Singh, J.P. Singh, B. Lal, M.J.K. Thomas, S. Bera: New $NiFe_{2-x}Cr_xO_4$ spinel films for O_2 evolution in alkaline solutions. Electrochim. Acta **51**(25), 5515–5523 (2006)
23. A. Becheri, M. Dürr, P.L. Nostro, P. Baglioni, Synthesis and characterization of zinc oxide nanoparticles: application to textiles as UV-absorbers. J. Nanoparticle Res. **10**(4), 679–689 (2008)
24. G.B. Jung, T.J. Huang, M.H. Huang, C.L. Chang, Preparation of samaria-doped ceria for solid-oxide fuel cell electrolyte by a modified sol-gel method. J. Mater. Sci. **36**(24), 5839–5844 (2001)
25. L.F. Li, A. Selloni, Mechanism and Activity of Water Oxidation on Selected Surfaces of Pure and Fe-Doped NiO_x. ACS Catal. **4**(4), 1148–1153 (2014)
26. C. Xiao, X. Lu, C. Zhao, Unusual synergistic effects upon incorporation of Fe and/or Ni into mesoporous Co_3O_4 for enhanced oxygen evolution. Chem. Commun. **50**(70), 10122–10125 (2014)
27. D. Friebel, M.W. Louie, M. Bajdich, K.E. Sanwald, Y. Cai, A.M. Wise, M.-J. Cheng, D. Sokaras, T.-C. Weng, R. Alonso-Mori, R.C. Davis, J.R. Bargar, J.K. Nørskov, A. Nilsson, A.T. Bell, Identification of Highly Active Fe Sites in (Ni, Fe)OOH for Electrocatalytic Water Splitting. J. Am. Chem. Soc. **137**(3), 1305–1313 (2015)

Chapter 6
Review on the Effect of Compensation Ions on Zeolite's Hydrogen Adsorption

Redouane Melouki and Youcef Boucheffa

Abstract The development of a safe and efficient storage method is a key to achieving hydrogen economy. The zeolites are crystalline and porous aluminosilicate. They are potential candidates for hydrogen storage. These materials are well known for their electrostatic fields due to the differences in electronegativity between the atoms of aluminum, silicon, oxygen and compensation cations. In addition to temperature and pressure, the adsorption of hydrogen on zeolites depends also on the crystal lattice topology and compensating cations. Several studies have illustrated the effect of exchange on the adsorption capacity of these materials. These properties promote the zeolite's surface energy change leading to an increase in the hydrogen's uptake capacity. Charge compensation ions in the zeolite's framework are considered as adsorption centers and the structure's oxygen bridges as minor adsorption sites. Indeed, the gained mass per unit of area reveals the effect of the ion's exchange in terms of cation's size and charge.

Keywords Hydrogen · Adsorption · Zeolite · Ions exchange

6.1 Introduction

The development of human civilization is closely linked to mastering energy. Originally, man had fire and his own physical strength as a source of energy. Later on, he added energies that today are known as renewable energies (wind, water). When the industrial revolution came on, it has completely disrupted this order and a new era of energy revolution sees the day.

R. Melouki (✉)
Ecole Militaire Polytechnique, Algiers, Algeria
e-mail: melouki.redouane@gmail.com

Y. Boucheffa
Laboratoire d'Etude Physico-Chimique des Matériaux et Application à l'Environnement,
Université des Sciences et de la Technologie Houari Boumediene, El-Alia, Bab-Ezzouar, Algiers, Algeria

© The Editor(s) (if applicable) and The Author(s), under exclusive license to Springer Nature Singapore Pte Ltd. 2021
A. Khellaf (ed.), *Advances in Renewable Hydrogen and Other Sustainable Energy Carriers*, Springer Proceedings in Energy,
https://doi.org/10.1007/978-981-15-6595-3_6

The energy consumption is increasing by about 1% in the developed countries and by 5% in the developing countries [1]. The world global consummation has doubled in 40 years, which is unprecedented in history: the world has consumed 5.52 billion tons of oil equivalent (Toe) in 1971 and 13.76 billion tons in 2016 [2]. The established scenarios by the International Energy Agency (IEA) for the upcoming decades are predicting an increase in the global energy demand of about 50% in 2040 [3].

Nowadays, the major problem in the twenty-first century, on the one hand, is the fossil energy sources depletion and, on the other hand, the environmental impact of high hydrocarbon consumption. This situation has placed renewable energies at the heart of the countries' challenges aiming to ensure their energy income while maintaining sustainable development.

The total investment in renewable energies was amounted to 279.8 billion dollars in 2017. Research and development in this field has a significant part to play [4].

As a matter of fact, the abundant hydrogen seems to be one of the most interesting and efficient energy carriers and its associated technologies are likely to significantly grow in the near future. It can be used either as a heat source in gas turbines or as a fuel in fuel cell devices.

The use of this energy carrier remains dependent on the development of safe and efficient storage methods. Currently, hydrogen is stored either in a gaseous form at high pressure or in a liquid form at low temperature. However, these storage forms are not totally convenient in terms of gravimetric and volumetric density. Different types of storage technology are studied, such as hydride metal storage and porous materials storage.

Although fuel cell engines are more efficient, the gap between the two fuels is very large. As good illustration, under the same conditions, a car tank filled with hydrogen provides 1/10 of the gasoline's storage. That fact leads to develop safe and cost-effective storage facilities [5].

Several laboratories are interested in adsorption storage technique in order to store the same amount of hydrogen; the use of porous materials reduces the required pressure to about 40%. Therefore, to store 4.1 kg of H_2 are stored in a 100 L' tank at 25 °C and 750 bar. Meanwhile, at a cryogenic temperature of -196 °C, only 150 bars are sufficient. Nevertheless, at the same temperature and using the same volume filled with activated carbon straws AX-21, the required pressure for storage was only 90 bar [6].

Over the past thirty years, a large number of studies was dedicated to the hydrogen storage possibilities in porous materials and has been remarkably increasing [7]. As a consequence, various porous materials have been tested: carbon nanotubes [8], MOFs (Metal Organic Frameworks) [9], graphene [10], activated carbons [11] and zeolites [12].

6.2 Hydrogen Storage in Zeolites

Zeolites are crystalline and porous aluminosilicates. They are potential candidates for hydrogen storage because of their low cost, high thermal stability and different types of porous structures [13]. The adsorption storage in zeolites is based on the fact that a gas can be adsorbed to the surface of a solid where it is retained by Van der Waals forces [14]. These materials are well known for their electrostatic fields due to the differences in electronegativity between the atoms of aluminum, silicon, oxygen and compensation cations.

The adsorption of hydrogen on zeolites depends on: the temperature, the pressure, the porous volume, the surface area, the pore size and the compensating cations [15]. Although the availability of large porous volume increases the adsorbed quantities, the surface's thermodynamic has significant contribution (or impact). Indeed, two materials with equivalent textural proprieties, under the same conditions of temperature and pressure, can adsorb unequal quantities of hydrogen since the surface's affinities are different [16]. The cations of the structure are considered to be the hydrogen adsorption sites [17], although Kazansky et al. have suggested that oxygen zeolite framework also contributes to the hydrogen's interactions [18].

Vitillo et al. have established a theoretical calculation of the maximum quantity that could be stored in several aluminosilicates zeotype's structures [19]. These quantities are intrinsically limited by geometric constraints and can reach a maximum value of 2.86 %wt for FAU and RHO zeolites.

In order to promote the adsorption phenomenon, the surface-gas interaction's energy must be high enough so it can retain the gas. However, a very high energy level makes the desorption process difficult to achieve. The zeolites have large specific surfaces, but surface-hydrogen interactions remain too low to retaining the desired amount of gas.

Different methods have been developed aiming the zeolite's modification in order to increase their storage capacity. Chung has reported that the amount of the adsorbed hydrogen on the mordenite type zeolites (MOR) has increased with increasing the Si/Al molar ratio [20]. Also, the ion exchange property of these structures gives them the ability to modulate both textural and thermodynamic of surface [21]. The nature of the extra-structure cations has a great influence on the adsorption capacities.

In order to optimize the storage process's cost, different conditions than standard temperature and pressure values are preferable. The adsorption is an exothermic phenomenon; the process is favored by a decrease in temperature. All room temperature adsorption tests revealed very low mass gains. Chung examined the storage capacity of five zeolites at 30 °C and 50 bar [20]. At this temperature, the largest amount was 0.4 %wt and it was recorded on a Y type zeolite which is very low to be considered as a good storage solution.

6.3 Adsorption at Low Pressure

The adsorption capacities of zeolites at low pressure remain small and insufficient for mobile applications. To our knowledge, the greatest adsorbed amount on these materials was reported by Musyoka et al. on a commercial X zeolite type and was around 1.6 %wt at 1 bar and −196 °C [22].

Testing the same zeolite type, at −196 °C and 1 bar, Li et al. have reported that the hydrogen adsorption capacity is highly dependent on the cation's structure radius and density [17]. The storage capacity decreases with the ion exchange capacity (Li+ > Na+ > K+ giving 1.50 %wt, 1.46 %wt and 1.33 %wt respectively). The interaction energies between cations and H2 follow the predicted order, on the ionic size basis. The oxygen bridges of the zeolite structure present a minor adsorption sites. Compared to natural zeolites, these quantities remain higher. ErdoğanAlver et al. have performed hydrogen adsorption tests at −196 °C and 1 bar on three natural zeolites of the clinoptilolite (CLN), mordenite (MOR) and chabazite (CHB) types, exchanged at the various cations [23]. The largest uptake was equal to 1.08 %wt recorded on Na-CHB (458 m^2/g), followed by 0.55 %wt on Li-CLN (41 m^2/g) and 0.34 %wt on Li-MOR (90 m^2/g). These results show the complexity of the competition phenomenon due to the influence of the cation's nature and size as well as the specific surface area.

6.4 Adsorption at High Pressure

In adsorption processes, the pressure's increase effect is well known. Higher pressure leads to more molecule's accumulation on the adsorbent surface. Several studies have reported that the increase in pressure enhances the adsorbed stored quantities. In the work of Rahmati et al. various simulations have been carried out considering several zeolite structures [24]. Under optimal conditions of −196 °C and 600 bar, hydrogen adsorption in RWY zeolite is 6.93 %wt. However, the experiments carried out on these materials show very low values compared to the simulation.

Du et al. have reported that the largest adsorbed mass of hydrogen was about 2.55 %wt, on NaX zeolite, at −196 °C and 70 bar [25]. However, with the increase in temperature to −78 °C and then 20 °C this quantity was reduced to 0.98 %wt and then to 0.4 %wt respectively. These results show that storage is strongly influenced by temperature. It remains to determine the optimal pressure of the process in order to avoid any energy loss. In this work, the maximum adsorbed quantity was equal to 2.55 %wt and has been reached at 40 bar, then the quantity remains constant during the pressure increases.

6.5 Conclusion

Hydrogen adsorption in zeolites is a complex function of zeolite temperature, pressure and structure. Storage capacities remain very low at normal temperature and pressure. When working under extreme physical conditions (low temperature and high pressure) the adsorbed quantities increase, but remain insufficient for an embedded storage. The zeolites' modification via the ion exchange influences the adsorption capacities. The nature of the exchanged cations, their size and density considerably influence the interaction energies between the hydrogen and the corresponding material's surface. Considering these class of porous material, all work show that their hydrogen uptake remains insufficient to be applied as a storage solution.

References

1. A.B. Stambouli, H. Koinuma, A primary study on a long-term vision and strategy for the realisation and the development of the Sahara Solar Breeder project in Algeria. Renew. Sustain. Energy Rev. **16**(1), 591–598 (2012)
2. World energy balances, *International Energy Agency* (OECD/IEA, Paris, 2018), p. 2018
3. World Energy Outlook, *International Energy Agency* (OECD/IEA, Paris, 2015), p. 2015
4. Global Trends in Renewable Energy Investment, Frankfurt School of Finance & Management gGmbH (2018)
5. D. Mori, K. Hirose, Recent challenges of hydrogen storage technologies for fuel cell vehicles. Int. J. Hydrogen Energy **34**(10), 4569–4574 (2009)
6. R. Paggiaro, P. Bénard, W. Polifke, Cryo-adsorptive hydrogen storage on activated carbon. I: Thermodynamic analysis of adsorption vessels and comparison with liquid and compressed gas hydrogen storage. Int. J. Hydrogen Energy **35**(2), 638–647 (2010)
7. D.P. Broom, Potential Storage Materials, in *Hydrogen Storage Materials*, ed. by D.P. Broom (Springer, London, 2011), pp. 19–59
8. A. Lueking, R.T. Yang, Hydrogen storage in carbon nanotubes: Residual metal content and pretreatment temperature. AIChE J. **49**(6), 1556–1568 (2003)
9. D. Lupu, O. Ardelean, G. Blanita, G. Borodi, M.D. Lazar, A.R. Biris, C. Ioan, M. Mihet, I. Misan, G. Popeneciu, Synthesis and hydrogen adsorption properties of a new iron based porous metal-organic framework. Int. J. Hydrogen Energy **36**(5), 3586–3592 (2011)
10. V. Tozzini, V. Pellegrini, Prospects for hydrogen storage in graphene. Phys. Chem. Chem. Phys. **15**(1), 80–89 (2013)
11. R. Melouki, P.L. Llewellyn, S. Tazibet, Y. Boucheffa, Hydrogen adsorption on activated carbons prepared from olive waste: effect of activation conditions on uptakes and adsorption energies. J. Porous Mater. **24**(1), 1–11 (2016)
12. H.W. Langmi, D. Book, A. Walton, S.R. Johnson, M.M. Al-Mamouri, J.D. Speight, P.P. Edwards, I.R. Harris, P.A. Anderson, Hydrogen storage in ion-exchanged zeolites. J. Alloys Compd **404–406**, 637–642 (2005)
13. J. Shi, J. Li, E. Wu, Adsorption of hydrogen and deuterium in MnO_2 modified NaX zeolites. Microporous Mesoporous Mater. **152**, 219–223 (2012)
14. J. Weitkamp, M. Fritz, S. Ernst, Zeolites as media for hydrogen storage. Int. J. Hydrogen Energy **20**(12), 967–970 (1995)
15. S.H. Jhung, J.W. Yoon, H.K. Kim, J.S. Chang, Low temperature adsorption of hydrogen on nanoporous materials. Bull. Korean Chem. Soc. **26**(7), 1075–1078 (2005)

16. A.W. van den Berg, C.O. Arean, Materials for hydrogen storage: current research trends and perspectives. Chem. Commun. (Camb.) **6**(6), 668–681 (2008)
17. Y. Li, R.T. Yang, Hydrogen storage in low silica type X zeolites. J. Phys. Chem. B **110**(34), 17175–17181 (2006)
18. V.B. Kazansky, F.C. Jentoft, H.G. Karge, First observation of vibration–rotation drift spectra of para- and ortho-hydrogen adsorbed at 77 K on LiX, NaX and CsX zeolites. J. Chem. Soc., Faraday Trans. **94**(9), 1347–1351 (1998)
19. J.G. Vitillo, G. Ricchiardi, G. Spoto, A. Zecchina, Theoretical maximal storage of hydrogen in zeolitic frameworks. Phys. Chem. Chem. Phys. **7**(23), 3948–3954 (2005)
20. K.H. Chung, High-pressure hydrogen storage on microporous zeolites with varying pore properties. Energy **35**(5), 2235–2241 (2010)
21. E. Garrone, B. Bonelli, C. Otero Areán, Enthalpy–entropy correlation for hydrogen adsorption on zeolites. Chem. Phys. Lett. **456**(1–3), 68–70 (2008)
22. N.M. Musyoka, J. Ren, H.W. Langmi, B.C. North, M. Mathe, A comparison of hydrogen storage capacity of commercial and fly ash-derived zeolite X together with their respective templated carbon derivatives. Int. J. Hydrogen Energy **40**(37), 12705–12712 (2015)
23. B. Erdoğan Alver, M. Sakızcı, Hydrogen (H_2) adsorption on natural and cation-exchanged clinoptilolite, mordenite and chabazite. Int. J. Hydrogen Energy **44**(13), 6748–6755 (2019)
24. M. Rahmati, H. Modarress, The effects of structural parameters of zeolite on the adsorption of hydrogen: a molecular simulation study. Mol. Simul. **38**(13), 1038–1047 (2012)
25. X.M. Du, Wu, E.d.: Physisorption of hydrogen in A, X and ZSM-5 types of zeolites at moderately high pressures. Chin. J. Chem. Phys. **19**(5):457–462 (2006)

Chapter 7
Catalytic Reforming of Methane Over Ni–La$_2$O$_3$ and Ni–CeO$_2$ Catalysts Prepared by Sol-Gel Method

Nora Yahi, **Kahina Kouachi**, **Hanane Akram**, **and Inmaculada Rodríguez-Ramos**

Abstract Lantanium and Cerium supported Nickel catalysts with Ni-loading close to 15 %wt were synthesized using sol-gel methods in order to design efficient catalysts for the dry reforming of methane to produce syngas (H$_2$ + CO). The catalytic test was performed after calcining the as-prepared samples at 700 °C and subsequent in situ reduction was performed under hydrogen flow at 600 °C. The resulting catalysts were characterized by X-ray diffraction (XRD), Temperature Programmed Reduction (TPR), transmission electron microscopy (TEM) and N$_2$ adsorption-desorption isotherm measurements. The investigation of the catalytic performances of Ni–CeO$_2$ and Ni–La$_2$O$_3$ catalysts prepared by sol gel method (SG), for a duration of 12 h, under a reaction CH$_4$/CO$_2$ shows that, for the same synthesis method, the efficiency varies with according to the support nature. Indeed, conversions and yields are higher in the presence of the lanthanum than with the cerium support (XCO$_2$ = 18% and YCo = 15% for Ni–La compared of XCO$_2$ = 8% and YCo = 6% for Ni–Ce. Comparing the catalysts stability, we can notice that it is 100% (no deactivation) in the presence of Ni–La$_2$O$_3$ catalyst compared to Ni–CeO$_2$, which showed a high deactivation (78% after 12 h of reaction). This difference in stability is probably related to the perovskite structure and to the strong interactions between the active phase and the support, reinforced by the basicity of lanthanum support which inhibits carbon deposition during the CH$_4$/CO$_2$ reaction.

Keywords Sol-gel · Methane · Dry reforming

N. Yahi (✉)
Faculty of Sciences, University of Saad Dahlab Blida 1, Blida, Algeria
e-mail: norayahi@yahoo.fr

K. Kouachi
Faculty of Science of the Nature and Life, University of Bejaia, Béjaïa, Algeria

H. Akram
Faculty of Sciences and Techniques of Tangier, UAE/L01FST, Laboratory LGCVR, Abdelmalek Essaadi University, B.P. 416, Tangier, Morocco

I. Rodríguez-Ramos
Instituto de Catálisis Y Petroleoquímica, CSIC, C/Marie Curie, 2 L10, Cantoblanco, 28049 Madrid, Spain

© The Editor(s) (if applicable) and The Author(s), under exclusive license to Springer Nature Singapore Pte Ltd. 2021
A. Khellaf (ed.), *Advances in Renewable Hydrogen and Other Sustainable Energy Carriers*, Springer Proceedings in Energy,
https://doi.org/10.1007/978-981-15-6595-3_7

7.1 Introduction

Carbon dioxide reforming of methane (CH_4/CO_2) has aroused considerable attentions in recent years thanks to its enormous advantages, not only on the environment protection by the reducing and recycling of greenhouse gases (CH_4 and CO_2) to valuable gas but as an efficient and sustainable technique for the valorization of natural gas.

Dry reforming of methane DRM allows obtaining syngas with a H_2/CO ratio near to one, which is suitable for the Fischer-Tropsch process of long chain hydrocarbons [1, 2]. Commonly, the used catalysts are based on noble metals (Pd, Rh, Ru, Pt and Ir) and transition metals (Ni, Co and Fe) [3, 4] deposited on various supports, such as alumina, silica and rare earths oxides [5]. Nevertheless, the main problem with the catalysts used for dry reforming of methane is their quick deactivation due to the strong carbon deposition. However, noble metal based catalysts have been found to be less sensitive for carbon deposition than other experienced catalysts [6], but their high costs represents a major drawback for their widespread use.

For the DRM, the most used catalysts are usually based on Ni thanks to their high activity and efficiency and low cost [7]. However, the problem with these catalysts is their easy deactivation because of the coke deposition and metal sintering which causes plugging of the reactor. The stability of this kind of catalysts is, generally, affected by the Ni particles size. In fact, catalysts with nanosized Ni particles show lower carbon deposition and, of course, better stability [8].

In our work, we prepared Ni/La_2O_3 and Ni/CeO_2 catalysts by sol-gel method [9]. We investigated the structures of Ni–La_2O_3 and Ni–CeO_2 catalysts by XRD, TEM, Brunauer-Emmett-Teller theory (BET) and TPR in order to correlate these structural properties with their catalytic activity in the dry reforming of methane process.

7.2 Experimental Procedure/Details

7.2.1 Catalyst Preparation

Two Ni-based catalysts were prepared using sol-gel method [9]; they are labeled Ni–CeO_2, Ni–La_2O_3 respectively. All the reagents were of analytical grade and purity. A mixture of adequate amounts of $Ni(NO_3)_2$ $6H_2O$ and $Ce(NO_3)_3$ $6H_2O$/La $(NO_3)_3$ $6H_2O$ was dissolved in 40 ml of distilled water. Subsequently, 20 ml of saturated solution of stearic acid $C_{18}H_{36}O_2$ (98%, Gpr Rectapur) was added to this mixture. The resulting solution was stirred at 80 °C for 5 h. The gel prepared was dried at room temperature for 6 h then at 110 °C for 24 h and finally calcined during 8 h in air at 700 °C using a ramp of 5 °C/min.

7.2.2 Characterization Methods

The BET surface areas of the catalysts were measured with N_2 adsorption at N_2 liquid temperature by using a micromeritics ASAP 2020 instrument. Prior to each measurement, the samples were degassed at 403 K (130 °C) in vacuum for 1 h.

Temperature-programmed reduction (TPR) was conducted with a BROOKS 5878 instrument in Materials Research Laboratory. About 200 mg of the samples were loaded in a quartz reactor and heated from room temperature to 750 °C at heating rate of 8 °C/min in stream of (5%) H_2 in Ar with total flow of 100 ml/min, the hydrogen consumption is determined as function of the temperature.

Power X-ray diffraction (XRD) patterns were recorded on a Xpert Pro X-ray diffractometer with Cu/Kα radiation ($\lambda = 0.1544$ nm) operating at 45 kV and 40 mA, ranging from 4 to 90°. The crystallite size was determined from the Scherrer-Warren equation [10].

7.2.3 Catalytic Test

The catalytic activity of the prepared samples was performed in a fixed-bed tubular reactor (with an inner diameter of 9.5 mm) that was heated in an electric furnace equipped with a programmable temperature controller. A fresh 100 mg of catalysts, with a particle size between 150 and 250 μm, were diluted with silicon carbide (SiC) to obtain 50 mm bed height and packed in the middle of the reactor. The temperature was monitored by a K-type thermocouple placed in the center of the catalyst bed. Before starting the reaction, the catalyst was reduced in situ at 600 °C (maximum operation temperature) for 2 h with a mixture of 25 vol. % H_2 in helium at a flow of 100 ml/min. After reduction, helium gas was used during 30 min to sweep the H_2 from the reactor. The feed steam gas mixture consisted on CH_4, CO_2 and He to balance 1:1:8. The total flow rate was 100 ml/min. The reactant gases were measured-controlled by mass flow meters (Brook). The catalytic activity was measured at 600 °C. Gas analyses of both reactants and products were carried out by on line GC (Varian 3400) equipped with a TCD detector. Porapaq Q and Chromosorb 102 columns were used to separate the sample gas (CH_4, H_2, CO and CO_2). Blank experiments were done to verify the absence of catalytic activity under the conditions used in this study, either with the reactor empty or filled with silicon carbide. To discard the presence of diffusion problems, the experiments performed were replicated with other particles sizes and significantly changing both the flow rate and the catalyst amount while keeping the mass/flow rate ratio constant. The results obtained suggest the absence of both internal and external mass transfer effects. The carbon balance was close to 100% in all cases. The conversions (X), the yields (Y) and the H_2/CO ratio are calculated as follows:

$$XCH_4(\%) = (\text{moles of } CH_4 \text{ converted}) * \frac{100}{\text{moles of } CH_4 \text{ fed}} \qquad (1)$$

$$XCO_2(\%) = (\text{moles of } CO_2 \text{ converted}) * \frac{100}{\text{moles of } CO_2 \text{ fed}} \quad (2)$$

$$YH_2(\%) = (\text{moles de } H_2 \text{ produced}) * \frac{100}{2 \text{moles of } CH_4 \text{ fed}} \quad (3)$$

$$YCO(\%) = (\text{moles de CO produced}) * \frac{100}{\text{moles of } CH_4 \text{ Fed} + \text{moles of } CO_2 \text{ fed}} \quad (4)$$

$$\frac{H_2}{CO} \text{Ratio} = (\text{moles de } H_2 \text{ produced}) * \frac{100}{\text{moles CO produced}} \quad (5)$$

To investigate the stability, of the Ni–CeO$_2$ and Ni–La$_2$O$_3$ catalysts, the deactivation after 12 h of reaction was calculated as follows:

$$D(\%) = \left(CH_4^0 - CH_4^{12}\right)/CH_4^0 \quad (6)$$

7.3 Results and Discussion

The BET surface areas (SBET) of the catalysts are shown in Table 7.1. The Ni–CeO$_2$ catalyst presents higher specific surface area (45 m^2/g) than that of Ni–La$_2$O$_3$ SG catalyst (11 m^2/g). For Ni-La$_2$O$_3$ catalyst, (SBET) is much higher to those previously reported for perovskite-type oxides air-calcined beyond this temperature [11, 12]. However with Ni–CeO$_2$ catalyst, the obtained (SBET) is very near of those obtained by S. Pangranich [13] (65 m^2/g) with Ni–CeO$_2$ prepared by impregnation.

The TPR profiles of H$_2$ consumption for Ni–CeO$_2$ and Ni–La$_2$O$_3$ catalysts are shown in Fig. 7.1. The TPR profiles of Ni–CeO$_2$ showed a main peak located at 360 °C for Ni–Ce SG. This peak can be assigned to the reduction of bulk NiO phase to Ni^0 [14, 15]. Moreover, small peaks are observed at 244 and 514 °C. The first one can be attributed to the reduction of non-stoichiometric species Ni^{+3} [16, 17] the presence of free NiO particles [18] or hydrogen spillover effect [19], while the second one, much broader, can be assigned to the reduction of NiO interacting with support [20] or CeO$_2$ reduction [21]. However the TPR profile of Ni–La$_2$O$_3$ catalyst shows two peaks at 364 and 521 °C. The first one is due to the first LaNiO$_3$ reduction step. The resulting reduced compound, represented by the general formula LaNiO$_{2.5}$,

Table 7.1 BET specific surface area of the prepared catalysts and their crystallite size obtained from the XRD patterns before TPR

Catalyst	S$_{BET}$ (m^2 g^{-1})	Particle size (DRX) (nm) Deby Scherer
Ni–La$_2$O$_3$	11	La$_2$O$_3$ (35), LaNiO$_3$(18)

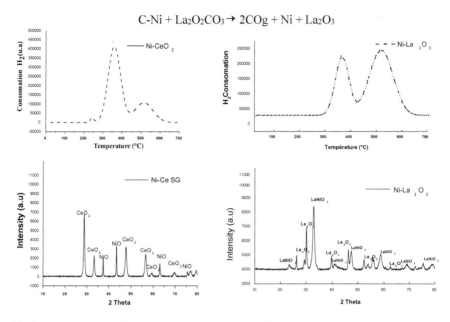

Fig. 7.1 XRD patterns (bottom) and TPR profiles (top) of the studied catalysts

is an oxygen deficient structure developed during the nickel reduction. This reaction is presumably accompanied by the reduction of nickel oxide NiO [22].

$$LaNiO_3 + 1/2H_2 \rightarrow LaNiO_{2.5} \text{ (1st step)} \qquad (7)$$

The second peak observed at 521 °C can be assigned to the second $LaNiO_3$ reduction step (Ni^{+2} to Ni^0) may be demonstrates the high interactions between the nickel and catalyst support.

$$LaNiO_{2.5} + H_2 \rightarrow 1/2La_2O_3 + Ni + H_2O \text{ (2nd step)} \qquad (8)$$

The X-ray diffraction patterns profiles of Ni–La_2O_3 and Ni–CeO_2 catalysts are depicted in Fig. 7.1 and Table 7.1. Ni–La_2O_3 catalyst prepared by SG technique exhibits, at the same time, perovskite structure rhombohedra (R) (00-034-1028, JCPDS), La_2O_3 hexagonal (01-083-1344, JCPDS). However, no NiO phase was detected by this technique because of its undersized particles likely less than 5 nm [23]. However, CeO_2 cubic system (00-034-0394, JCPDS) (Fm-3m, a = 5.41 Å) and NiONiO monoclinic (C2/m) (03-065-6920, JCPDS) are observed with Ni–CeO_2 [9].

The size of the different phases detected in the as-studied Ni–La_2O_3 catalysts is obtained by Scherer equation and are gathered in Table 7.1. In fact, crystallite size is a key factor for a high reforming activity as well as the coke deposition degree [24, 25]. For the Ni–La catalyst prepared by Sol-Gel technique, small particles $LaNiO_3$

Table 7.2 Conversion of CH_4, CO_2, yield of CO and H_2 ($CH_4:CO_2:He$) = (1:1:8), reduction temperature (Tr = 600 °C), reaction time (tr = 12 h) obtained from Ni–CeO_2 and Ni–La_2O_3 catalysts

Catalysts	XCH_4 (%)	XCO_2 (%)	YCO (%)	YH_2 (%)	CO/H_2 (%)	D
Ni–La SG	12	18	15	06	2.5	–
Ni–Ce SG	05	08	06	03	2.0	78

(18 nm) and relatively big La_2O_3 crystallite (35 nm) are presented simultaneously. However a particles size of CeO_2 and NiO calculated with Ni-CeO_2 catalyst is 11 nm [9].

The studied catalysts were tested for the dry reforming of methane reaction (DRM), with a feed of ($CH_4/CO_2/He$ = 1/1/8, volume ratio) at 600 °C during 12 h. The conversion of CH_4, CO_2 and CO, H_2 yields are presented in Table 7.2. The results, indicate that the CO_2 conversion is higher than that of CH_4 for all the studied catalysts, suggesting that, besides methane dry reforming, reverse water–gas shift reaction is also occurring at low temperature [26]. The reaction 9: RWGS before Table 7.2.

$$(RWGS) \; (CO_2 + H_2 \leftrightarrow CO + H_2O) \tag{9}$$

The catalytic performances of Ni–CeO_2 and Ni–La_2O_3 catalysts prepared by SG, for a duration of 12 h, under a reaction CH_4/CO_2 shows that, conversions and yields are higher in the presence of the lanthanum than with the cerium support (XCO_2 = 18% and YCO = 15% for Ni–La_2O_3 compared of XCO_2 = 8% and YCo = 6% for Ni-CeO_2 [9]. Ni–La_2O_3 is more stable than Ni–CeO_2, This difference in stability is probably related to the perovskite structure and to the strong interactions between the active phase and the support in the perovskite $LaNiO_3$ [27], reinforced by the basicity of lanthanum support which inhibits carbon deposition during the CH_4/CO_2 reaction [28].

The tested catalysts were characterized by XRD, after 12 h of reaction in order to show their structures and its possible relationship with their deactivation cause. In fact, the deactivation of Ni–CeO_2 can result from carbone deposition confirmed by DRX analysis (Fig. 7.2), However, the stability of Ni-La_2O_3 can be due to formation of $La_2O_2CO_3$ (Fig. 7.2).

Indeed, it has been confirmed in literature that La_2O_3 reacts in situ with CO_2 to give $La_2O_2CO_3$, which positively contributes to the stability of the catalyst by preventing the carbon deposition that causes the deactivation of the active sites according to the reaction:

$$C-Ni + La_2O_2CO_3 \circledR 2COg + Ni + La_2O_3 \tag{10}$$

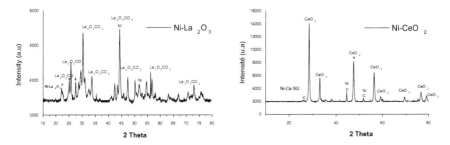

Fig. 7.2 XRD patterns after reaction of the studied Ni–La$_2$O$_3$ (left) and Ni–CeO$_2$ (right) catalysts

7.4 Conclusion

In the present work, we have prepared Ni–CeO$_2$ and Ni–La$_2$O$_3$ catalysts using sol-gel method. The resulting catalysts were characterized by XRD, BET and TPR analysis and tested in catalytic reforming of methane. The XRD analysis shows a good crystallization of all catalysts, with peaks assigned to CeO$_2$ and NiO in the case of Ni–CeO$_2$, La$_2$O$_3$ and LaNiO$_3$ for Ni–La$_2$O$_3$. Although Ni–CeO$_2$ catalyst presents a high surface area (45 m^2/g) but Ni–La$_2$O$_3$ is found to be more active in the methane reforming test, the catalytic tests show that the conversions of CH$_4$, CO$_2$ are near to 12%, 18%, Yields of H$_2$ and CO yields are of 15%, 06% respectively, with CO/H$_2$ ratio 2.5 (Table 7.2). After 12 h of reaction, no deactivation was observed, which can be due to the presence of single LaNiO$_3$ perovskite phase (18 nm) and the presence of stable La$_2$O$_2$CO$_3$ compound during the reaction.

References

1. H. Atashi, Kinetic study of Fischer-Tropsch process on titania-supported cobalt manganese catalyst. J. Ind. Eng. Chem. **16**(6), 952–961 (2010)
2. R. Sheldon, *Chemicals from Synthesis Gas* (D. Reidel Publishing Company, Dordrecht, 1983)
3. Y. Khani, High catalytic activity and stability of ZnLaAlO$_4$ supported Ni, Pt and Ru nanocatalysts applied in the dry, steam and combined dry-steam reforming of methane. Chem. Eng. J. **299**, 353–366 (2016)
4. D. Pakhare, A review of dry (CO$_2$) reforming of methane over noble metal catalysts. J. Chem. Soc. Rev. **43**(22), 7813–7837 (2014)
5. K. Nakagawa, Partial oxidation of methane to synthesis gas over supported iridium catalysts. Appl. Catal. **169**(2), 281–290 (1998)
6. M.C.J. Bradford, CO$_2$ reforming of CH$_4$. Catal. Rev. **41**(1), 1–42 (1999)
7. Y. Benguerba, Modelling of methane dry reforming over Ni/Al$_2$O$_3$ catalyst in a fixed-bed catalytic reactor. React. Kinet. Mech. Catal. **114**(1), 109–119 (2015)
8. J. Kim, Effect of metal particle size on coking during reforming of CH$_4$ over Ni–alumina aerogel catalysts. Appl. Catal. A **197**(2), 191–200 (2000)
9. Yahi, N., Dry reforming of methane over Ni/CeO$_2$ catalysts prepared by three different methods. Green Process. Synth. **4**(6), 479–486 (2015)
10. H.P. Klug, L.E. Alexander: X-ray Diffraction Procedures For Polycrystalline And Amorphous Materials. A Wiley-Interscience Publication (1974)

11. J. Requies, Partial oxidation of methane to syngas over Ni/MgO and Ni/La$_2$O$_3$ catalysts. Appl. Catal. A **289**(2), 214–223 (2005)
12. K. Rida, Effect of calcination temperature on structural properties and catalytic activity in oxidation reactions of LaNiO$_3$ perovskite prepared by Pechini method. J. Rare Earths **30**(3), 210–216 (2012)
13. S. Pangranich, Methane partial oxidation over Ni/CeO$_2$–ZrO$_2$ mixed oxide solid solution catalysts, Catal. Today 93–95, 95–105 (2004)
14. K. Junichiro, Effects of nanocrystalline CeO$_2$ supports on the properties and performance of Ni-Rh bimetallic catalyst for oxidative steam reforming of ethanol. J. Catal. **238**(2), 430–440 (2006)
15. A.M. Diskin, The oxidative chemistry of methane over supported nickel catalysts. Catal. Today **46**(2–3), 147–154 (1998)
16. F. Arena, The role of Ni^{2+} diffusion on the reducibility of NiO/MgO system: a combined TRP-XPS study. Catal. Lett. **6**, 139–149 (1990)
17. F. Frusteri, Steam and auto-thermal reforming of bio-ethanol over MgO and CeO$_2$ Ni supported catalysts. Int. J. Hydrogen Energy **31**(15), 2193–2199 (2006)
18. H.S. Roh: Highly active and stable Ni/Ce–ZrO$_2$ catalyst for H$_2$ production from methane. J. Mol. Catal. A Chem. **181**(1–2), 137–142 (2002)
19. S. Pengpanich, Catalytic oxidation of methane over CeO$_2$-ZrO$_2$ mixed oxide solid solution catalysts prepared via urea hydrolysis. Appl. Catal. A **234**(1–4), 221–233 (2002)
20. Q. Shi, Hydrogen production from steam reforming of ethanol over Ni/MgO-CeO$_2$ catalyst at low temperatur. J. Rare Earths **27**(6), 948–954 (2009)
21. B.S. Liu, Sol–Gel-generated La$_2$NiO$_4$ for CH$_4$/CO$_2$ Reforming. Catal. Lett. **85**(3), 165–170 (2003)
22. D.G. Cheng, Carbon dioxide reforming of methane over Ni/Al$_2$O$_3$ treated with glow discharge plasma. Catal. Today **115**(1–4), 205–210 (2006)
23. S.M. Lima, Evaluation of the performance of Ni/La$_2$O$_3$ catalyst prepared from LaNiO$_3$ perovskite-type oxides for the production of hydrogen through steam reforming and oxidative steam reforming of ethanol. Appl. Catal. A **377**(1), 181–190 (2010)
24. M.A. Goula, Characterization of carbonaceous species formed during reforming of CH$_4$ with CO$_2$ over Ni/CaO–Al$_2$O$_3$ catalysts studied by various transient techniques. J. Catal. **161**(2), 626–640 (1996)
25. Y. Matsumura, Steam reforming of methane over nickel catalysts at low reaction temperature. Appl. Catal. A **258**(1), 107–114 (2004)
26. A. Iulianelli, H$_2$ production by low pressure methane steam reforming in a Pd–Ag membrane reactor over a Ni-based catalyst: experimental and modeling. Int. J. Hydrogen Energy **35**(20), 11514–11524 (2010)
27. S. Tomiyama, Preparation of Ni/SiO$_2$ catalyst with high thermal stability for CO$_2$-reforming of CH$_4$. Appl. Catal. **241**(1–2), 349–361 (2003)
28. A.F. Lucrédio, Methane conversion reactions on Ni catalysts promoted with Rh: influence of support. Appl. Catal. A **400**, 156–165 (2011)

Chapter 8
Hydrogen Effect on Soot Formation in Ethylene-Syngas Mixture Opposed Jet Diffusion Flame in Non-conventional Combustion Regime

Amar Hadef, Selsabil Boussetla, Abdelbaki Mameri, and Z. Aouachria

Abstract The influence of the addition of hydrogen on the formation of soot in an ethylene-gas/air mixture of a counter-current laminar diffusion flame in the flameless regime and at atmospheric pressure has been studied. A detailed gas phase reaction mechanism, including aromatic chemistry up to four cycles and complex thermal and transport properties, was used. The soot is modeled by the moments method. The interactions between soot and gas phase chemistry have been taken into account. Losses by thermal radiation (from CO_2, CO, H_2O from CH_4 and soot) modeled by a thin body. Adding hydrogen to the fuel eliminates the formation of soot. The calculations further suggest that the effect of the addition of hydrogen on soot formation is due to the absence of the concentration of hydrogen atoms in the surface growth regions of soot (stagnation plane) and at a higher concentration of molecular hydrogen in the flame zone. It also reduces the concentrations of C_3H_3, C_6H_6 as well as PAHs (for example, pyrene) which all suppress the process of soot formation.

Keywords Soot · Opposed jets flame · Radiation · Hydrogen addition

---

A. Hadef (✉) · S. Boussetla · A. Mameri
Laboratory (LCMSMTF), Department of Mechanical Engineering, University of Oum El Bouaghi, Oum El Bouaghi, Algeria
e-mail: hadef_am@yahoo.fr

S. Boussetla
e-mail: selsabil.boussetla@outlook.com

A. Mameri
e-mail: mameriabdelbaki@yahoo.fr

Z. Aouachria
Applied Physical Energetics Laboratory, University of Batna1, 05000 Batna, Algeria

© The Editor(s) (if applicable) and The Author(s), under exclusive license to Springer Nature Singapore Pte Ltd. 2021
A. Khellaf (ed.), *Advances in Renewable Hydrogen and Other Sustainable Energy Carriers*, Springer Proceedings in Energy,
https://doi.org/10.1007/978-981-15-6595-3_8

8.1 Introduction

The use of alternative fuels gives very attractive solution to depletion of fossil fuel energy resources. Among these fuels syngas (synthesis gas) which comes from the transformation of various raw materials, both solid and carbonated or liquid, which are rich in carbon. It can also be produced from coal, natural gas, organic waste or biomass. It is composed of hydrogen (H_2) and carbon monoxide (CO). However, depending on production techniques, it may also contain methane (CH_4), carbon dioxide (CO_2) or nitrogen (N_2) [1].

For the combustion process, a new flameless regime (MILD) is discovered. It is based on low oxygen content with preheating of oxidizer, furthermore difference between maximum temperature during combustion and temperature of reactive mixture injection should be lower than the self-ignition temperature of mixture [2].

The temperature reduction in MILD combustion reduces polluting species significantly. Among these species, conversion of fuel carbon in a series of chemical reactions to CO_2 and CO. When combustion takes place with air excess, solid carbonaceous particles known as soot appear in flue gases zone. Soot are detrimental to combustion efficiency and human health, [3, 4]. They participate in global warming because they have the ability not only to absorb some of the solar radiation but also to reflect some of the infrared radiation reemitted by the earth and cause an intensification of the greenhouse effect. This effect is all the more marked because they are the most absorbing and their global warming potential is about 680 times greater than that of carbon dioxide [5].

8.2 Configuration and Computations Strategy

The flame in question is represented by Fig. 8.1, it consists of two circular opposed jets with a separating distance D = 2.9 cm, one injects a mixture of ethylene/syngas

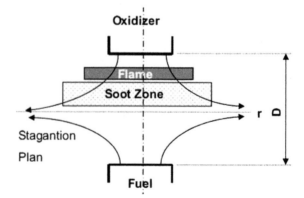

Fig. 8.1 Opposed jets geometry

Table 8.1 Operating conditions

Fuel	M1	M2	M3	M4
Mole fraction of C_2H_4	0.70	0.70	0.70	0.70
Mole fraction of H_2	0.05	0.10	0.15	0.20
Mole fraction of CO	0.25	0.20	0.15	0.10
Temperature (K)	300	300	300	300
Strain rate s^{-1}	50	50	50	50
Velocity (cm/s)	25.23	25.65	26.08	26.55
Stagnation plan (cm)	1.87	1.85	1.83	1.81
Maximum temperature (cm)	2.29	2.29	2.27	2.27
Oxidizer				
Mole fraction of N_2	0.79			
Mole fraction of O_2	0.06			
Mole fraction of CO_2	0.15			
Temperature (K)	1200			
Velocity (cm/s)	25.23	25.65	26.08	26.55

and the other preheated oxidant of a temperature of 1200 K. A low oxygen content of oxidizer is considered (Table 8.1) with constant global strain rate [6]. Unidimensional conservation equations of mass, momentum, energy, and species [7] are solved with their corresponding boundary conditions.

The moments method [8] is adopted for soot modelling, it is widely used for models adapted to aerosols for its flexibility and low-cost calculation. In this method, the first moment corresponds to total concentration of particles and the second one permits the computation of soot volume fraction Fv. For chemical kinetics, the reaction mechanism composed by 101 species and 543 elementary reactions and which is developed by Appel et al. [9] is adopted in this study. This mechanism includes growth reactions from PAHs to pyrene.

8.3 Results and Discussions

8.3.1 Hydrogen Effect on Flame Structure

As shown in Fig. 8.2, increase of hydrogen did not significantly influence variations in maximum combustion temperatures, since it substitutes CO species which has nearly the same volumetric heat value. The addition of 20% of H_2 in mixture (M4) increases temperature by 24 K relatively to the mixture (M1), with a small displacement of maximum to the fuel side. The results clearly show that molar fraction of OH species increases with hydrogen addition, which increases the efficiency of the mixture with

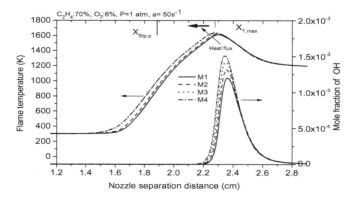

Fig. 8.2 Flame temperature and OH species evolution in function of hydrogen addition

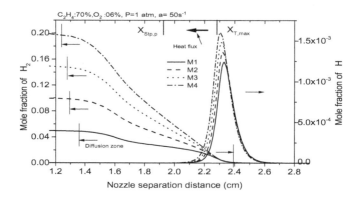

Fig. 8.3 Distributions of species H_2 and H in function of hydrogen addition

its high diffusivity (Fig. 8.2) as well as its dissociation and which increases the production of the radical H (Fig. 8.3).

8.3.2 Effect of Hydrogen Addition on the Soot Particles Formation

Soot production is a very complex phenomenon involving many chemical reactions and physical mechanisms [10]. It is widely accepted that acetylene (C_2H_2) is considered as the potential precursor of the propargyl radical (C_3H_3) which is itself the major source of benzene (propargyl radical recombination reaction) [11]. The positions of flame front represented by maximum temperature (T_{max}) position and stagnation plane of particles ($X_{Stp.p}$) by axial velocity vanish position are indicated in the figures.

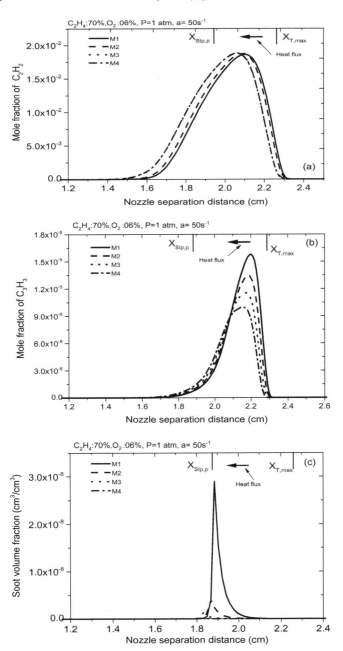

Fig. 8.4 Mole fractions of: **a** acetylen, **b** propargyl, and **c** soot volume fraction

As a soot growth promoter, acetylene only appears on the fuel side in the four flames (M1–M4), indicating that it is mainly produced from the pyrolysis of the fuel. It is not really affected by hydrogen (Fig. 8.4a), unlike C_3H_3 which is a radical, where the effect of the addition of H_2 plays an important role in its reduction (Fig. 8.4b), the volumetric fraction of soot (Fig. 8.4c) is highly induced by the addition of hydrogen.

8.4 Conclusion

In this study, the effect of hydrogen concentration in ethylene/syn-gas/air mixture on formation of soot precursors was studied in a flameless combustion regime. The increase of H_2 in the H_2/CO ratio does not vary the flame temperature significantly, but promotes the production of OH radical. The amount of C_3H_3 formed decreases continuously with increase of H_2 concentration in the mixture, as well as the volume fraction of soot which is greatly affected by addition of hydrogen.

References

1. H.C. Lee, A.A. Mohamad, L.Y. Jiang, Comprehensive comparison of chemical kinetics mechanisms for syngas/biogas mixtures. Energy Fuels **29**, 6126–6145 (2015)
2. C. Antonio, J. Mara, Mild combustion. Prog. Energy Combust. Sci. **30**, 329–366 (2004)
3. C.A. Pope III, R.T. Burnet, M.J. Thun, E.E. Calle, D. Krewski, K. Ito, G.D. Thurston, Lung cancer, cardiopulmonary mortality, and long-term exposure to fine particulate air pollution. JAMA-J. Am. Med. Assoc. **6**, 1132–1141 (2002)
4. J. Hansen, L. Nazarenko, Soot climate forcing via snow and ice albedos. Proc. Natl. Acad. Sci. U.S.A. **101**, 423–428 (2004)
5. N. Mac Carty, D. Ogle, D. Still, T. Bond, C. Roden, A laboratory comparison of the global warming impact of five major types of biomass cooking stoves. Energy Sustain. Dev. **12**, 56–65 (2008)
6. E.M. Fisher, B.A. Williams, J.W. Fleming, Determination of the strain in counter flow diffusion flames from flow conditions, in *Proceedings of the Eastern States Section of the Combustion Institute*, 191–194 (1999)
7. A. Cuoci, A. Frassoldati, T. Faravelli, E. Ranzi, Formation of soot and nitrogen oxides in unsteady counter flow diffusion flames. Combust. Flame **156**, 2010–2022 (2009)
8. A. Kazakov, M. Frenklach, Dynamic modeling of soot particle coagulation and aggregation: implementation with the method of moments and application to high-pressure laminar premixed flames. Combust. Flame **114**, 484–501 (1998)
9. R.S. Barlow, A.N. Karpetis, J.H. Frank, J.Y. Chen, Scalar profiles and NO formation in laminar opposed-flow partially premixed methane/air flames. Combust. Flame **127**, 2102–2118 (2001)
10. W. Grosshandler, RadCal: a narrow band model for radiation calculations in a combustion environment. NIST technical note TN 1402 (1993). https://nvlpubs.nist.gov/nistpubs/Legacy/TN/nbstechnicalnote1402.pdf. Last accessed 10 Nov 2018
11. J. Appel, H. Bockhorn, M. Frenklach, Kinetic modeling of soot formation with detailed chemistry and physics: laminar premixed flames of C_2 hydrocarbons. Combust. Flame **121**, 122–136 (2000)

Chapter 9
A Two-Dimensional Simulation of Opposed Jet Turbulent Diffusion Flame of the Mixture Biogas-Syngas

Abdelbaki Mameri, Selsabil Boussetla, and Amar Hadef

Abstract To reduce combustion harmful emissions such as CH_4, CO_2 and NO, low calorific renewable biofuels are found to be the best candidate. Combustion of biofuels gives a weak unstable flame which can be enhanced by hydrogen addition. The latter can also be obtained from a renewable source like syngas. In this context the mixture biogas-syngas is studied to overcome combustion emissions issues. A two-dimensional numerical procedure has been used to investigate the opposed jet turbulent flame structure of biogas-syngas mixture. The standard k-ε model is adopted for turbulence modeling and the Steady Laminar Flammelettes Model (SLFM) to handle turbulent combustion. The combustion kinetics is modeled by the detailed Glarborg's N-mechanism. Equimolar biogas-syngas mixture is considered, namely: biogas $0.25CH_4 + 0.25CO_2$ and syngas $0.25H_2 + 0.25CO$. Injection velocity is equal for both jets and it is varied from 3 to 12 m/sec. It has been noticed that when injection velocity increases, the flow residence time and non-equilibrium effects are enhanced. This improves incomplete reaction species production and reduces final combustion products volume, temperature and NO species emission. As a summary, the NO emission can be avoided by using important injection velocities.

Keywords Turbulent 2D opposed jets · Biogas-syngas mixture · SLF modelisation

---

A. Mameri (✉) · S. Boussetla · A. Hadef
LCMASMTF, Department of Mechanical Engineering, Science and Applied Science Faculty, Oum El Bouaghi University, Oum El Bouaghi, Algeria
e-mail: mameriabdelbaki@yahoo.fr

S. Boussetla
e-mail: selsabil.boussetla@outlook.com

A. Hadef
e-mail: hadef_am@yahoo.fr

© The Editor(s) (if applicable) and The Author(s), under exclusive license to Springer Nature Singapore Pte Ltd. 2021
A. Khellaf (ed.), *Advances in Renewable Hydrogen and Other Sustainable Energy Carriers*, Springer Proceedings in Energy,
https://doi.org/10.1007/978-981-15-6595-3_9

9.1 Introduction

Renewability and sustainability of biofuels make them the favorite in feeding numerous applications such as households, furnaces, IC engines and gas turbines. Biofuels are always contaminated by diluent species like carbon dioxide and water vapor; which makes them Low Calorific Value fuels (LCV) with weak flames [1]. The latter cannot sustain high injection velocity of reactants which enhances non-equilibrium effects and decrease temperatures to the extinction [2]. To overcome these issues biofuels are blended by hydrogen which can also be recovered in syngas from renewable sources by gasification [3, 4]. Hydrogen addition to biogas improves the flame resistance to injection velocity (strain rate) and inhibits NO species formation [5]. In the biogas-syngas blends, it was found that hydrogen increases the flame burning velocity and consequently enhances combustion stability [6]. Besides, it was revealed that the NO species levels are lower than those in methane. Their primary formation routes in biogas are both thermal and prompt; however, in syngas the only route is the thermal one [7]. For the mixtures biogas-syngas, it was found that the results obtained by the NUIG2013 mechanism were in best match with the measured laminar flame burning velocity and ignition delay [8]. This mechanism did not include NO_x sub-mechanisms and cannot be used in this work. The up to date mechanism that include the NO_x sub-mechanisms is the Glarborg's one which is formed by 1397 elementary reactions linking 151 species [9]. This numerical simulation purposes is to investigate the effect of injection velocity (strain rate) on turbulent flame structure and NO emission. The equimolar B25S25 (biogas $0.25CH_4 + 0.25CO_2$ and syngas $0.25H_2 + 0.25CO$) is considered and injection velocity is varied from 3 to 12 m/s. Temperature and pressure are constant and equal to 300 K and 1 atm respectively. The resolution of the phenomenon equations is achieved by Ansys Fluent program [10].

9.2 Governing Equations and Numerical Procedure

The geometry is composed by two opposed jets streaming fuel mixture from the left side and air from the right side (Fig. 9.1). Each jet is formed by a circular duct of

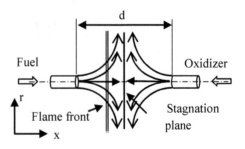

Fig. 9.1 Geometry of the flame

9 A Two-Dimensional Simulation of Opposed Jet Turbulent ...

inner diameter of 2 cm, the separating distance between the ducts is 2 cm Fig. 9.1. A diffusion toroidal flame is formed between the jets which simplifies the analysis to axisymmetric two-dimensional configuration.

The governing equations for two-dimensional, axisymmetric, turbulent flames are:

The continuity equation is:

$$\frac{\partial}{\partial x}(\rho u) + \frac{1}{r}\frac{\partial}{\partial r}(r\rho v) = 0 \tag{9.1}$$

The axial and radial momentum conservation equations are:

$$\frac{\partial}{\partial x}(\rho u u) + \frac{1}{r}\frac{\partial}{\partial r}(r\rho v u) = -\frac{\partial p}{\partial x} + \frac{\partial}{\partial x}\left[2\mu_{eff}\frac{\partial u}{\partial x} - \frac{2}{3}(\mu_{eff}\nabla V + \rho k)\right]$$
$$+ \frac{1}{r}\frac{\partial}{\partial r}\left[r\mu_{eff}\left(\frac{\partial u}{\partial r} + \frac{\partial v}{\partial x}\right)\right] \tag{9.2}$$

$$\frac{\partial}{\partial x}(\rho u v) + \frac{1}{r}\frac{\partial}{\partial r}(r\rho v v) = -\frac{\partial p}{\partial r} + \frac{\partial}{\partial x}\left[\mu_{eff}\left(\frac{\partial u}{\partial r} + \frac{\partial v}{\partial x}\right)\right]$$
$$+ \frac{1}{r}\frac{\partial}{\partial r}\left[r\left(2\mu_{eff}\frac{\partial v}{\partial r} - \frac{2}{3}(\mu_{eff}\nabla V + \rho k)\right)\right]$$
$$- \frac{2}{r}\left(\mu_{eff}\frac{v}{r} + \frac{1}{3}(\mu_{eff}\nabla V + \rho k)\right) \tag{9.3}$$

Energy conservation equation is:

$$\nabla(V(\rho E + p)) = \nabla\left(\lambda_{eff}\nabla T - \sum_{i=1}^{N} h_i V_i + (\bar{\bar{\tau}} V)\right) + \sum_{i=1}^{N} \frac{h_i^0}{W_i}\dot{\omega}_i \tag{9.4}$$

With

$$E = h - \frac{p}{\rho} + \frac{V^2}{2}$$

$$h = \sum_{i=1}^{N} Y_i \int_{T_1}^{T_2} Cp_i dT$$

$$\bar{\bar{\tau}} = \mu\left[2S - \frac{2}{3}\delta\nabla.V\right] \text{ and } S = \frac{1}{2}(\nabla V + \nabla V^T)$$

And species equation is:

$$\frac{\partial \rho u Y_i}{\partial x} + \frac{1}{r}\frac{\partial}{\partial r}(r\rho v Y_i) + \frac{\partial}{\partial x}(\rho Y_i V_i) + \frac{1}{r}\frac{\partial}{\partial r}(r\rho V_i) = \dot{\omega}_i \tag{9.5}$$

Turbulence is modeled by the well-known k-ε model [11] and turbulent combustion by the Steady Laminar Flamelet Model (SLFM) [11].

An equimolar composition is considered with 0.25 mol for each fuel components i.e. biogas $0.25CH_4 + 0.25CO_2$ and syngas $0.25H_2 + 0.25CO$. Oxidizer is air composed by $0.21O_2 + 0.79N_2$. Injection temperature and pressure are constant for both fuel and oxidizer and equal to 300 K and 1 atm respectively. Injection velocity guarantees the turbulent regime and it is varied from 3 to 12 m/s.

9.3 Numerical Results and Interpretation

The fuel mixture and oxidizer are injected by two opposed ducts at the same velocity which forms a toroidal diffusion flame between the nozzles. To characterize the flame structure, two sections are operated through the flame center. The first one is longitudinal (denoted L) and begins from the axis and ends at the exit plane following the radial direction, whereas the second section is transverse (denoted T) and follows the axial direction (Fig. 9.2). Here, it should be noticed that the sections are operated according to the temperature field which represents the flame position.

Globally, it has been observed that injection velocity increase enhances kinetic energy Fig. 9.3a. Kinetic energy is maximal on the axis at the impinging point, since velocity gradients induced by injection are very high. Moreover, kinetic energy increases at flame front since gas expansion increase velocity gradients Fig. 9.3b–d show that temperature increases from the surroundings to reach its maximum at the flame front. The latter is slightly shifted to the radial direction and towards the fuel side with injection velocity. Obviously, when the flame residence time is reduced, the flame position is lifted downstream, also the heavier oxidizer stream shifts the light fuel one.

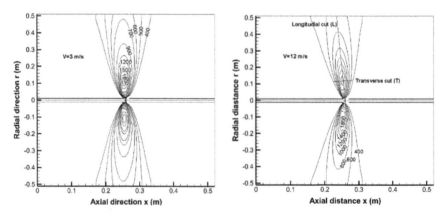

Fig. 9.2 Temperature filed and positions of longitudinal and transverse cuts

9 A Two-Dimensional Simulation of Opposed Jet Turbulent ...

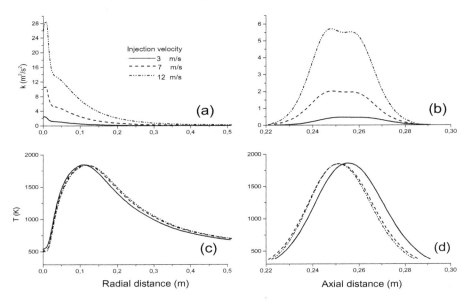

Fig. 9.3 Effect of injection velocity on kinetic energy and flame temperature

The global flame shift which can be seen on Fig. 9.2 and others is combined effect of flame residence time reduction and stream mass flow rate.

Figure 9.4 shows variations of the fuel mixture compounds (CH_4, CO_2, H_2 and CO) and the H_2O produced from the flame. It can be seen that methane and hydrogen are quickly consumed compared to CO and CO_2 which is a stable combustion product.

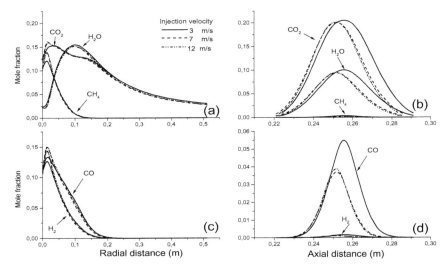

Fig. 9.4 Effect of injection velocity on major species

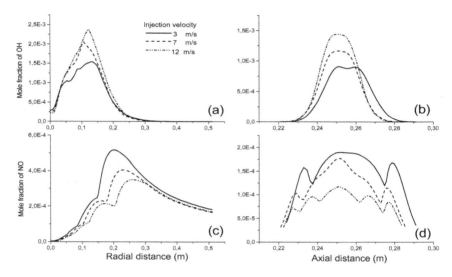

Fig. 9.5 Effect of injection velocity on minor species

Of course, when injection velocity increases, more fuel is injected and the mole fraction is increased at the injectors Fig. 9.4a, c. At the flame transverse cross section Fig. 9.4b, d, it can be seen that all species are reduced by injection velocity augmentation. Near or at the flame front, the velocity increase makes the flame more reactive and consumes more fuel such as methane, hydrogen and CO. On the other hand, CO_2 and H_2O are also reduced by injection velocity increase Fig. 9.4b. This is due to the non-equilibrium effects which enhances incomplete reaction products and decreases complete reaction species.

The OH species which is an indicator of the flame strength is shown by Fig. 9.5a, b. It can be seen that when injection velocity increases, the OH species production is enhanced. The OH is an incomplete reaction product which is augmented by non-equilibrium effects which are amplified velocity increase. The NO species is a harmful combustion product which should be avoided in any combustion system. It can be noticed that NO is produced after the flame front Fig. 9.5c and reduced by injection velocity augmentation.

9.4 Conclusion

A two-dimensional procedure is used to investigate the effect of fuel and oxidizer injection velocity on the flame structure and NO emission. When injection velocity increases:

- Flame is shifted towards the fuel duct and its volume is reduced.

- Final combustion products are reduced and intermediate species are increased since non equilibrium effect are enhanced.
- The NO species is significantly reduced.

References

1. D. Wanneng, Q. Chaokui, C. Zhiguang, T. Chao, L. Pengjun, Experimental studies of flame stability limits of biogas flame. Energy Convers. Manag. **63**, 157–161 (2012)
2. M.C. Drake, R.J. Blint, Structure of laminar opposed-flow diffusion flames With $CO/H_2/N_2$ fuel. Combust. Sci. Technol. **61**, 187–224 (1988)
3. P. Sabia, G. Sorrentino, P. Bozza, G. Ceriello, R. Ragucci, M. de Joannon, Fuel and thermal load flexibility of a MILD burner. Proc. Combust. Inst. **37**(4), 4547–4554 (2019)
4. G. Sorrentino, P. Sabia, M. de Joannon, et al., The effect of diluent on the sustainability of MILD combustion in a cyclonic burner. Flow Turbulence Combust. **96**(2), 449–468 (2016)
5. A. Mameri, F. Tabet, Numerical investigation of counter-flow diffusion flame of biogas–hydrogen blends: Effects of biogas composition, hydrogen enrichment and scalar dissipation rate on flame structure and emissions. Int. J. Hydrogen Energy **41**(3), 2011–2022 (2016)
6. R. Selfarski, J. Sacha, P. Grzymislavski, Combustion of mixtures of biogases and syngases with methane in strong swirl flow, in *6th European Combustion Meeting* (Lund, Sweden), pp. 25–28 (June 2013)
7. G.M. Watson, J.D. Munzar, J.M. Bergthorson, NO formation in model syngas and biogas blends. Fuel **124**, 113–124 (2014)
8. H.C. Lee, A. Mohamad, L. Jiang, Comprehensive comparison of chemical kinetics mechanisms for syngas/biogas mixtures. Energy Fuels **29**(9), 6126–6145 (2015)
9. P. Glarborg, J.A. Miller, B. Ruscic, S.J. Klippenstein, Modeling nitrogen chemistry in combustion. Prog. Energy Combust. Sci. **67**, 31–68 (2018)
10. R.J. Kee, F.M. Rupley, J.A. Miller, M.E. Coltrin, J.F. Grcar, E. Meeks, H.K. Moffat, A.E. Lutz, L.G. Dixon, M.D. Smooke, J. Warnatz, G.H. Evans, R.S. Larson, R.E. Mitchell, L.R. Petzold, W.C. Reynolds, M. Caracotsios, W.E. Stewart, P. Glarborg, C. Wang, O. Adigun, CHEMKIN collection, Release 3.6, Reaction Design, Inc., San Diego, CA (2000)
11. https://www.ansys.com/products/fluids/ansys-fluent

Chapter 10
Numerical Evaluation of NO Production Routes in the MILD Combustion of the Biogas-Syngas Mixture

Selsabil Boussetla, Abdelbaki Mameri, and Amar Hadef

Abstract In the present study, the effect of composition, pressure and inlet temperature on NO production in non-premixed MILD (Moderate or Intense Low-oxygen Dilution) combustion of biogas-syngas mixture is evaluated numerically. Several pressures, compositions and inlet temperatures are considered. Results showed that NO formation increase highly with increasing pressure and oxidizer injection temperature (T_{ox}). It is found that production of NO is more sensitive to increasing T_{ox} than decreasing pressure. Moreover, effect of pressure is reduced as volume of hydrogen decreases in mixture. Also, increase of maximum NO mole fraction, as a function of injection temperature is important for the B50S25 mixture. Furthermore, effect of pressure and injection temperature on different NO production routes (thermal, prompt, N_2O and NO_2) were evaluated. It is noticed that at low pressure, NO is mainly produced by prompt path, whereas NO_2 mechanism prevails when pressure is increased. When oxidizer temperature increases, most important mechanism that produces NO is prompt one followed by thermal route and finally NO_2 path. It is noticed that N_2O production route is trivial for all values of pressure and T_{ox}. The reaction path diagram was presented to show the formation routes of NO for different pressures.

Keywords MILD combustion · NO emission · Opposed-jet flames · Biogas-syngas mixture

---

S. Boussetla (✉) · A. Mameri · A. Hadef
Laboratory (LCMSMTF), Department of Mechanical Engineering FSSA, University Larbi Ben MHIDI Oum El Bouaghi, Oum El Bouaghi, Algeria
e-mail: selsabil.boussetla@outlook.com

A. Mameri
e-mail: mameriabdelbaki@yahoo.fr

A. Hadef
e-mail: hadef_am@yahoo.fr

© The Editor(s) (if applicable) and The Author(s), under exclusive license to Springer Nature Singapore Pte Ltd. 2021
A. Khellaf (ed.), *Advances in Renewable Hydrogen and Other Sustainable Energy Carriers*, Springer Proceedings in Energy,
https://doi.org/10.1007/978-981-15-6595-3_10

10.1 Introduction

Climate change induced by human activities is the most serious environmental problems caused by energy consumption. Combustion installations are key sources of greenhouse gas emissions and many different types of air pollutants, such as carbon monoxide (CO) which can cause harmful effects on human health, nitric oxides (NO), which is toxic at low amounts and may combine with water to form nitric acid (HNO_3), responsible of acid rain [1]. Solutions to these problems comes from the use of clean and renewable energy sources. The latter are produced from elements found in nature, like biomass, as an alternative to traditional energy sources and which is an inevitable trend in future [2].

Control of polluting emissions is crucial in design of new combustion systems. Flameless or MILD (Moderate or Intense Low-oxygen Dilution) combustion technology has capacity to emit very small amounts of pollutants, in particular nitrogen oxides, carbon monoxide and unburnt hydrocarbons [3]. In conventional combustion systems using diffusion flames, it is found that NO is formed according to NO_2 and prompt routes, whereas others important pathways such as thermal one is inhibited [4]. Few studies [5–7] examined the effect of operating conditions such as pressure, carbon dioxide dilution, hydrogen addition, nitrogen concentration and preheated air temperature on NO formation in MILD combustion. It has been demonstrated that in MILD combustion regimes, thermal NO is implicitly reduced by lowering flame peak temperature. The other pathways may also have crucial role in formation of NO, namely: Fenimore-NO, also called prompt NO, which is activated at temperatures below 1800 K, the N_2O-intermediate and the NO_2 pathway.

The objective of present study is to evaluate numerically NO formation routes of laminar diffusion combustion in MILD regime of biogas ($CH_4 + CO_2$)-syngas ($H_2 + CO$) mixture, taking into account compositions, pressure and inlet temperature effects. Several pressures ranging from 1 to 10 atm, compositions (B00S25, B25S00, B25S25, B25S50 and B50S25) and inlet temperatures (from 900 to 1600 K) were considered, strain rate (injection velocity) is kept constant.

10.2 Governing Equations and Numerical Strategy

The counter flow geometry is adopted, distance between the two nozzles is fixed at d = 2 cm as shown in Fig. 10.1. The mixture biogas-syngas is injected at the first side while the oxidizer composed by oxygen and nitrogen ($0.04O_2 + 0.96N_2$) is injected at the second one. This configuration simplifies on one-hand flow equations, which become unidimensional, and on other hand produces a stagnation plane flame for characterizing the structure.

Fig. 10.1 Schematic of counter flow non-premixed flame configuration

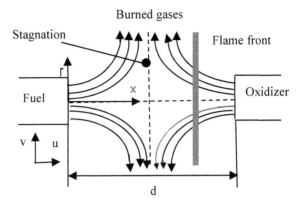

10.2.1 Governing Equations

Mathematical model adopted for this geometry is that developed by Kee et al. [8]:
Conservation of mass in axisymmetric coordinates writes as:

$$\frac{\partial}{\partial x}(r\rho u) + \frac{\partial}{\partial r}(\rho v r) = 0 \tag{10.1}$$

Equation of momentum in axial direction x is:

$$H - 2\frac{d}{dx}\left(\frac{UG}{\rho}\right) + \frac{3G^2}{\rho} + \frac{d}{dx}\left[\mu \frac{d}{dx}\left(\frac{G}{\rho}\right)\right] = 0 \tag{10.2}$$

With $H = \frac{1}{r}\frac{\partial P}{\partial r} = Cste$, $G(x) = \frac{dU}{dx}$, $U(x) = \frac{\rho u}{2}$,
Energy and species conservation equations are:

$$\rho u \frac{dT}{dx} - \frac{1}{Cp}\frac{d}{dx}\left(\lambda \frac{dT}{dx}\right) + \frac{\rho}{Cp}\sum_k Cp_k V_k Y_k \frac{dT}{dx}$$
$$+ \frac{1}{Cp}\sum_k h_k \dot{\omega}_k + \frac{1}{Cp}\left[4\sigma p \sum_j X_j a_i (T^4 - T_f^4)\right] = 0 \tag{10.3}$$

$$\rho u \frac{dY_k}{dx} + \frac{d}{dx}(\rho Y_k V_k) - \dot{\omega}_k W_k = 0 \tag{10.4}$$

Diffusion velocities are given by multi-component formulation:

$$V_k = \frac{1}{X_K \bar{W}}\sum_{j\neq k}^{k} W_j D_{kj}\frac{dX_j}{dx} - \frac{D_k^T}{\rho Y_k}\frac{1}{T}\frac{dT}{dx} \tag{10.5}$$

The Gri-mech 3.0 mechanism [9], formed by 53 species linking 325 elemental reactions (and 40 NO associated reactions), is used to describe chemical kinetics and NO formation. Contribution of the four NO formation mechanisms (thermal, prompt, N_2O and NO_2) in NO production was evaluated [10].

10.2.2 Boundary Conditions

For all studied cases, injection temperature of fuel is 300 K and strain rate is kept constant and equals to 200 s^{-1}. Several compositions explicitly: B00S25, B25S00, B25S25, B25S50 and B50S25 were considered. Mixture composition is written in the form BαSβ, where α and β represent volume of methane and hydrogen respectively in mixture.

The Chemkin software [11] is used for solving with their appropriated boundary conditions.

10.3 Results and Discussion

10.3.1 Composition, Pressure and Inlet Temperature Effects on Maximum NO Mole Fraction

Effect of pressure and inlet temperature on maximum NO mole fraction for different compositions is illustrated in Fig. 10.2. Maximums variation of molar fraction of NO is shown by Fig. 10.2a. It is noticed that in the interval 1–3 atm increase in pressure

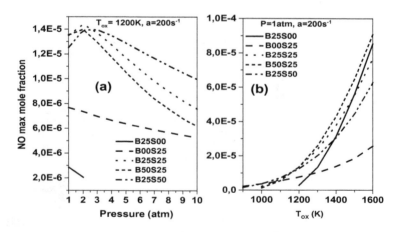

Fig. 10.2 Effect of pressure and inlet temperature on maximum NO mole fraction

increases the NO emission. The most sensitive composition to pressure in terms of NO production is B25S50, which contains maximum volume of hydrogen (50%). Whereas, the least sensitive one is B25S00 which contains minimum of hydrogen volume. In the interval 3–10 atm a reduction in NO production is observed with maximum under 15 ppm for all compositions. For B25S00, it can be noticed that from P = 2 atm NO is null.

Figure 10.2b shows maximum mole fraction of NO which is significantly enhanced by oxidizer injection temperature increase and methane volume augmentation. For oxidizer temperature $T_{ox} = 1600$ K, maximum NO production increases from 26 ppm for B00S25 to 91 ppm for B50S25. While for B25S00, it can be noticed that no NO species is produced until $T_{ox} = 1200$ K.

10.3.2 Pressure and Inlet Temperature Effects on Net Rate of NO Production

Effects of pressure and injection temperature on net rate of production for each NO route and flame temperature are presented in Fig. 10.3. Three values of pressure (1, 5 and 10 atm) and injection temperature (1000, 1200 and 1600 K) were considered for the composition B25S25 at a strain rate a = 200 s^{-1}. It can be observed that when pressure increases, reaction zone thickness is reduced and combustion region is shifted to fuel side. From Fig. 10.3a, at pressure P = 1 atm, it can be observed that most important mechanism that produces NO is the prompt one followed by thermal and finally NO_2 mechanism. When pressure increases to 5 atm (Fig. 10.3b), the NO_2 route becomes more important than thermal one. At a pressure of 10 atm (Fig. 10.3c), the NO_2 mechanism become the most important with a maximum net rate of NO production of 5.8 mol cm^{-3}s^{-1} reached in lean fuel region at x = 1.36 cm. The production through N_2O mechanism is very small for all values of pressure. Moreover, it can be seen that maximum NO production position for NO_2 mechanism is shifted to oxidizer side compared to other routes. Variation of temperature along direction x is shown by Fig. 10.3d, it can be seen that when pressure increases, maximum flame temperature is increased from 1572 K for P = 1 atm to 1620 K for P = 5 atm and then to 1629 K when pressure increases to 10 atm. Reaction zone is significantly reduced and shifted to the fuel side especially in the range P = 1–5 atm.

To elucidate effect of injection temperature on NO production, Fig. 10.3e–g show distribution of NO net rate production for each path. It can be seen that all NO production routes are enhanced by oxidizer injection temperature. Furthermore, prompt mechanism dominates NO production despite of oxidizer temperature. Relatively to the other mechanisms, NO_2 route decreases when oxidizer temperature augments. The variation of temperature in function of distance x is shown in Fig. 10.3h for different oxidizer injection temperatures. It can be seen that when oxidizer temperature increases, the maximum flame temperature and thickness are increased and reaction zone is shifted to the oxidizer side.

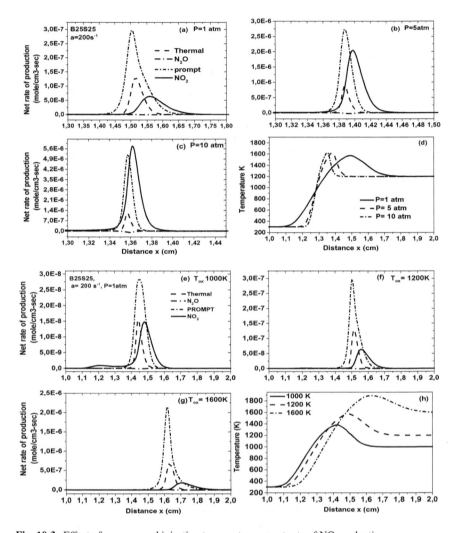

Fig. 10.3 Effect of pressure and injection temperature on net rate of NO production

10.3.3 Effect of Pressure on the NO Formation Paths

In Fig. 10.4 are presented reaction paths for NO production through thermal and prompt mechanisms for different pressures P = 1, 5 and 10 atm. Arrows represent reaction paths while theirs thickness (six levels) shows the maximum production rate. Globally, it can be seen that pressure has significant influence on the rate of NO formation, also thermal and prompt mechanisms are strongly coupled via reactions $N + O_2 = NO + O$ and $N + NO = N_2 + O$.

When pressure increases, NO produced by Fenimore route (prompt) is more important than that produced by thermal path. For an ambient pressure of 1 atm,

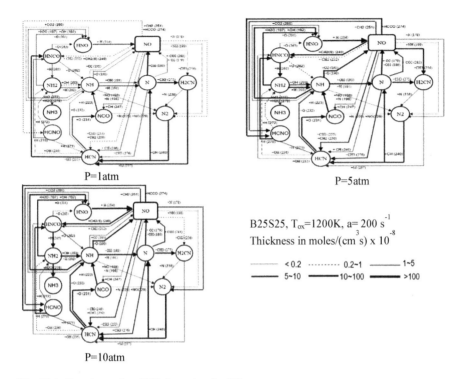

Fig. 10.4 Reaction paths of NO formation for different pressures

the maximum rate of NO formation by reaction $HNO + H = NO + H_2$ is 7×10^{-8} mol/(cm^{3}s), it reaches 2×10^{-6} mol/(cm^{3}s) when P increases to 5 atm and for P = 10 atm, it increases to 2.8×10^{-6} mol/(cm^{3}s). Whereas, for reaction $N + OH = NO + H$, maximum rate of NO formation is 1.2×10^{-7} mol/(cm^{3}s) for P = 1 atm, it increases to 5.5×10^{-7} mol/(cm^{3}s) for both pressures P = 5 atm and P = 10 atm.

10.4 Conclusion

Effect of compositions, pressure and injection temperature on NO production are considered. In summary, results show that:

- Production of NO is more sensitive to increasing oxidizer injection temperature T_{ox} than decreasing pressure.
- For the mean mixture B25S25 and an ambient pressure of P = 1 atm, the main source of NO formation is the prompt path followed by thermal and finally NO_2 mechanism. When pressure increases, NO_2 mechanism become dominant. If inlet temperature increases the principal source of NO is the Fenimore mechanism.

- The N_2O production route is trivial for all values of pressure and oxidizer injection temperature T_{ox}.

References

1. I.A. Reşitoğlu, K. Altinişik, A. Keskin, The pollutant emissions from diesel-engine vehicles and exhaust after treatment systems. Clean Technol. Environ. Policy **17**(1), 15–27 (2015)
2. S.E. Hosseini, M.A. Wahid, Feasibility study of biogas production and utilization as a source of renewable energy in Malaysia. Renew. Sustain. Energy Rev. **19**, 454–462 (2013)
3. S.E. Hosseini, M.A. Wahid, A.A.A. Abduelnuor, Biogas flameless combustion: a review. Appl. Mech. Mater. **388**, 273–279 (2013)
4. A. Mameri, S. Boussetla, R. Belalmi, Z. Aouachria, Combustion characterization of the mixtures biogas-syngas, strain rate and ambient pressure effects. Int. J. Hydrogen Energy **44**(39), 22478–22491 (2019)
5. A. Mameri, F. Tabet, A. Hadef, MILD combustion of hydrogenated biogas under several operating conditions in an opposed jet configuration. Int. J. Hydrogen Energy **43**(6), 3566–3576 (2018)
6. M. Mehregan, M. Moghiman, A numerical investigation of preheated diluted oxidizer influence on NOx emission of biogas flameless combustion using Taguchi approach. Fuel **227**, 1–5 (2018)
7. A. Hadef, A. Mameri, F. Tabet, Z. Aouachria, Effect of the addition of H_2 and H_2O on the polluting species in a counter-flow diffusion flame of biogas in flameless regime. Int. J. Hydrogen Energy **43**(6), 3475–3481 (2018)
8. R.J. Kee, J.A. Miller, G.H. Evans, G. Dixon-Lewis, A computational model of the structure and extinction of strained, opposed flow, premixed methane-air flames, in *proceeding of the 22ND Symposium (International) on Combustion*, The Combustion Institute, Pittsburgh, Pennsylvania, pp. 1479–1494 (1988)
9. Chemical-kinetic mechanisms for combustion applications. http://web.eng.ucsd.edu/mae/groups/combustion/mechanism.html. Mechanical and Aerospace Engineering (Combustion Research), University of California at San Diego. http://combustion.ucsd.edu. (2011)
10. M. Nishioka, S. Nakagawa, Y. Ishikawa, T. Takeno, NO emission characteristics of methane-air double flame. Combust. Flame **98**(1–2), 127–138 (1994)
11. Chemkin-Pro 15131. Reaction Design, San Diego (2013)

Chapter 11
Effect of H_2/CO Ratio and Air N_2 Substitution by CO_2 on CH_4/Syngas Flameless Combustion

Amar Hadef, Abdelbaki Mameri, and Z. Aouachria

Abstract Energy conversion through combustion is one of main energy production processes. The research related to combustion has progressed enormously, leading to an improvement in energy performance in industrial combustion systems. Relative to strong growth in global energy demand, gradual depletion of fossil resources is reported. The latter generates an improvement in combustion processes and makes energy optimization a major challenge for researchers. On the other hand, the combustion emissions to the atmosphere are considered as one of main environmental and climate change concerns. This work analyzes the non-premixed flameless combustion of CH_4-syngaz mixture in opposed jet configuration. Kinetics of combustion is described by Grimech 3.0 mechanism, calculations are achieved by Chemkin code. Several ratio values were considered, fuel injection temperature is T = 300 K, that of oxidant is 1200 K with a constant pressure equals to one atmosphere. It has been found that increasing H_2/CO ratio does not affect combustion temperature and consequently NO emission. Furthermore, a large reduction in NO_2 has been depicted. The replacement of N_2 by CO_2 in the oxidizer has a significant impact on the thermal field and species.

Keywords Flameless combustion · CH_4/syngaz mixture · Strain rate

---

A. Hadef (✉) · A. Mameri
Laboratory (LCMSMTF), Department of Mechanical Engineering, University of Oum El Bouaghi, Oum El Bouaghi, Algeria
e-mail: hadef_am@yahoo.fr

A. Mameri
e-mail: mameriabdelbaki@yahoo.fr

Z. Aouachria
Applied Physical Energetics Laboratory, University of Batna1, 05000 Batna, Algeria

© The Editor(s) (if applicable) and The Author(s), under exclusive license to Springer Nature Singapore Pte Ltd. 2021
A. Khellaf (ed.), *Advances in Renewable Hydrogen and Other Sustainable Energy Carriers*, Springer Proceedings in Energy,
https://doi.org/10.1007/978-981-15-6595-3_11

11.1 Introduction

Energy demand is steadily increasing since it is one of main drivers of global economic growth, while global fossil fuel energy reserves are declining. The increase in global energy need leads to an increase in polluting emissions. To reduce pollution and energy dependency, developed countries begin implementing alternative renewable biofuels [1].

Environmental laws and regulations have evolved and are leading, or at least attempting to support, the transition from pollution-based to prevention-based approaches, and finding fuels that are more respectful of the environment and more sustainable to reduce the effect of climate change and reduce toxic pollutants.

Hydrogen-based fuels are currently an important alternative to fossil fuels; they do not contribute to exacerbating certain global environmental impacts such as greenhouse effect since they emit less carbon dioxide [2].

Synthesis gas (syngas) is a non-primary fuel such as methane, it can be produced from a large number of raw materials containing carbon and hydrogen. It does not necessarily depend on a raw material of fossil origin; it can also be produced from waste or biomass through gasification. Among its advantages, it is renewable, it can be used in energy generation systems and cogeneration systems and it emits less pollution than conventional fuel [3].

The objective of this work is to compare the response of flameless combustion characteristics of syngas/methane mixtures to the changes in hydrogen content, and the replacement of air nitrogen by CO_2 with oxygen content of 6% in the oxidizer and a constant strain rate a = 200 s^{-1}.

11.2 Geometry and Operating Conditions

The flame in question is represented by Fig. 11.1, it consists of two opposing round jets separated by a distance D = 1.4 cm, one injecting a mixture of methane-syngaz (Table 11.1) and the other a low oxygen containing preheated oxidizer at a temperature of 1200 K (Table 11.2). In order to ensure that the stagnation plane is in the center, the momentum ratio of the species with respect to the opposing nozzles is the same and with a deformation rate a = 200 s^{-1} [4].

The governing equations of the opposed flow flame model are mass conservation, conservation of momentum, energy conservation, and species conservation [5]. Chemkin Oppdiff program established by Kee et al. [6] is used for solving equations. For chemical kinetics, the Grimech 3.0 reaction mechanism, which is formed by 53 species and 325 elemental reactions, is adopted.

Fig. 11.1 Configuration of the opposed jet flame

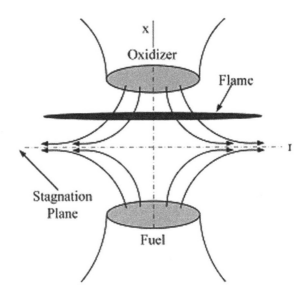

Table 11.1 Fuel mixture composition

CH$_4$ (%)	40	40	40	40
H$_2$ (%)	20	30	36	40
CO (%)	40	30	24	20
Ratio H$_2$/CO	0.5	1	1.5	2
LHV (MJ/m^3)	7.21	7.02	6.90	6.45

Table 11.2 Oxidizer composition

O$_2$ (%)	N$_2$ (%)	CO$_2$ (%)
6	79	15

11.3 Validation of the Computation Procedure

The opposed jets configuration has been used in several studies with different fuels and oxidizers, it was introduced by Sung et al. [7]. Figure 11.2 represents a comparison of the results calculated by the Chemkin program and values derived from the experiment which are temperature and major species CH$_4$, O$_2$, H$_2$O, CO$_2$ and CO. Good agreement is shown. Furthermore, the maximum flame temperature computes accurately which shows that radiation heat losses taken into account are well modeled (Fig. 11.2).

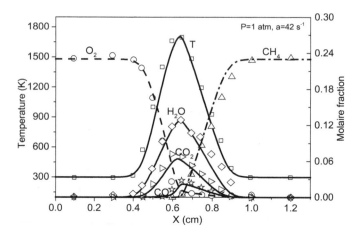

Fig. 11.2 Comparison between calculated and measured values: temperature and major species profiles (symbols represent experimental results)

11.4 Results and Discussion

11.4.1 Effect of H_2/CO Ratio

The Fig. 11.3a, b show the effects of change in syngaz composition according to four H_2/CO ratio values R1, R2, R3 and R4 on temperature and NO formation. As initial heat input for different ratios is nearly the same (Table 11.1), the temperature does not vary significantly, except for R4 (Fig. 11.3a) in which reduction is important, since lower heating value of H_2 is 10.76 MJ/m^3 while that of CO is 12.64 MJ/m^3. The mole fraction of thermal NO is dominated by temperature, it is represented by Fig. 11.3b, it follows the same trend as temperature.

The nitrogen oxides generated by combustion are designated by notation NO_x, mainly NO, NO_2, N_2O, their formation is due on one hand to nitrogen of the air, and on other hand to the presence of a nitrogenous body in the fuel. Figure 11.4a shows the variation of molar fraction of NO_2, where its reduction is greatly affected by increase in molar fraction of hydrogen in the mixture.

Nitrous oxide, also known as nitrous oxide, is a pollutant found in most flames and is represented by Fig. 11.4b. It is produced in oxidizer before flame front and then completely disappear in the zone where the temperature is higher. On the other hand, the variation of the H_2/CO ratio does not have a significant influence on the variation of this species which is not stable.

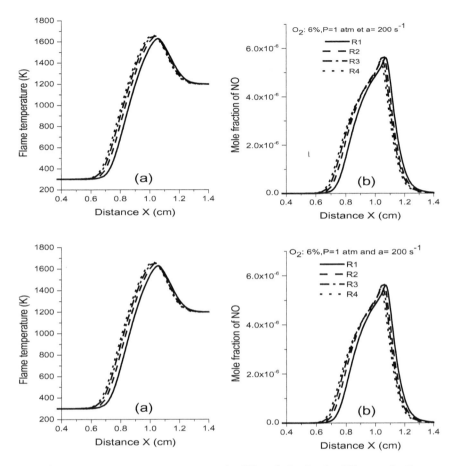

Fig. 11.3 Axial profiles: **a**—flame temperature **b**—NO mole fraction for different ratios R

11.4.2 Effect of Replacement of N_2 by CO_2 in the Air

Figure 11.5 illustrates the variation of maximum temperature as a function of the amount of CO_2 added to the oxidizer which substitutes N_2 contained in the air for different ratios H_2/CO. Oxygen volume in the oxidizer is equal to 6% and strain rate $a = 200\ s^{-1}$. It is noticed that carbon dioxide dilution reduces significantly maximum temperature by 12% for all four ratios. The maximum molar fraction of NO is represented by Fig. 11.5b, it is greatly affected by amount of CO_2 added, and it is reduced with respect to the increase of CO_2.

Le maximum de la fraction molaire de NO_2 est très induit par l'ajout du CO_2 jusqu'à une valeur de égale à 40% (Fig. 11.6a); L'effet de l'addition de CO_2 réduit le maximum de fraction molaire de N_2O progressivement et relativement au pourcentage de d'addition (Fig. 11.6b).

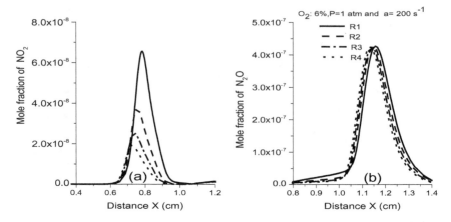

Fig. 11.4 Axial profiles of **a**—molar fraction of NO_2, **b**—molar fraction of N_2O according to the different ratios

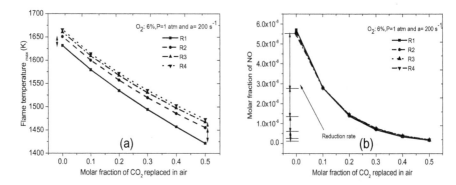

Fig. 11.5 Variation of maximum of: **a**—flame temperature, **b**—NO molar fraction as a function of CO_2 replaced in the oxidant

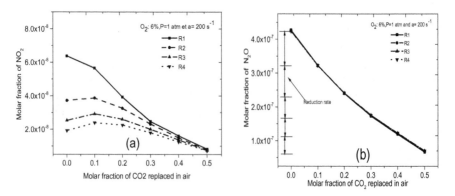

Fig. 11.6 Variation of the maximum of: **a**—NO_2 molar fraction, **b**—N_2O molar fraction as a function of CO_2 replaced in the oxidizer

Figure 11.6a shows that maximum of molar fraction of NO_2 is very induced by addition of CO_2 in the interval 0–40% of CO_2. The effect of CO_2 addition reduces maximum N_2O molar fraction gradually and relative to percentage of CO_2 addition (Fig. 11.6b).

11.5 Conclusion

The conclusions that can be drawn: The temperature is not affected by the variation of H_2/CO ratio, since the initial heat input is nearly constant. The formation of thermal NO is proportional to temperature variation. Increasing hydrogen in the fuel reduces the formation of NO_2. Adding CO_2 to the air greatly reduces temperature and NO species; NO_2 and N_2O.

References

1. K. Jeffrey, Information communication technology development and energy demand in African countries. Energy **189**, 116–192 (2019)
2. A. Mameri, S. Boussetla, R. Belalmi, Z. Aouachria, Combustion characterization of the mixtures biogas-syngas, strain rate and ambient pressure effects. Int. J. Hydrogen Energy **44**(39), 22478–22491 (2019)
3. M. Safer, F. Tabetb, A. Ouadha, K. Safera, A numerical investigation of structure and emissions of oxygen-enriched syngas flame in counter-flow configuration. Int. J. Hydrogen Energy **40**(6), 2890–2898 (2015)
4. E.M. Fisher, B.A. Williams, J.W. Fleming, Determination of the strain in counter flow diffusion flames from flow conditions, in *Proceedings of the Eastern States Section of the Combustion Institute*, pp. 191–194 (1997)
5. A.E. Lutz, R.J. Kee, J.F. Grcar, et al., OPPDIF: a fortran program for computing opposed-flow diffusion flames. Sandia National Laboratories, Livermore, Calif: Report SAND96–8243 (1997)
6. R.J. Kee, J.A. Miller, G.H. Evans, Computational model of the structure and extinction of straind, opposed flow, premixed methane-air flames, in *Twenty-Second Symposium (International) on Combustion*, The Combustion Institute, pp. 1479–1494 (1988)
7. C.J. Sung, J.B. Liu, C.K. Law, Structural response of counter flow diffusion flames to strain rate variations. Combust. Flame **102**, 481–492 (1995)

Chapter 12
Chaotic Bacterial Foraging Optimization Algorithm with Multi-cross Learning Mechanism for Energy Management of a Standalone PV/Wind with Fuel Cell

Issam Abadlia, Mohamed Adjabi, and Hamza Bouzeria

Abstract Variability and intermittency are some of the features renewable energies (REs). Due to their intermittent nature, it is very difficult to predict energy production, which requires either additional supply plants or new storage and control technologies. This work presents the sustainable development of a RE production chain. Reinforcement and optimization of the chain are also considered. At the same time, an energy management strategy (EMS) for a standalone photovoltaic (PV) and wind system integrated with fuel cell is presented. The EMS is aimed to coordinate the power flow of the system components while satisfying load demand and other constraints. System optimization and EMS are combined such that it is unusual to discuss them individually from a system-level design perspective. Therefore, optimization by a Chaotic Bacterial Foraging Optimization (CBFO) algorithm based on multi-cross learning (M-CL) mechanism is proposed to ensure an EMS of the system. The performance of the proposed system is validated by simulation and obtained results prove the efficacy and the feasibility of the proposed approach.

Keywords Standalone multisource · Energy management · Chaotic bacterial foraging optimization algorithm with multi-cross learning

---

I. Abadlia (✉)
Division Énergie Solaire Photovoltaïque, Centre de Développement des Énergies Renouvelables, B.P. 62 Route de l'Observatoire, 1600 Bouzaréah, Algeries, Algeria
e-mail: i_abadlia@yahoo.fr

M. Adjabi
Department of Electrical Engineering, University of Badji Mokhtar Annaba, B.P. 12 Annaba, 23000 Annaba, Algeria

H. Bouzeria
Laboratory LITE, Department of Transport Engineering, University of Mentouri Constantine 1, 25000 Constantine, Algeria

© The Editor(s) (if applicable) and The Author(s), under exclusive license to Springer Nature Singapore Pte Ltd. 2021
A. Khellaf (ed.), *Advances in Renewable Hydrogen and Other Sustainable Energy Carriers*, Springer Proceedings in Energy,
https://doi.org/10.1007/978-981-15-6595-3_12

12.1 Introduction

The population of the planet is increased and its mobility needs to the more and more of the electrical energy. How do to respond without accentuating the greenhouse effect or accelerating the scarcity of fossil resources? RE is one of the possible responses to this major challenge. REs, also called green energies, are not new, although they are becoming increasingly popular in the context of energy transition.

RE describes a collection of energy technologies i.e. wind, solar, geothermal derived from sources that are never-ending to produce the electrical energy. Variability and intermittency are the major drawbacks of REs. However, PV and wind generation are the most promising technologies for supplying load in remote and rural regions. In this case, the possibility of the integrating of diesel generators and storage systems are approaches to overcome reliability problem. The combination of computer technologies, energy sources and storage devices smoothes and reduces the most costly peak production capacities with increase network security.

Energy management is very important for multisource systems. In order to present a preferment algorithm, many conditions and constraints have considerate. The principle of the "intelligent" power grid is presented as being able to reduce the emissions of greenhouse gases and thus fight against the global warming. Finally, it is one of the components for the concept "smart grid". On the other hand, control and management of energy systems aim to ensure the balance of energy flows between suppliers and consumers through the match between production, distribution and consumption. They improved the security of power supply and the overall energy efficiency (minimizing losses and overconsumption), while limiting the environmental impact as CO_2 emission [1–3]. Generally, controlling of storage systems, powers disturbance and equipment costs are the mean task of any EMS. Intelligent algorithm offered always a big success in the EMS applications. It permitted to overcome the nonlinearity problems that also represent a good solution to defeat the problem of modelling and the complexity of the conventional EM algorithms.

In this paper, an optimization and energy management of a multi-power standalone generation system based on CBFO technique with M-CL is proposed. Generation system is composed of renewable PV and wind sources with hydrogen fuel. Where CBFO optimize the system and ensure an optimal power management. Hence, M-CL approach of the algorithm adds more performances. The main contributions of the work are: Investigate CBFO performance to perform a good EMS for a standalone system; Using M-CL technique to perform the CBFO; Increasing the efficiency of the hydrogen storage in fuel system; Maximizing the use of the renewable sources.

12.2 System Description and Problem Formulation

12.2.1 Study of Standalone System

Standalone architecture considered for the application as shown in Fig. 12.1 consists of controllable and uncontrollable micro-sources, where PV generator and wind turbine (WT) are the uncontrollable micro-sources, and storage hydrogen fuel cell production system (H_2FC) constitutes the controllable micro-sources. There are two types of charge DC and AC. Power converters convert power with maximizing function of wind and PV sources. Intelligent switches operating all power system. The standalone conditions operating presented as follows: All loads powered directly from renewable sources, In case of surplus of renewable production, electrolyzer is activated in order to produce the H_2. The case of the demand is more than the production, the energy stored in hydrogen as add, intelligent switches activate the operations of sources and loads. Then, H_2 tank level is controlled at maximum and minimum states.

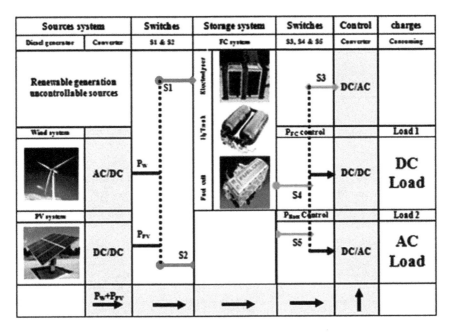

Fig. 12.1 Standalone configuration [1]

12.2.2 Problem Formulation

The standalone optimization model is introduced using the energy flow between the different generators and devices loads. To provide the optimization model, its objective functions are firstly introduced. The total power, denoted as P_t^{AC}, represents the optimized target, which defined as follows [1]:

$$P_t^{AC} = P_t^{WT} + P_t^{PV} + P_t^{FC} - \frac{P_t^{RC}}{bn_{RC}} \quad (12.1)$$

where: P_t^{WT} is the output power from wind turbine at instant t, bn_{RC} are the efficiencies of the converters, P_t^{PV}, P_t^{FC} and P_t^{RC} indicate respectively the power of the photovoltaic generator, the fuel cell generation and the power inverter/rectifier at time t.

The role of the H_2 reservoir is to store H_2 produced by electrolysis of water using alkaline electrolyser. The number of H_2 mol in the tank (NH_2) is given by the difference between the H_2 level generated (NH_2; in) and the H_2 level consumed by the NH_2 FC; out [1]. It is expressed as follows:

$$C-H_2 = \frac{NH_2}{NH_2 Max} \times 100 \quad (12.2)$$

where: $NH_2 Max$ is the number of moles of H_2 in the tank is in plain position. H_2 content in the reservoir given as a percentage using the following equation:

$$NH_2 = \int (NH_2, in - NH_2, out) \, dt \quad (12.3)$$

In our case, we find to: Minimize balanced power P_t^{AC} considering mean square error (MSE) bases. Maximize the injecting renewable powers via intelligent switches. Controlling the C–H_2. Minimize power converters absorbing.

To do, the C–H_2 represents a circumstances optimization problem and variables of EMS unit as defined following.

Ucontrollable sources		Controlable sourses	Power converters
P_W	P_{PV}	P_{FC}	p^{RC}
–	–	C–H_2	bn_{RC}
Total MG power: $P_t^{AC} = P_t^{WT} + P_t^{PV} + P_t^{FC} - \frac{P_t^{RC}}{bn_{RC}}$			

Figure 12.2 Schows the unite of power flow management and switches control that select states of optimaized power system respecting all conditions operation.

Fig. 12.2 EMS control unit of the standalone system

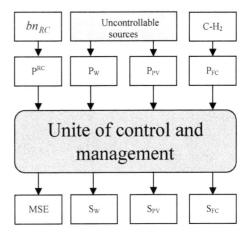

12.3 Energy Management System

In this section, EMS operating standalone system based on BFOA win MC-L mechanism is proposed. BFOA has been widely accepted as a global optimization algorithm of current interest for distributed optimization and control. BFOA [4] inspired by the social foraging behavior of Escherichia coli. BFOA has already drawn the attention of researchers because of its efficiency in solving real-world optimization problems arising in several application domains. The steps of the BFOA applied to reduce the power loss are as follows [5]: Step 1: Initialization of the algorithm parameters. Step 2: Elimination/dispersal loop. Step 3: Reproduction loop. Step 4: Chemotaxis loop. Step 5: Reproduction. Step 6: Elimination/dispersal. Step 7: Otherwise end. The offspring's of individuals are built in using the test vector and parent based on the chaotic procedure shown in [6, 7]. These values are also obtained from the learning mechanism of [7] is also a randomly chosen index.

12.4 Results

Table 12.1 shows the main characteristics of the simulated model. Performance of the Standalone power system under the proposed intelligent management algorithm over a typical uncontrollable sources conversion for one day (Fig. 12.3a) has been evaluated. Demand side is given by Fig. 12.3b. Figure 12.3c shows the different power system that can see the complementarily of the injecting power considering the controllable sources (Fig. 12.3d) via C–H_2 in the tank of gas. Figure 12.3e shows the MSE of the optimization model that prove a small error compared with the max of the produced power.

Table 12.1 Simulation parameters

Fuel cell generator		Electrolyzer	
Type	Nexa ballard 310-0027	Electrolyte	Alkaline
Rating	10,000 W	Electrolyte section	300 cm^2
Number of cells	40	Distance between electrodes	3.0×10^{-4}
Rated current	45	Temperature of the electrolyzer	52 °C
Rated voltage	28	Cathode transfer coefficient	0.5
–	–	Anode transfer coefficient	0.3
PMSG wind turbine		*PV system*	
Rated power	10000 W	Type	ELR-615
Rated voltage	220 V	Rating	10000 W
Rated frequency	50 Hz	Series modules	50
Pole-pairs	2	Parallel modules	4
Stator resistance	2.9 Ω	S-circuit current	3.28 A
Rotor resistance	1.52 Ω	O-circuit voltage	21.6 V
Stator inductance	223 mH	Maxi power/cell	50 W
Rotor inductance	229 mH	–	–
Moment of inertia	0.0048 kg m^2	–	–

12.5 Conclusion

An efficient Chaotic Bacterial Foraging Optimization algorithm with multi-cross learning mechanism for EMS was presented for a standalone power system. Under the proposed scheme, the power plant is able to operate, effectively under both normal climatic conditions. In addition, the scheme provides a dynamic during abnormal conditions, thus making it compatible with the modern power generation.

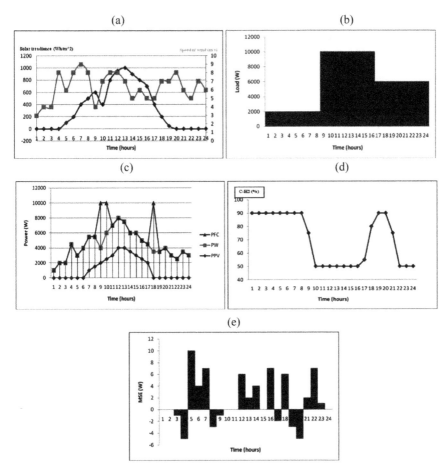

Fig. 12.3 a: Uncontrollable sources (PV + Wind) data, b: load (AC + DC) profile, c: different power system, d: C–H_2 tank, e: MSE of optimization problem

References

1. A. Issam, B. Hamza, Energy management strategy based on fuzzy logic for compound RES/ESS used in stand-alone application. Int. J. Hydrogen Energy **41**(38), 16705–16717 (2016)
2. A. Issam, A. Mohamed, B. Hamza, Sliding mode based power control of grid connected photovoltaic-hydrogen hybrid system. Int. J. Hydrogen Energy **42**(40), 28171–28182 (2017)
3. A. Issam, B. Hamza, Active and reactive power neurocontroller for grid-connected photovoltaic generation system. J. Electr. Syst. **12**(1), 146–157 (2016)
4. K.M. Passino, Biomimicry of bacterial foraging for distributed optimization and control. IEEE Trans. Control Syst. **22**(3), 52–67 (2002)
5. D. Sambarta, D. Swagatam, Adaptive computational chemotaxis in bacterial foraging optimization: an analysis. IEEE Trans. Evol. Comput. **13**(4), 919–941 (2009)

6. J. Zhang, Arthur C: JADE: adaptive differential evolution with optional external archive. IEEE Trans. Evol. Comput. **13**(5), 945–958 (2009)
7. M. Hemmati, A. Nima, System modeling and optimization for islanded micro-grid using multi-cross learning-based chaotic differential evolution algorithm. Int. J. Electr. Power Energy Syst. **56**, 349–360 (2014)

Chapter 13
Mechanical Properties of the Tetragonal $CH_3NH_3PbI_3$ Structure

Kamel Benyelloul, Smain Bekhechi, Abdelkader Djellouli, Youcef Bouhadda, Khadidja Khodja, and Hafid Aourag

Abstract The hybrid organic-inorganic halide perovskite solar cells based on $CH_3NH_3PbI_3$ have attracted enormous attention in the last few years due to their rapid improvement and high certified efficiencies over 20%. In this paper, the tetragonal structure of $CH_3NH_3PbI_3$ hydrides was investigated by first-principles calculations based on density functional theory (DFT) with the generalized gradient approximation (GGA). The elastic constants for the tetragonal $CH_3NH_3PbI_3$ hydride are successfully obtained from the stress-strain relationship calculations. The obtained equilibrium lattice parameters and elastic constant (C_{ij}) are in good agreement with other theoretical values and experimental data available in the literature. The shear modulus (G), bulk modulus (B), Young's modulus (E), Poisson's ratio (ν) and the ratios (B/G) are also determined. The calculated bulk modulus and the ductility factor (B/G) shows that the $CH_3NH_3PbI_3$ hydrides which indicates that this material is ductile behaviour. Also, the universal Poisson's ratio (ν) is higher than the critical value.

Keywords Perovskite structure · First-principles calculation · Elastic properties

---

K. Benyelloul (✉) · Y. Bouhadda · K. Khodja
Unité de Recherche Appliquée en Energies Renouvelables, URAER, Centre de Développement des Energies Renouvelables, CDER, 47133 Ghardaïa, Algeria
e-mail: benyelloul_kamel@yahoo.fr

Y. Bouhadda
e-mail: bouhadda@yahoo.com

S. Bekhechi · H. Aourag
Departement de Physique, Université de Tlemcen, 13000 Tlemcen, Algeria

A. Djellouli
Departement de Physique, Université de Tiaret, 14000 Tiaret, Algeria

© The Editor(s) (if applicable) and The Author(s), under exclusive license to Springer Nature Singapore Pte Ltd. 2021
A. Khellaf (ed.), *Advances in Renewable Hydrogen and Other Sustainable Energy Carriers*, Springer Proceedings in Energy,
https://doi.org/10.1007/978-981-15-6595-3_13

13.1 Introduction

The Methylammounium triiodideplumbate $CH_3NH_3PbI_3$ takes a structural phase transition, cubic, tetragonal and orthorhombic depending on the temperature [1]. Kazim [2], Snaith [3], Park [4] and Kim et al. [5] have discussed on the physical and electronic structure of lead halide perovskites with their photovoltaic application. In addition, Feng [1] have been determined the elastic constants, mechanical properties such bulk, shear, Young's modulus, anisotropic indices, sound velocity and Debye temperature of $CH_3NH_3BX_3$ (B = Sn, Pb; X = Br, I), obtained by the DFT calculations.

$MAPbI_3$, presents different phase transition as a function of temperature, above 327.4 K it adopts a cubic phase (space group Pm-3 m), between 162 and 327.4 K it is tetragonal phase (space group I4/*mcm*, Z = 4) and below 162.2 it is orthorhombic phase (*Pna*21) [6].

The tetragonal phase (165 K < T < 328 K), was confirmed, which is the same space group with the tetragonal phase of $SrTiO_3$.

It is well known that the elastic properties, optical properties and electronic structure of complex hydrides play an important role in many applications related to the mechanical properties of materials and optoelectronic technology. In this paper, the first principle was carried out to investigate the structural and elastic properties of tetragonal Methylammounium triiodideplumbate $CH_3NH_3PbI_3$ compounds. The structural and elastic properties are compared comprehensively to the available theoretical results [1].

The paper is organized as follows. In Sect. 13.2, details of the theory and computational are presented. The most relevant results obtained are presented and discussed in Sect. 13.3. Finally, a conclusion of the present work is given in Sect. 13.4.

13.2 Computational Methods

The organic-inorganic halide perovskite solar cells based on $CH_3NH_3PbI_3$ have been studied by the density functional theory (DFT). The projector augmented wave (PAW) method [7] as implemented in the Vienna Ab initio Simulation Package (VASP) code [8, 9] was employed. Also, the exchange-correlation interaction was treated and described using the generalized gradient approximation correction GGA [10]. The convergence is achieved when the cut-off energy is 800 eV. For Brillouin zone interaction we use the Monkhorst-Pack scheme [11] and k-point was sampled by 5 × 5 × 4. For tetragonal structure organic-inorganic halide perovskite $CH_3NH_3PbI_3$ with space group I4/mcm [6]. The orthogonal structure of $CH_3NH_3PbI_3$ is illustrated in Fig. 13.1 [1].

The elastic constants are computed by applying the strain energy-strain curve method [12]. There is six independent elastic constants for tetragonal structure. The

13 Mechanical Properties of the Tetragonal ...

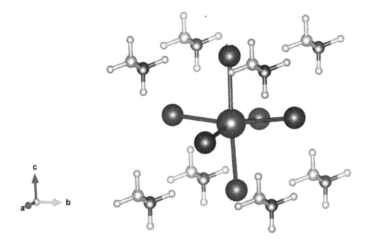

Fig. 13.1 The tetragonal structure for $CH_3NH_3PbI_3$

elastic constants were deduced from the variation of total energy calculations by applying five small dimensionless strains δ with ($\delta = 0$, ± 0.01, ± 0.02).

13.3 Results and Discussion

The atomic positions and lattice parameters were relaxed; this relaxation lowered the total energy. The calculated equilibrium volume (V_0) and bulk modulus (B) are extracted by fitting the total energy as a function of volume, the plot of volume versus energy is shown in Fig. 13.2. The optimized lattice parameters (a and c) are determined. The results are reported in Table 13.1.

Our relaxed structures are in good agreement with reported structures from experimental. For the compounds we have obtained a value of 8.765 and 12.635 Å for the lattice a and c, respectively. This value are in good agreement with experimental values of 8.800 Å [1] and 12.685 Å [1], for a and c respectively, the difference is less then 0.39% and 0.56% for a and c, respectively. To get the bulk modulus (B) we employed the four-parameter Birch-Murnaghan equation of state (EOS) [13]. Consequently, the obtained bulk modulus is 41.52 GPa.

From Table 13.2, it can be seen that the calculated results on this hybrid halide perovskite satisfy the elastic stability conditions, such $C_{11} > 0$, $C_{33} > 0$, $C_{44} > 0$, $C_{66} > 0$, $C_{11} - C_{12} > 0$, $C_{11} + C_{33} - 2C_{13} > 0$, $2C_{11} + C_{33} + 2C_{12} + 4C_{13} > 0$. For tetragonal $CH_3NH_3PbI_3$, it can be seen that $C_{11} > C_{33}$, that means the atomic bonding between neighbours along (100) plane is stronger than along (001) plane.

In case of polycrystalline bulk modulus (B), shear modulus (G), Young's modulus (E) and Poisson ratio are determined. The general expressions for the Voigt and Reuss approaches are represented in Eqs. (13.1)–(13.4) [14] for bulk modulus and shear

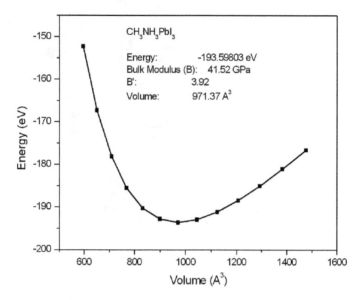

Fig. 13.2 Energy as a function of volume for CH$_3$NH$_3$PbI$_3$

Table 13.1 Structural parameters of CH$_3$NH$_3$PbI$_3$

	a (Å)	c (Å)	V$_0$ (Å^3)	B (GPa)	B'
This work	8.765	12.635	971.37	41.52	3.92
Exp.	8.800[a]	12.685[a]	982.3		

[a]Ref. [1]

Table 13.2 Elastic constants of CH$_3$NH$_3$PbI$_3$ perovskite compounds by calculations. The unit is GPa

System	C$_{11}$	C$_{12}$	C$_{13}$	C$_{33}$	C$_{44}$	C$_{66}$
CH$_3$NH$_3$PbI$_3$	88.51	22.25	10.5	21.8	5.6	21.48
	20.1[a]	14.6[a]	6.8[a]	17.9[a]	1.6[a]	9.2[a]

[a]Ref. [1]

modulus as follows. In Eqs. (13.3) and (13.4) S_{ij}'s represent the elastic compliance constants. The elastic moduli of polycrystalline CH$_3$NH$_3$PbI$_3$ aggregates can then be approximated by Hill's average [14]. $B_H = (B_R + B_V)/2$ and $G_H = (G_R + G_V)/2$.

$$B_V = \frac{1}{9}(C_{11} + C_{22} + C_{33}) + \frac{2}{9}(C_{12} + C_{23} + C_{13}) \quad (13.1)$$

Table 13.3 The bulk modulus, shear modulus, Young's modulus, Poisson ratio and B/G ratio of $CH_3NH_3PbI_3$. The unit is GPa

B_V	B_R	B_H	G_V	G_R	G_H	E	ν	B_H/G_H
31.7	19.52	25.61	14.91	8.91	11.91	30.94	0.298	2.15
		12.2[a]			3.7[a]	12.8[a]	0.33[a]	2.52[a]

[a]Ref. [1]

$$G_V = \frac{1}{15}(C_{11} + C_{22} + C_{33}) - \frac{1}{15}(C_{12} + C_{23} + C_{13}) + \frac{1}{5}(C_{44} + C_{55} + C_{66}) \quad (13.2)$$

$$\frac{1}{B_R} = (S_{11} + S_{22} + S_{33}) + 2(S_{12} + S_{23} + S_{13}) \quad (13.3)$$

$$\frac{1}{G_R} = \frac{4}{15}(S_{11} + S_{22} + S_{33}) - \frac{4}{15}(S_{12} + S_{23} + S_{13}) + \frac{1}{5}(S_{44} + S_{55} + S_{66}) \quad (13.4)$$

Also, the Young's modulus, Poisson's ratio, can be estimated by the following equations [14].

$$E = 9BG/3B + G \quad (13.5)$$

$$N = (3B - 2G)/2(3B + G) \quad (13.6)$$

All the calculated results based on Eqs. (13.1)–(13.6), the results are summarized in Table 13.3.

From Table 13.3, in order to predict the brittleness or the ductility behaviour of our studied compounds, we calculated the ratio between bulk modulus and shear modulus (B/G) as introduced by Pugh [15]. Our calculation shows the ratio B/G is 2.15 is greater than the critical value (1.75), which indicates that this material is ductile behaviour. The universal Poisson's ratio (ν) can also be applied to characterize materials: a material is ductile if ν is higher than 0.26, otherwise the material is brittle. From our results listed in Table 13.3, we can notice that the value of Poisson's ratio for $CH_3NH_3PbI_3$ are larger than 0.26, which is consistent with the results found earlier using the ratio B/G. Also, we can notice that the obtained values for B_H, G_H and E, our results are great than calculated by Feng [1], that is due to the different methods calculations.

13.4 Conclusion

In this paper, ab initio was performed in order to predict the elastic properties of hybrid organic-inorganic halide perovskite $CH_3NH_3PbI_3$. The calculated lattice parameters are in excellent agreement with the experimental data. Elastic properties are calculated by first principle calculation within the generalized gradient approximation (GGA) based on density functional theory (DFT). The single crystal elastic constants for the tetragonal $CH_3NH_3PbI_3$ compounds are successfully obtained from the stress-strain relationship calculations. The obtained elastic constant (C_{ij}) are in good agreement with other theoretical values in the literature. The shear modulus, bulk modulus, Young's modulus, Poisson's ratio and the shear anisotropic factors are also calculated. Analysis of single-crystal and polycrystalline elastic parameters revealed that the hybrid organic-inorganic halide is mechanically stable. The calculated ratio B_H/G_H shows that the compound is ductile. The best of our knowledge, there are, till now, no available experimental data about the elastic constant of tetragonal structure $CH_3NH_3PbI_3$. This research work and obtained results could provide a useful for future studies.

References

1. J. Feng, Mechanical properties of hybrid organic-inorganic $CH_3NH_3BX_3$ (B = Sn, Pb; X = Br, I) perovskites for solar cell absorbers. APL Mater. **2**, 081801 (2014)
2. S. Kazim, M.K. Nazeeruddin, M. Grätzel, S. Ahmad, Perovskite as light harvester: a game changer in photovoltaics. Angew. Chem. Int. Ed. Engl. **53**(11), 2812–2824 (2014)
3. H.J. Snaith, Perovskites: the emergence of a new era for low-cost, high-efficiency solar cells. J. Phys. Chem. Lett. **4**(21), 3623–3630 (2013)
4. N.G. Park, Organometal perovskite light absorbers toward a 20% efficiency low-cost solid-state mesoscopic solar cell. J. Phys. Chem. Lett. **4**(15), 2423–2429 (2013)
5. H.S. Kim, S.H. Im, N.G. Park, Organolead halide perovskite: new horizons in solar cell research. J. Phys. Chem. C **118**(11), 5615–5625 (2014)
6. Y. Kawamura, H. Mashiyama, K. Hasebe, Structural study on cubic–tetragonal transition of $CH_3NH_3PbI_3$. J. Phys. Soc. Jpn. **71**(7), 1694–1697 (2002)
7. G. Kresse, D. Joubert, From ultrasoft pseudopotentials to the projector augmented-wave method. Phys. Rev. B **59**, 1175–1758 (1999)
8. G. Kresse, J. Furthmuller, Efficient iterative schemes for ab initio total-energy calculations using a plane-wave basis set. Phys. Rev. B **54**, 11169–11186 (1996)
9. G. Kresse, J. Hafner, Ab initio molecular dynamics for liquid metals. Phys. Rev. B **47**, 558–561 (1993)
10. J.P. Perdew, K. Burke, M. Ernzehof, Generalized gradient approximation made simple. Phys. Lett. **77**, 3865 (1996)
11. H.J. Monkhorst, J.D. Pack, Special points for Brillouin-zone integrations. Phys. Rev. B **13**, 5188–5192 (1976)
12. K. Benyelloul, H. Aourag, Elastic constants of austenitic stainless steel: investigation by the first-principles calculations and the artificial neural network approach. Comp. Mater. Sci. **67**, 353–358 (2013)
13. F. Birch, J. Geophys. Res. Solid Earth **83**, 1257 (1978)
14. E. Schneider, D.L. Anderson, N. Soga, *Elastic constants and their measurement* (MGraw-Hill, New York, 1974)
15. S.F. Pugh, Relation between elastic moduli and plastic properties of polycrystalline pure metals. Phylos. Mag. **45**, 823–843 (1954)

Chapter 14
Hydrogen Production by the *Enterobacter cloacae* Strain

Azri Yamina Mounia, Tou Insaf, and Sadi Meriem

Abstract Today, alternative and sustainable solutions are proposed to replace fossil fuels. Hydrogen production from renewable energy sources has gained special attention in recent years, especially biohydrogen production from biomass resources which is accepted as a sustainable ecofriendly approach, it is a colorless gas, tasteless, odorless, light and non-toxic. When this gas is used as fuel, air pollution will not be produced, only water is produced and considered as end-product. This work aims to develop a microbial biohydrogen electrosynthesis using *Enterobacter cloacae* strain bacteria isolated from Beni-Messous wastewater. Very little informations are available on the hydrogen production using *E. cloacae*, which is a gram negative, motile, facultative anaerobe. The present work deals an electrochemical study from linear and cyclic voltammetry tests examining the hydrogen evolution reduction in biocathode in order to optimize the biohydrogen production. The acetate fermentation, the proton H^+ accumulation and reduction had allowed an increase in current density from -19 to -96.6 mA/m^2 with the *E. cloacae* strain.

Keywords Biohydrogen · *E. cloacae* · Linear sweep voltammetry first section

14.1 Introduction

The exponential growth of the world's population requires potential sources of clean and low cost energy, most of energy production processes used to date are known for their pollution and greenhouse gases emissions. The renewable energies [1, 2] (wind, solar, bioenergy, etc …) has emerged as hope for humanity even if their applications remain limited. A new bioenergy can be produced by biomass fermentation that is the bio-hydrogen [2], contrary to chemical hydrogen production which is an energy intensive process. Biologically, hydrogen can be produced both by photosynthetic and fermentation pathway [3]. In the fermentation, the hydrogen production rates by *Clostridium butyricum* and *Enterobacter aerogen* were 7.3 and 17 mmol/(g dry cell

---

A. Y. Mounia (✉) · T. Insaf · S. Meriem
Centre de Développement des Energies Renouvelables (CDER), Bouzaréah, Algiers, Algeria
e-mail: y.azri@cder.dz

© The Editor(s) (if applicable) and The Author(s), under exclusive license to Springer Nature Singapore Pte Ltd. 2021
A. Khellaf (ed.), *Advances in Renewable Hydrogen and Other Sustainable Energy Carriers*, Springer Proceedings in Energy,
https://doi.org/10.1007/978-981-15-6595-3_14

per h) respectively, these rates are higher than 0.3 mmol/(g dry cell h) given by the blue green algae *Oscillatoria.* sp via photosynthetic process. The fermentative hydrogen production has been reported from Gram negative and facultative anaerobes, this later can be grown more easily than strict anaerobic bacteria [3]. The *Enterobacter* species studied for hydrogen production include *Enterobacter asburiae* [4], *Enterobacter cloacae* [5] *Enterobacter aerogenes* [6] are recognized as a promising biological tool for biohydrogen production because of their rapid growth in aerobic conditions and their easy adapting to a broad range of substrates as well as the possibility of combining energy generation with simultaneous biodegradable waste materials treatment [7].

The present study concerns bio-hydrogen production from *E. cloacae* strain isolated from a wastewater biofilm. *E. cloacae* is a Gram-negative bacterium with facultative anaerobic metabolism. The bio-hydrogen produced was detected by linear sweep voltammetry on graphite electrode during the fermentation process using acetate substrate according to the following reaction:

$$C_2H_4O_2 + 2H_2O \rightarrow 2CO_2 + 8e^- + 8H^+ \quad \text{bacterial oxidation} \quad (14.1)$$

$$8H^+ + 8e^- \rightarrow 4H_2 \quad \text{reduction on graphite cathode} \quad (14.2)$$

The cathodic reduction of proton H^+ allows the production of biohydrogen and detects it without going through technics often tedious and expensive, this technic remain qualitative and allows to quickly get a result of biohydrogen production.

14.2 Materials and Methods

14.2.1 Microorganism and Culture Conditions

Enterobacter cloacae was isolated from biofilm of Beni-Messous wastewater formed on the working electrode surface under imposed potential +0.155 V/CSE. The bacterial identification was carried out by biochemical method using the API 20E [8]. A culture medium composed of macro and micro nutrients: Acetate 4 g/L, NH_4Cl 0.5 g/L, KH_2PO_4 0.2 g/L, K_2HPO_4 0.2 g/L, $MgCl_2$ 0.25 g/L, $CoCl_2$ 20 mg/L, $ZnCl_2$ 10 mg/L, 10 mg/L $MnCl_2$ regulated at pH 7, was used for bacterial growth. The acetate was used as an energy source and as the sole e-donor for electrochemical experiments. The culture medium is mixed with sodium sulfate to increase the conductivity.

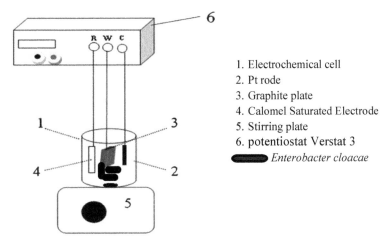

Fig. 14.1 Experimental setup

14.2.2 Electrochemical Cell and Measurements

The first step is to form a cathodic biofilm by chronoamperometric method at 0.155 V/CSE using a three electrode electrochemical system measured by potentiostat Verstat 3 AMTEK Princeton applied research. A sealed electrochemical cell contain-ing a graphite plate 2.5 × 2.5 cm Fig. 14.1. The counter electrode was Platine rod. All potential was quoted versus CSE. The media in both cells were flushed with N_2 for 20 min to create anaerobiosis.

Cyclic voltammetry (CV) was carried at the beginning and after chronoamperometry to compile the information of catalytic responses. Start and end potential were −1 to +1 V with scan rate 0.01 mV/s.

Linear sweep voltammetry (LSV) from +0.2 to −1 V (vc CSE) was performed to evaluate the electroactivity of the cathode biohydrogen evolution reduction at different times of the chronoamperometric kinetic during the cathodic biofilm formation on about 10 days. The LSV was practiced every 2 days during chroamperometry kinetic.

Scanning electron microscopy analysis of biocathode was carried out using HITACHI model TM 1000 at an acceleration voltage of 10 kV with a variable pres-sure system (University of Boubker Belkaid, Department of Physics, Tlemcen).

14.3 Results and Discussion

In the present study, the cathode was modified and so performed by cultivating the *E. cloacae* collected from bioanode effluent wastewater operated in electrochemical at +0.155 V during 10 days. Figure 14.2 shows the monitored current during

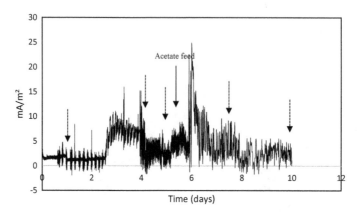

Fig. 14.2 Biofilm formation with *E. cloacae* using chronoamperometric method under + 0.155 V/CSE, linear sweep voltammetry

the *E. cloacae* biofilm formation at potential +0.155 V/SCE. A current increase after 2 days and remains stable until the 4th day, after that a current decrease was gradually registered. With the substrate consumption, the current density increased significantly and thus demonstrated maximum bioelectricity activity indicating electroactive biofilm formation. In fact, the Gram$^-$ bacteria group to which *E. cloacae* belongs is known to generate a higher current compared to Gram$^+$ bacteria [9]. It was also previously studied that the current density obtained with the *E. cloacae* bacteria studied separately was very low compared to a results given with bacteria consortium containing *E. cloaceae* [10]. Kumar and Das have already reported the *E. cloacae* for biohydrogen production [5]. According to reaction 1, the substrate was consumed by *E. cloacae* in anaerobic condition by releasing protons H^+ and electrons e^-. Thus, this study will help the biocathode formation for hydrogen production by recording the hydrogen evolution reduction through the linear sweep voltammetry at different points.

From the linear sweep voltammetry, it can be demonstrated that biocathode led to higher current production compared to the current produced with electrode without inoculum Fig. 14.3. The hydrogen evolution in the biocathode was −0.6 V, indeed the current obtained increases with the proton H^+ reduction according to the reaction 2. It can be demonstrated that the biocathode formed with *E. cloacae* biofilm shows a higher current when accumulation of H^+ protons and their reduction allows the current increase of about −19 mA/m^2 at 1st day to −96.6 mA/m^2 at 10th days under −0.6 V.

Figure 14.4 shows the catalytic activity between biocathode (with *E. cloacae*) and control (without *E. cloacae*) under the potential range +1 to −1 V, a significant reduction activity was observed from −0.6 V and no oxidation peak is noticed. However, an important oxidation current was observed when the voltammetry was scanned from 1 to 0 V, this current density increase was attributed to hydrogen oxidation reaction where the generated hydrogen (at −0.6 V) was re-oxidized under

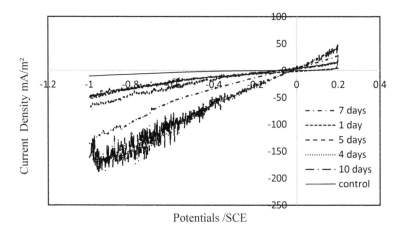

Fig. 14.3 Linear sweep voltammetry of hydrogen evolution

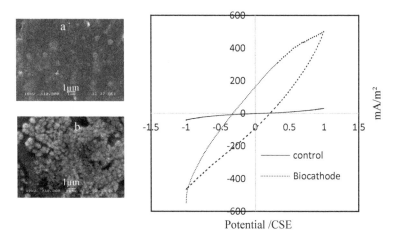

Fig. 14.4 Cyclic voltammetry of the *E. cloacae* biocathode and SEM surfaces images of the **a** control graphite, and **b** *E. cloacae* biocathode

the extra-membrane hydrogenases effect [11]. Otherwise, the control showed only reduction activity at -0.6 V with current of -4.16 mA/m^2 which was significantly lower than biocathode current of -96.6 mA/m^2. The catalytic properties proved bacterial growth on the electrode surface [11]. Data suggest that hydrogen production was significant after cathodic potentials -0.6 V. The cyclic and linear sweep voltammetries do not show an action on the hydrogen evolution reduction (HER) on the biocathode but only on the current increase indicating a high concentration of biohydrogen produced.

It can be demonstrated that the biocathode formed with *E. cloacae* shows a higher current compared to the control electrode (without *E. cloacae*). The accumulation of H$^+$ proton and their reduction allowed to increase the current from -19 to -96.6 mA/m^2. According to this study, the electrochemical system tension can be fixed at -0.6 V in the biocathode of *E.cloacae* for hydrogen production.

14.4 Conclusion

The biocathode formed with *E. cloacae* led to higher performance compared to the untreated electrode, it allows to produce H$^+$ protons through the acetate fermentation. The linear sweep voltammetry corresponding to the reduction of accumulated protons and cyclic voltammetry confirms the electrode modification by significant current density product. The linear sweep voltammetry measurements corresponded to the accumulated protons reduction.

References

1. A. Benkaraache, Production de l'hydrogène par un procédé d'électrosynthèse microbienne (ESM) (2017)
2. L. Hong, S. Grot, B. Logan, Electrochemically assisted microbial production of hydrogen from acetate. Environ. Sci. Technol. **39**(11), 4317–4320 (2005)
3. N. Kumar, D. Das, Enhancement of hydrogen production by *Enterobacter cloacae*. Process. Biochem. **35**(6), 589–593 (2000)
4. S. Jong-Hwan, Y. Jong Hyun, A. Eun Kyoung, K. Mi-Sun, S. SangJun, P. Tai Hyun, Fermentative hydrogen production by the newly isolated *Enterobacter asburiae* SNU-1 32. Int. J. Hydrogen Energy **32**(2), 192–199 (2007)
5. K. Nath, A. Kumar, D. Das, Effect of some environmental parameters on fermentative hydrogen production by *Enterobacter cloacae* DM11. Can. J. Microbiol. **52**(6), 525–532 (2006)
6. S. Tanisho, Hydrogen production by facultative anaerobe *Enterobacter aerogenes*. Biohydrogen 273–279 (1998)
7. D. Levin, L. Pitt, M. Love, Biohydrogen production: prospects and limitations to practical application. Int. J. Hydrogen Energy **29**(2), 173–185 (2004)
8. L. Prescott, J. Harley, D. Klein, J. Willey, Microbiologie. De Boeck Supérieur (2003)
9. C.E. Milliken, H.D. Ma, Sustained generation of electricity by the spore-forming, gram-positive, *Desulfitobacterium hafniense* strain DCB2. Appl. Microbiol. Biotechnol. **73**, 1180–1189 (2007)
10. S. Ishii, K. Watanabe, S. Yabuki, B. Logan, Y. Sekiguchi, Comparison of electrode reduction activities of *Geobacter sulfurreducens*. Appl. Environ. Microbiol. **73**(23), 7348–7355 (2008)
11. F. Aulenta, L. Catapano, L. Snip, M. Villano, M. Majone, Linking bacterial metabolism to graphite cathodes: electrochemical insights into the H$_2$-producing capability of Desulfovibrio sp. Chem. Sus. Chem. **5**, 1080–1085 (2012)

Chapter 15
Visible Light Hydrogen Production on the Novel Ferrite CuFe$_2$O$_4$

S. Attia, N. Helaïli, Y. Bessekhouad, and Mohamed Trari

Abstract The spinel type compounds represent a new family of photocatalysts that can be used as photoelectrodes capable to produce hydrogen under visible light. In the current study, the CuFe$_2$O$_4$ spinel, which is prepared by the sole-gel method, is investigated as a possible candidate, and the structure, opto-electronic, electrochemical and photoactive properties are characterized. CuFe$_2$O$_4$ exhibits a p-type semiconductor; the conduction mechanism is assigned to the typical small polaron hopping of semiconductor conduction band 'd'. Determining the potential of the flat strip electrochemically has established an energy digraph to predict the duality between the physico-chemical properties and photocatalytic activity. The production of hydrogen under visible light was selected to evaluate the photoactivity. Sensors holes: Na$_2$S, Na$_2$SO$_3$ and oxalic acid were studied. The best performance was obtained by Na$_2$S in basic medium.

Keywords CuFe$_2$O$_4$ · Spinel · Physical properties · Hydrogen production

15.1 Introduction

The spinels oxides are important materials for a wide range of applications such as electronics, materials with magnetic properties, as catalysts and photo-catalysts [1–7]. The ternary oxides (Cu^{2+}Fe$_2^{3+}$O$_4$ → A $^{+\text{II}}$B$_2^{+\text{III}}$O$_4$) of spinel structure are very promising semiconductors for the production of H$_2$ as well as for the oxidation of minerals such as S^{2-} or SO$_3^{2-}$ under visible light [8–10]. These compounds offer an interesting compromise between optical and transport properties, which is in contrast with that observed for simple oxides such as TiO$_2$ or ZnO [11, 12]. These

---

S. Attia · N. Helaïli (✉) · M. Trari
Laboratory of Storage and Valorization of Renewable Energies, Faculty of Chemistry, U.S.T.H.B, BP 32, 16111 El-Alia, Algiers, Algeria
e-mail: nassimahelaili@yahoo.fr

Y. Bessekhouad
National Veterinary High School, BP 161, El Harrach, Algiers, Algeria

are considered as wide band gap materials and suffer from an insulative nature [13–15]. These oxides have the advantage of being chemically stable with a suitable Vbp flat band potential. This fact makes it possible to induce desired oxidation-reduction reactions without polarization [16]. However, such oxides absorb only a small portion of sunlight and have low on version efficiency. This behavior appears to result from the anionic nature of the valence band that flows from the deep O^{2-}: 2p6 orbital [17].

15.2 Experimental Procedure

The spinel CuF_2O_4 was prepared by a classical Sol-Gel method. A certain amount of $Cu(NO_3)_2$, $6H_2O$ and $Fe(NO_3)_3$, $9H_2O$ with purity greater than 99% were dissolved in 100 mL of water containing Agar-Agar (1 mg mL^{-1}). The mixture is heated at 70 °C for 3 h before to be completely dehydrated at 100 °C. A black precipitate is generally obtained which becomes homogeneous by grinding in agate mortar. This powder was calcined in static air condition at 850 °C for 12 h. This calcination was repeated three times. The photoactivity is studied in a double walled Pyrex cell connected with a cooling system; the temperature was fixed at 50 °C ± 1 °C. 100 mg of the CuF_2O_4 powder was dispersed in 200 mL of solutions (Na_2SO_4, 0.5 M) containing Na_2SO_3, Na_2S and oxalic acid (10^{-3} M) at pH 7 under stirring (210 rpm). The nitrogen was bubbled in the solution (10 mL mL^{-1}) during 35 min before the launch of the test. The tree tungsten lamps (3 × 200 W) were used for as a source of light photocatalytic system. Hydrogen gas was identified by gas chromatography (Agilent Technology 7890A) using a thermal conductivity detector. The volume was collected in a water manometer at standard temperature and pressure (STP).

The crystal structure of the spinel phase was confirmed by X-ray diffraction measurements (XRD). The patterns are recorded with a Philips diffractometer equipped with Ni-filtered CuK_α radiation.SEM Analysis For this analysis, MEP-EDX Quanta TM 250 with tungsten filament was used by the company FEI. It is a domain microscope to explore the properties of nanoscale materials and to characterize their structure. The Fourier transform infrared reflectance spectroscopy is recorded in the range of 400–4000 cm^{-1} using the Bio-Rad-FTS 3000MX system. The morphology of the oxide is obtained using a HITACHI S2500 scanning electronmicroscope. The Cary 500 UV-VIS-NIR spectrophotometer is used for the study of optical properties. On the other hand, the transport properties are studied by means of the thermal variation of both conductivity (s) and thermopower S = (DV/DT). The intensity potential J(E) curves and capacitance plots were drawn with apotentiostat/galvanostat (PGZ 301 Radiometer), Pt-electrode and a saturated calomel electrode (SCE) were employed as emergency and reference electrodes respectively. The photoactivity is studied in a double walled Pyrex cell connected with a cooling system; the temperature was fixed at 50 °C ± 1 °C. 100 mg of the CuF_2O_4 powder was dispersed in **200 mL** of solutions (Na_2SO_4, 0.5 M) containing Na_2SO_3, Na_2S and oxalic acid (10^{-3} M) at pH 7 under stirring (210 rpm). The nitrogen was bubbled in the solution (10 mL mL^{-1}) during 35 min before the launch of the test. The tree tungsten lamps (3 × 200 W) were used for as a source of light photocatalytic system. Hydrogen gas was identified by gas chromatography (Agilent Technology 7890A) using a thermal conductivity detector. The volume was collected in a water manometer at standard temperature and pressure (STP).

15.3 Results and Discussion

The XRD Patterns of the crystalline $CuFe_2O_4$ synthesized by the sol-gel method is presented in (Fig. 15.1). All the picks of the spinel are indexed in tetragonal unit cell with the space group I41/amd, in concordance with the JCPDS cards 34-0425. The lattices parameters of $CuFe_2O_4$ a = 0.5836 and b = 0.8632 nm (Table 15.1), obtained by Rietveld refinement of the XRD pattern (see Fig. 15.1).

The FTIR spectra show the same vibration bands for all samples (see Fig. 15.2) and are similar to those reported in the literature [18, 19].

The large vibration band in the ranges (3000–4000 cm^{-1}) and (1600 and 1700 cm^{-1}) are ascribed respectively to the stretching and bending of OH groups of adsorbed water. It is well known that spinel type oxides show two characteristic vibration bands that are attributed to tetrahedral and octahedral sites [20, 21]. In our case, the vibration bands observed in the 510–535 cm^{-1} range are attributed to the atoms in the tetrahedral sites while those observed in the 580–680 cm^{-1} domain are

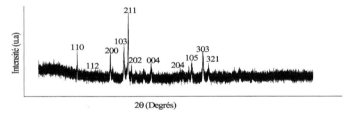

Fig. 15.1 XRD patterns of $CuFe_2O_4$ prepared via sol–gel at 850 °C

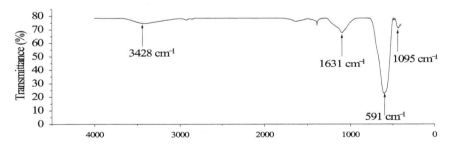

Fig. 15.2 The infrared spectrum of $CuFe_2O_4$ elaborated by sol–gel at 850 °C

Fig. 15.3 MEB analysis of CuFe$_2$O$_4$ treated for 12 h at 850 °C

attributed to the atoms in octahedral sites. The surface morphology of the CuFe$_2$O$_4$ sample is obtained by scanning electron microscopy (see Fig. 15.3).

It reveals the formation of pseudo-spherical grains of micrometric size. The grains are connected to each other and form blocks.

The variation of diffuse reflectance (%R) shows an intense reflection in the visible region (see Fig. 15.4).

The Pankov relation (15.2) is used to determine the optical gap energy (E$_g$) [15]:

$$\alpha h\upsilon = \text{const} * \left(h\upsilon - E_g\right) \quad (15.1)$$

The exponent n indicates the nature the transition, 0.5 or 2 respectively for direct or indirect transition. The direct transition (m = 2), was obtained from the plot $(\alpha h\upsilon)^n$ versus the photon energy (hυ). The extrapolation of the straight line $(\alpha h\upsilon)^2 = 0$ gives a gap E$_g$ of 1.47 eV (see Fig. 15.4). The E$_g$ values of these materials were between 1.3 and 1.58 eV (Table 15.2). It can be seen that each of these ferrites have E$_g$ lower than 2 eV, making them efficient photo-catalysts under visible light irradiation. It is also important to note that these ferrites have a smaller gap compared to other photo-catalysts such as Fe$_2$O$_3$ (~2 eV), CdS (2.4 eV) and TiO$_2$ (~3 eV) [22, 23]. Consequently, their ability to absorb visible light was enhanced compared to the

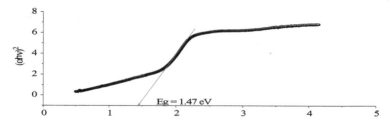

Fig. 15.4 Optical gap of CuFe$_2$O$_4$ prepared by sol-gel route

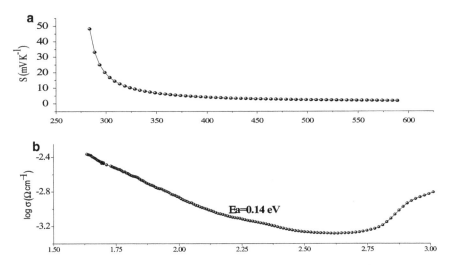

Fig. 15.5 a The thermal variation of the thermo-power (S) of $CuFe_2O_4$, **b** the logarithm of the electrical conductivity as function of $10^3/T$ for $CuFe_2O_4$

photocatalysts that absorb in the UV and visible regions. The dependence of the Seebeck (S) coefficient on the temperature is shown in (see Fig. 15.5). The positive values $S_{300\,K} = 589$ (μV/K) all samples close to room temperature indicate that the conductivity is that of a p-type material, i.e., the holes represent the majority carriers. The conduction mechanism is believed to take place through small polaron-hoppings since such materials present an incomplete 'd' orbital that constitute the conduction band.

The plot of the variation of the logarithm of the electrical conductivity (σ) as a function of the inverse of the temperature exhibits a linear trend (see Fig. 15.5b).

The increase of σ with temperature confirms the semiconductor character of spinel $\sigma_{300K} = 4.58 \times 10^{-5}$ (ohm.cm^{-1}). This indicates that the conduction mechanism occurs by thermal activation of the charge carriers and that the Arrhenius model can be applied for the determination of the activation energy $\Delta E\sigma$. This leads us to deduce an activation energy $\Delta E\sigma$ of $CuFe_2O_4$ is 0.14 eV. The photocurrent voltage (J-E) characteristics for the ferrite materials in the dark and under illumination were investigated in the (Na_2SO_4, 0.5 M) with a scanning speed rate fixed at 10 mV/s. The photocurrent (Jph) increased along the positive potential confirming that the SC material behaves like p-type semi-conductor. The flat band potential V_{fb} (−0.041 V) and the holes density NA (4.52×10^{19} cm^{-3}) are provided, respectively, from the intercept of the potential axis at $C^{-2} = 0$. To understand how the light is converted into chemical energy, we have drawn the energy diagram that establishes the relationship between the semiconductor and the charge transfer [20]. The electrons excited by visible light should have a potential enough negative to reduce water molecules i.e. more cathodic than HER. S_2O_3, Na_2S and oxalic oxide were used to scavenge the

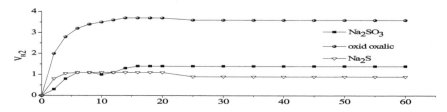

Fig. 15.6 Hydrogen evolution of CuFe$_2$O$_4$ in KOH electrolyte as function for time

photo hole, the half electrochemical reactions occur simultaneously on the opposite sides of the crystallite:

Cathodic pole:

$$\mathrm{CuFe_2O_4} + h\upsilon \rightarrow \mathrm{CB}(e^-) + V_B(h^+) \tag{15.2}$$

$$\mathrm{CuFe_2O_4} - \mathrm{CB}(2e^-) + 2\mathrm{H_2O} \rightarrow \mathrm{H_2} + 2\mathrm{OH}^- \tag{15.3}$$

Anodic pole:

$$\mathrm{S_2O_3^{2-}} + 4h^+ + 6\mathrm{OH}^- \rightarrow 2\mathrm{SO_3^{2-}} + 3\mathrm{H_2O} \tag{15.4}$$

The determination of the optical, electrochemical and electrochemical optical characteristics makes it possible to locate the CuFe$_2$O$_4$ valence and conduction bands on the energy diagram with respect to the vacuum and with respect to the saturated calomel electrode and to predict thus the photocatalytic reactions. The CuFe$_2$O$_4$ conduction band is positioned at a more negative potential than the H$_2$O/H$_2$ level. This causes a spontaneous evolution of H$_2$ under illumination. The hydrogen evolution is see in Fig. 15.6.

The quantity of H$_2$ is measured volumetrically and it was identified by gas chromatography. A good activity is observed on oxalic acid after 20 min of irradiation corresponding to 3.4 mmol h^{-1} g^{-1} in oxalic acid.

15.4 Conclusion

The photoelectrochemical performance of the pure tetragonal phase of the CuFe$_2$O$_4$ was successfully prepared by the sol-gel method. Their structural and optical properties point to a direct allowed transition. Under visible illumination, the photocatalytic results indicated that the show a much higher activity in oxalic oxide to water photo-reduction. This can be ascribed to a narrower band gap, a higher electrical conductivity and amore facile flow of charge due to enhanced (e$^-$/h$^+$) separation at the catalyst surface. The conduction band is clearly negative as the potential of H$_2$O/H$_2$ couple allowing unbiased hydrogen liberation. Three holes scavengers in

KoH medium were studied: Na_2S, Na_2SO_3 and oxalic acid, the best amount to H_2 production attains 3.4 mmol h^{-1} g^{-1} in oxalic acid under ideal conditions.

References

1. K. Gurunathan, J.O. Baeg, S.M. Lee, E. Subramanian, S. Moon, K.J. Kong, Visible light active pristine and Fe^{3+} doped $CuGa_2O_4$ spinel photocatalysts for solar hydrogen production. Int. J. Hydrogen Energy **33**(11), 2646–2652 (2008)
2. P. Cheng, C. Deng, Gu. Mingyuan, W. Shangguan, Visible-light responsive zinc ferrite doped titania photocatalyst for methyl orange degradation. J. Mater. Sci. **42**(22), 9239–9244 (2007)
3. S. Khangembam, M. Victory, W. Surchandra, S. Phanjoubam, Electrical and structural properties of zinc substituted nickel ferrites synthesized by sol-gel tech. J. Adv. Mater. Proc. **2**(3), 162–166 (2016)
4. D.P. Lapham, A.C.C. Tseung, The effect of firing temperature, preparation technique and composition on the electrical properties of the nickel cobalt oxide series $Ni_xCo_{1-x}O_y$. J. Mater. Sci. **39**, 251–264 (2004)
5. T. Nakamura, Y. Ogawa, N. Sengoku, Y. Yamada, Magnetic susceptibility study on Li-Mn spinel oxides. J. Mater. Sci. **38**, 4597–4601 (2003)
6. X. Yan, J. Wang, L. Zhang, J. L. Cong, Photocatalytic activity of magnetically separable La-doped $TiO_2/CoFe_2O_4$ nanofibers prepared by two-spinneret electrospinning. J. Mater. Sci. **47**, 465–472 (2012)
7. R. Sharma, S. Singhal, Structural magnetic and electrical properties of zinc doped nickel ferrite and their application in photo catalytic degradation of methylene blue. J. Physica B **414**, 83–90(2013)
8. V. Preethi, S. Kanmani, Photocatalytic hydrogen production over $CuGa_{2-x}Fe_xO_4$ spinel. Int. J. Hydrogen Energy **37**(24), 18740–18746 (2012)
9. Y. Bessekhouad, M. Trari, Photocatalytic hydrogen production from suspension of spinel powders AMn_2O_4 (A = Cu and Zn). Int. J. Hydrogen Energy **4**(27), 357–478 (2002)
10. A. Kezzim, N. Nasrallah, A. Abdi, M. Trari, Visible light induced hydrogen on the novel hetero-system $CuFe_2O_4/TiO_2$. J. Energy Convers. Manag. **52**(8), 2800–2806 (2011)
11. W. Chen, L. Siyao, L. Changyan, L. Xiaoyan, C. Chuansheng, Photocatalytic performance of single crystal ZnO nanorods and ZnO nanorods films under natural sunlight. J. Inorg. Chem. Commun. **114**, 107842 (2020)
12. A. Fujishima, T.N. Rao, D.A. Tryk, Synthesis and characterization of efficient TiO_2 mesoporous photocatalysts. J. Mater. Today **11**(4), 11526–11533 (2017)
13. S.H. Deng, M.Y. Duan, M. Xub, L. Hea, Structural, electronic and optical properties of copper-doped $SrTiO_3$ perovskite. J. Physica B: Condens. Matter **11**(406), 2055–2322 (2011)
14. H. Sato, K. Ono, T. Sasaki, A. Yamagishi, Semiconductor nanosheet crystallites of quasi-TiO_2 and their optical properties. J. Phys. Chem. B **36**(107), 9824–9828 (2003)
15. R. Abe, K. Sayama, H. Arakawa, Significant effect of iodide addition on water splitting into H_2 and O_2 over Pt-loaded TiO_2 photocatalyst: suppression of backward reaction. J. Chem. Phys. Lett. **3–4**(371), 360–364 (2003)
16. N. Helaili, USTHB. Doctoral thesis (2014)
17. R. Brahimi, Y. Bessekhouad, M. Trari, Effect of S-doping toward the optical properties of WO_3. J. Physica B **18**(407), 4689–3924
18. S. Chitra, P. Kalyani, T. Mohan, M. Massot, S. Ziolkiewicz, R. Gangandhanran, M. Eddrief, C. Julien, Sol. Sta. Ion. **4**, 8 (1998)
19. S. Chitra, P. Kalyani, T. Mohan, R. Gangadharan, B. Yebka, S. Castro-Garcia, M. Massot, C. Julien, M. Eddrief, Characterization and electrochemical studies of $LiMn_2O_4$ cathode materials prepared by combustion method. J. Electroceram. **3**, 433–441 (1999)
20. F. Galosso, *Structure and properties of inorganic solids*, 1st edn. (Pergamom, New York, 1970)

21. G. Rekhila, Y. Gabes, Y. Bessekhouad, M. Trari, Hydrogen production under visible illumination on the spinel $NiMn_2O_4$ prepared by sol gel. J. Sol. Energy **166**, 220–225 (2018)
22. R. Bagtache, I. Sebai, K. Abdmeziem, M. Trari, Visible light induced H_2 evolution on the heterojunction Ag/NiO prepared by nitrate Route. J. Sol. Energy. **177**, 652–656 (2019)
23. I. Sebai, N. Salhi, G. Rekhila, M. Trari, Visible light induced H_2 evolution on the spinel $NiAl_2O_4$ prepared by nitrate route. Int. J. Hydrogen Energy **43**(42), 26652–26658 (2017)

Chapter 16
Prediction of New Hydrogen Storage Materials: Structural Stability of SrAlH$_3$ from First Principle Calculation

Youcef Bouhadda, Kamel Benyelloul, N. Fenineche, and M. Bououdina

Abstract The structural stability of the SrAlH$_3$ has been investigated using the density-functional theory with the generalized-gradient approximation, pseudo-potential and plane wave method, for hydrogen storage application. Indeed, we have predicted structures for hypothetical compounds SrAlH$_3$ by considering 24 different guess structures possible and have been attained through energy minimization and force relaxation, for the above guess structures. The most stable arrangement being the tetragonal PCF$_3$-type (tP40) structure, which contains distorted octahedral. At higher pressure this phase transform into the orthorhombic GdFeO$_3$-type (oP20) at 31.8 GPa. We have fitted the total energy as function of cell volume using the so-called universal equation of state and we have found that SrAlH$_3$ have a bulk modulus B equal to 37.11 and 44.88 Gpa for P4$_2$/nmc (space group n° 137) and Pnma (space group 62) respectively. The formation energy, the electronic density of states, charge density, charge transfer, electron-localization function and Born effective charge are investigated and discussed.

Keywords Hydrogen storage · Ab initio · SrAlH$_3$

Y. Bouhadda (✉) · K. Benyelloul
Unité de Recherche Appliquée en Energies Renouvelables, URAER, Centre de Développment des énergies Renouvelables, CDER, 47133 Ghardaïa, Algeria
e-mail: bouhadda@yahoo.com

K. Benyelloul
e-mail: benyelloul_kamel@yahoo.fr

N. Fenineche
FR FCLAB, UTBM bât. F, Rue Thierry Mieg, 90010 Belfort Cedex, France

M. Bououdina
Department of Physics, College of Science, University of Bahrain, 32038, Zallaq, Bahrain

© The Editor(s) (if applicable) and The Author(s), under exclusive license to Springer Nature Singapore Pte Ltd. 2021
A. Khellaf (ed.), *Advances in Renewable Hydrogen and Other Sustainable Energy Carriers*, Springer Proceedings in Energy,
https://doi.org/10.1007/978-981-15-6595-3_16

16.1 Introduction

The recent meteorological and ecological reports showed an anomalous climate change compared to old studies [1]. This can be explained by pollution and the green house effect due to the progressive consumption of fossil energy especially in the transport field.

Using clean energy as hydrogen can be a solution of the green house effect and climate change problem. Indeed, hydrogen can be burned in the presence of oxygen and produce only energy and water with zero CO_2 gas [1]. However, the widespread of hydrogen as fuel in transportation is penalized by its storage in vehicle. Many methods are used to hydrogen in vehicle but the solid-state storage is considered as a promising technique. In fact, numerous materials can store hydrogen in solid state and can be used under specific conditions (temperature and pressure) in Fuel cell car. Among them, ABH_3 family has been intensively studied due to their attractive perovskite crystal structure. The perovskite hydrides are of interest for hydrogen storage and can have cubic, orthorhombic, hexagonal or tetragonal phases depending on pressure and temperature. The known perovskite hydrides with the ideal *Pm3m* structure (cubic) are $LiBeH_3$, $SrLiH_3$, $CsCaH_3$, $BaLiH_3$, $KMgH_3$, and $RbCaH_3$ [2–6].

We have previously studied the electronic, dynamic, thermodynamic and mechanic properties of both $NaMgH_3$ (orthorhombic) and $KMgH_3$ (cubic) [7–10]. Indeed, we have investigated the optical response and the theoretical optical spectrum [7] of $NaMgH_3$. However, the study of the thermodynamic functions of $NaMgH_3$ was done for the first time [8] and show that $NaMgH_3$ is dynamically stable.

We have also investigated the mechanical properties of $NaMgH_3$ and we have found that this compound is mechanically stable at ambient pressure and can be classified as brittle material [9]. Also, we have found that the calculated linear bulk moduli are in good agreement with the theoretical value reported in the literature. We have determined Debye temperature (648 K°) using theoretical elastic constants. This value is higher compared to that of $KMgH_3$, which reveals that $NaMgH_3$ is harder than $KMgH_3$. Moreover, we have calculated and presented the dynamical and thermodynamic properties of $KMgH_3$, such as thermodynamic functions and formation energy which was calculated for different possible reaction pathways [10]. These results encouraged us to extend our research activity to find and predict unexplored ABH_3 hydrides.

In this work, we investigate and predict the structural stability of unknown $SrAlH_3$ phases. In the best of our knowledge, no experimental or theoretical investigation has been reported in the literature on $SrAlH_3$ hydride.

16.2 Computational Methods

In this study, we have used for all electronic structure computations the ABINIT code [11], which based on plane-waves and pseudopotentials within density functional theory (DFT) [12]. Indeed, an efficient fast Fourier transform algorithm [13] is employed for the conversion of wave functions between real and reciprocal space. In addition, we used an adaptation to a fixed potential of the band-by-band conjugate-gradient method [14] with a potential-based conjugate-gradient algorithm for the determination of the self-consistent potential [15]. To represent atomic cores we used the Generalized Gradient Approximation (GGA–PBE) [16] to describe exchange energy with Fritz–Haber–Institute GGA pseudopotentials [17].

The atoms are progressively relaxed to equilibrium until the Hellmann–Feynman forces on all atoms were less than 0.005 meV/Å. The electronic wave functions were extended in plane waves up to a kinetic energy of cut-off 60 hartree. Integrals over the Brillouin zone were done with similar k-grid densities in all calculations: 6 × 6 × 6 mesh of special k-point. A tolerance in the total energy of 0.01 meV/atom was chosen for the self-consistency.

16.3 Results and Discussion

In order to determine the crystal structure of $SrAlH_3$ stable phase, we have used as inputs in the structural optimization calculations, twenty-four potential structure types. The concerned structure types are (Pearson structure classification notation in parenthesis): α-$CsMgH_3$ (Pmmm), $BaRuO_3$ (hR45), $SrZrO_3$ (oC40), $GdFeO_3$ [$NaCoF_3$(oP20)], $BaTiO_3$ [$RbNiF_3$ (hP30)], $CsCoF_3$ (hR45), PCF_3 (tP40), $KCuF_3$ (tP5), $KCaF_3$ (mP40), $NaCuF_3$ (aP20), $SnTlF_3$ (mC80), $KCaF_3$ (mB40), $LiTaO_3$ (hR30), $KCuF_3$ (oP40), $CaCO_3$ (mP20), $KNbO_3$ (tP5), $KNbO_3$ (oA10), $KNbO_3$ (hR5), $LaNiO_3$ (hR30), $CaTiO_3$ (oC10), $FeTiO_3$ (hR30), $KCuF_3$ (tI20), $CaTiO_3$ [$CsHgF_3$ (cP5)], and $KMnF_3$ (tP20) [18].

We have performed full geometry optimization with both force and stress minimization, starting with the above chosen structure.

It should be noted that, during the full geometry optimization, some of the initial structures are physically instable or converted into another structure type. We report in Table 16.1 and Fig. 16.1 the atomic positions and crystal structure of only the stable structures with lowest total energy as drown in Fig. 16.2. We must note that the crystal structure of $SrAlH_3$ is up to now not experimentally or theoretically determined.

Among the considered structures, a PCF_3-type (tP40) ($P4_2/nmc$ space group n° 137) atomic position arrangements (Table 16.1, Fig. 16.1a) occurs at the lowest total energy (Fig. 16.2).

In this tetragonal structure, $SrAlH_3$ has 40 atoms (Fig. 16.1) and include two occupation sites of hydrogen (8 g, 16 h) and two occupation sites for Sr (8 g) and Al

Table 16.1 Optimized equilibrium structural parameters for SrAlH$_3$ predicted structure

	Unit cell (Å)	Space group	Positional parameters
KMnF$_3$ (tP20), *P4/mbm*	a = 6.02298182 b = 6.02298182 c = 8.53803400	127	Sr (4f) 0.0, 0.5, 0.247928 Al (2a) 0.0, 0.0, 0.0 Al (2b) 0.0, 0.0, 0.5 H (4e) 0.0, 0.0, 0.249889 H (4h) 0.22581, 0.27419, 0.5 H (4g) 0.249916, 0.749916, 0.0
GdFeO$_3$ [NaCoF$_3$(oP20)], *Pnma*	a = 5.611 b = 7.669 c = 5.349	62	Sr (4c) 0.43716, 1/4, 0.01556 Al (4a) 0, 0, 0 H (8d) 0.1984, 0.0506, 0.3043 H (4c) 0.5328, 1/4, 0.6005
KCuF$_3$ (tI20), *I4/mcm*	a = 5.8604 b = 5.8604 c = 7.8528	140	Sr (4a) 0.0, 0.0, 0.25. Al (4d) 0.0, 0.5, 0.0. H (8h) 0.223, 0.723, 0.0 H (4b) 0.0, 0.5, 0.25
CaTiO$_3$ [CsHgF$_3$ (cP5)], *Pm-3m*	a = 3.8	221	Sr (a1) 0,0,0 Al (b1) 0.5,0.5,0.5 H (d3) 1/2,1/2,0
PCF$_3$ (tP40), *P4$_2$/nmc*	a = 10.1 c = 6.397	137	Sr (8g) 1/4, 0.6025, 0.3029 Al (8g) 1/4, 0.0036, 0.0641 H (8g) 1/4, 0.5557, 0.8794 H (16h) 0.1452, 0.0804, 0.0649
KCuF$_3$ (tP5), *P4/mmm*	a = 3.923 c = 3.8963	123	Sr(1a) 0,0,0 Al (1d) 1/2, 1/2, 1/2 H(1b) 0,0, 1/2 H (2f) 0, 1/2, 0
KNbO$_3$ (tP5) *P4mm*	a = 4.02866548 b = 4.02866548 c = 4.21874013	99	Sr (1b): 0.5, 0.5, 0.505211 Al (1a): 0.0, 0.0, 0.021747 O (2c): 0.0, 0.5, 0.972551 O (1a): 0.0, 0.0, 0.462741
FeTiO$_3$ (hR30) *R-3*	a = 5.0881 c = 14.091	148	Sr (6c): 0.0, 0.0, 0.35543 Al (6c): 0.0, 0.0, 0.14643 H (18f) 0.3495, 0.03967, 0.08835

(8 g) respectively. However, when we employ pressure the stable P4$_2$/nmc structure of SrAlH$_3$ transform into orthorhombic *Pnma* (space group n° 62) structure at 38 Gpa.

Indeed, in an orthorhombic structure SrAlH$_3$ has 20 atoms (Fig. 16.1) and contains two occupation sites of hydrogen (4c, 8d) and that Sr (4a) and Al (4c) atoms, surround both of them. Reliable with the ABX$_3$ perovskite structure, each Sr (A-site) cation is bounded by 12 H anions, while each Al (B-site) cation is coordinated with 6H anions.

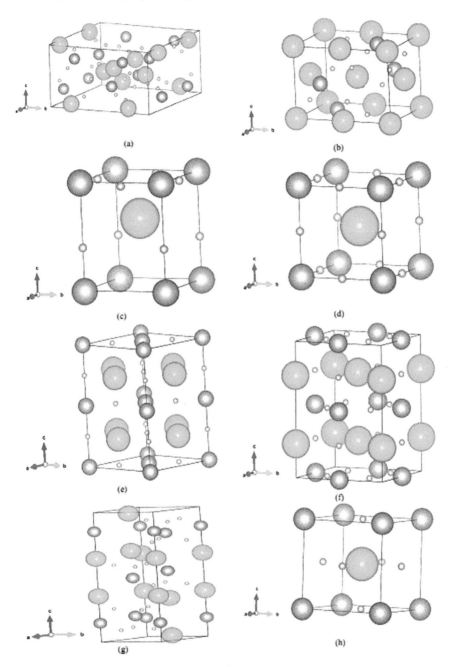

Fig. 16.1 Predicted crystal structure of SrAlH$_3$: **a** PCF$_3$ type, $P4_2/nmc$, **b** GdFeO$_3$ type, $Pnma$, **c** KNbO$_3$ type, $P4mm$; **d** KCuF$_3$ type, $P4/mmm$; **e** KMnF$_3$ type, $P4/mbm$; **f** KCuF$_3$ type, $I4/mcm$; **g** FeTiO$_3$ type R-3; **h** CaTiO$_3$ type, Pm-3 m. Sr, Al and H atoms are shown green (big), blue (medium) and pink (small) spheres, respectively

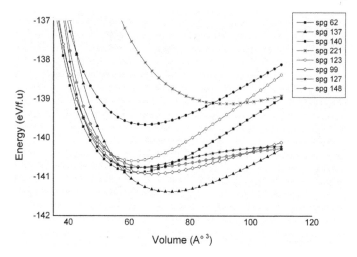

Fig. 16.2 Calculated cell volume versus total energy curves for SrAlH$_3$ in different possible arrangements (space group number indicated on the figure)

Therefore, each H anion is bonded with two Al and four Sr cations. Therefore, SrAlH$_3$ returns to a distorted perovskite structure similar to the NaMgH$_3$ hydride [7].

We have fitted the total energy as function of cell volume using the so-called universal equation of state and we have found that SrAlH$_3$ have a bulk modulus B equal to 37.11 Gpa and 44.88 Gpa for P4$_2$/nmc (space group n° 137) and Pnma (space group 62) respectively.

Finally, we must note that the other crystal structures are less stable than the above prototype structures for SrAlH$_3$.

16.4 Conclusion

In this work, we have used density functional calculation to predict the crystal structure of the unknown phases from the guess-structure approach for SrAlH$_3$ hydride. Indeed, the ground-state structures have been identified from a full optimization with both force and stress minimizations. The crystal structure of SrAlH$_3$ has been predicted to be of the PCF$_3$-type (tP40) (P4$_2$/nmc space group n° 137) at 0 K and atmospheric pressure. However, at high pressure this phase transforms into orthormbic *Pnma* (space group n° 62) structure at 38 Gpa. Using the equation of state, we have found that SrAlH$_3$ have a bulk modulus B equal to 37.11 Gpa and 44.88 Gpa for P4$_2$/nmc (space group n° 137) and Pnma (space group 62) respectively.

References

1. V. Masson-Delmotte, An IPCC special report on the impacts of global warming of 1.5 °C above pre-industrial levels and related global greenhouse gas emission pathways, in the context of strengthening the global response to the threat of climate change, sustainable development, and efforts to eradicate poverty (eds). World Meteorological Organization, Geneva, Switzerland (2018)
2. H.H. Park, M. Pezat, B. Darriet, A new ternary hydride: $CsCaH_3$. Rev. Chim. Min. **23**(3), 226–323 (1986)
3. H.H. Park, M. Pezat, B. Darriet, Deux nouveaux hydrures: $RbCaH_3$ et Rb_2CaH_4. C. R. Acad. Sci. **306**(14), 963–966 (1988)
4. R. Schuhmacher, A. Weiss, $KMgH_3$ single crystals by synthesis from the elements. J. Less Common Met. **163**(1), 179–183 (1990)
5. C.E. Messer, J.C. Eastman, R.G. Mers, A.J. Maeland, Ternary perovskite phases in systems of lithium hydride with barium, strontium, and calcium hydrides. Inorg. Chem. **3**(5), 776–778 (1964)
6. A.W. Overhauser, Crystal structure of lithium beryllium hydride. Phys. Rev. B **35**(1), 411–414 (1987)
7. Y. Bouhadda, Y. Boudouma, N. Fenineche, A. Bentabet, Ab initio calculations study of the electronic, optical and thermodynamic properties of $NaMgH_3$, for hydrogen storage. J. Phys. Chem. Solids **71**(9), 1264–1268 (2010)
8. Y. Bouhadda, N. Fenineche, Y. Boudouma, Hydrogen storage lattice dynamics of orthorhombic $NaMgH_3$. Phys. B **406**(4), 1000–1003 (2011)
9. Y. Bouhadda, M. Bououdina, N. Fenineche, Y. Boudouma, Elastic properties of perovskite-type hydride $NaMgH_3$ for hydrogen storage. Int. J. Hydrogen Energy **38**(3), 1484–1489 (2013)
10. Y. Bouhadda, N. Kheloufi, A. Bentabet, Y. Boudouma, N. Fenineche, K. Benyalloul, Thermodynamic functions from lattice dynamic of $KMgH_3$ for hydrogen storage applications. J. Alloy. Compd. **509**(37), 8994–8998 (2011)
11. X. Gonze, J.M. Beuken, R. Caracas, F. Detraux, M. Fuchs, G.M. Rignanese, L. Sindic, M. Verstraete, G. Zerah, F. Jollet, M. Torrent, A. Roy, M. Mikami, Ph Ghosez, J.Y. Raty, D.C. Allan, First principles computation of material properties: the ABINIT software project. Comput. Mater. Sci. **25**(3), 478–492 (2002)
12. W. Kohn, L.J. Sham, Self-consistent equations including exchange and correlation effects. Phys. Rev. A **140**(4A), 1133–1138 (1965)
13. S. Goedecker, Fast radix 2, 3, 4, and 5 kernels for fast fourier transformations on computers with overlapping multiply-add instructions. SIAM J. Sci. Comput. **18**(6), 1605–1611 (1997)
14. M.C. Payne, M.P. Teter, D.C. Allan, T.A. Arias, J.D. Joannopoulos, Iterative minimization techniques for Ab initio total-energy calculations: molecular dynamics and conjugate gradients. Rev. Mod. Phys. **64**(4), 1045–1097 (1992)
15. X. Gonze, Towards a potential-based conjugate gradient algorithm for order-N self-consistent total energy calculations. Phys. Rev. B **54**(7), 4383–4386 (1996)
16. J.P. Perdew, K. Burke, M. Ernzerhof, Generalized gradient approximation made simple. Phys. Rev. Lett. **77**(18), 3865–3868 (1996)
17. M. Fuchs, M. Scheffler, Ab initio pseudopotentials for electronic structure calculations of poly-atomic systems using density-functional theory. Comput. Phys. Commun. **119**(1), 67–98 (1999)
18. J. Daams, R. Gladyshevskii, O. Shcherban, V. Dubenskyy, V. Kuprysyuk, O. Pavlyuk, I. Savysyuk, S. Stoyko, R. Zaremba, Group III: condensed matter volume 43 crystal structures of inorganic compounds, Subvolume A, structure types, Landolt-Börnstein numerical data and functional relationships in science and technology new series Germany (2013)

Chapter 17
CFD Analysis of the Metal Foam Insertion Effects on SMR Reaction Over Ni/Al$_2$O$_3$ Catalyst

Ali Cherif, Rachid Nebbali, and Lyes Nasseri

Abstract This work presents a numerical study of hydrogen production via steam methane reforming (SMR) process in a monolithic reactor. Two parallel coated catalytic layers of order of micrometers consist of Nickel (Ni) supported on α- Al$_2$O$_3$ and structured in a rectangular channel are used to achieve more methane conversion and high levels of compactness. In order to find the influence of the copper foam, two configurations were numerically studied and compared. In the first case, the feed was introduced directly without using the metal foam. While in the second configuration, the reactor was fitted with copper foam. Indeed, SMR reaction is a high endothermic process requiring a great energy to achieve the activation condition. Although the Ni catalyst reduces this energy, it is recommended to increase the thermal efficiency improvement, thus the methane conversion, the hydrogen yield, and as a result pollution effects are minimized. It has been shown that using copper metal foam promotes the heat conduction, which is very small in gas phases. In addition, it has been found that the use of the copper foam enhances the hydrogen yield by 6.42%, and the catalyst length can be reduced by 9.67% while maintaining the same results. Therefore, the metal foam can positively impact the SMR reaction and reduces the process costs.

Keywords Catalyst · Reforming · Porous media

17.1 Introduction

Hydrogen is considered as a promising substitute of hydrocarbons. It uses as an energy source without producing any pollution. Even though it is the most abundant element in the universe, there is no process for mining it in primary form. Therefore, industrial processes have been developed for hydrogen production from

---

A. Cherif (✉) · R. Nebbali · L. Nasseri
Laboratory of Multiphase Flow and Porous Media 'LTPMP', Faculty of Mechanical Engineering and Process Engineering, University of Sciences and Technologies Houari Boumediene, Algiers, Algeria
e-mail: ali.cherif@usthb.dz

© The Editor(s) (if applicable) and The Author(s), under exclusive license to Springer Nature Singapore Pte Ltd. 2021
A. Khellaf (ed.), *Advances in Renewable Hydrogen and Other Sustainable Energy Carriers*, Springer Proceedings in Energy,
https://doi.org/10.1007/978-981-15-6595-3_17

natural gas such as Partial Oxidation of methane (POX) [1], Autothermal methane reforming (ATR) and methane steam reforming (SMR). The later developed in 1920 [1] became the principal method for the syngas production. Almost 50% of the hydrogen produced in the worldwide is obtained through this process [2].

In this work we studied the production of hydrogen via SMR process which consists mainly in a catalytic reaction of SMR, followed by WGS reaction.

As shown in Eq. (17.1), the SMR is proportionally a strong endothermic reaction. Thus, a high temperature range (980–1373 K) is required to activate it. The use of catalysts reduces significantly the amount of energy needed. There are different transient materials that can be used as catalysts in (SMR). One can cite the Platinum (Pt), Ruthenium (Ru), palladium (Pb) and the Nickel (Ni). The last one is the most used due to its low cost compared to the other named active elements [3].

Different methods are mainly used to use the catalyst in a reactor. The fluidized bed, which doesn't require a support to introduce it into operation [4], because it can be occurred within fluid flow. Another method widely used in the industry is the packed bed reactor which is more efficient in small-scale reactors, proportionally, the increasing of reactor section reduces the conversion of methane [5]. The coated bed reactor is the most common process in SMR activation, for its simplicity to put on the support and its efficiency compared to other deposition methods [6]. Reaction rate is an intrinsic parameter that links the energy and species governing equations and describes the reactive process evolvement. There are three different approaches to determine the reaction rate expression. The first one is the general model Langmuir-Hinshelwood kinetics [7], the first-order reaction with respect to methane is the fastest way in the numerical simulations and the power-law expression which is derived using data fitting results [8]. The latter methodology depends highly on the catalyst, reactor and boundary condition parameters. However, it has a high consistence with the experimental results and can be adjusted using a pre-exponential factor [9]. In our previous work, we achieved a decrement of temperature of 45% for the autothermal reformer compared by changing the arrangement of the catalyst compared to the traditional autothermal reactor [10].

In this work, we analyzed the influence of the metal foam (MF) insertion on the distribution of the species along the reactor, the methane conversion and the temperature field. The copper foam has a high thermal conductivity which can affect the thermal behaviors of the process as well as the improvement of the reactor compactness. Aiming to study the SMR reaction, two configurations have been established numerically. In the first configuration the inlet component occurs directly over the reactor. While, in the second configuration, the MF is inserted.

17.2 Numerical Modeling

SMR is the only considered reaction in this work due to its high activity compared to WGS reaction and the other involved reactions. The SMR reaction over a rectangular channel of 14 cm length and 0.2 cm height was integrated in a numerical model. The

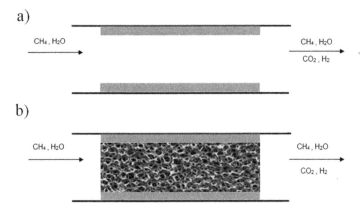

Fig. 17.1 SMR reformer schema, **a** without MF, **b** with MF

Ni/Al$_2$O$_3$ wash-coated catalyst is used to enhance the reaction activity with a length of 12 cm and a height of 0.01 cm. Figure 17.1 shows schematically the studied process in this paper.

The heat transfer process and the flow field inside the reformer is described by continuity, momentum, energy and spices governing equations. Consequently, the transport equations of the averaged physical properties for laminar flows, were used. The governing equations for the calculation setup were the continuity, the momentum, the energy conservation and the species equation. The governing equations are as follows:

$$\text{Continuity}: \frac{\partial \rho_e u}{\partial x} + \frac{\partial \rho_e v}{\partial y} = 0 \qquad (17.1)$$

$$\text{Momentum according to } x: \left(u\frac{\partial \rho_e u}{\partial x} + v\frac{\partial \rho_e v}{\partial y}\right) = -\frac{\partial p}{\partial x}$$
$$+ \left(\frac{\partial}{\partial x}\left(\mu\frac{\partial u}{\partial x}\right) + \frac{\partial}{\partial y}\left(\mu\frac{\partial u}{\partial y}\right)\right) - \frac{\mu}{K_p}u - \frac{\rho_f}{\sqrt{K_p}}u\sqrt{u^2+v^2} \qquad (17.2)$$

$$\text{Momentum according to } x: \left(u\frac{\partial \rho_e v}{\partial x} + v\frac{\partial \rho_e v}{\partial y}\right) = -\frac{\partial p}{\partial y}$$
$$+ \left(\frac{\partial}{\partial x}\left(\mu\frac{\partial v}{\partial x}\right) + \frac{\partial}{\partial y}\left(\mu\frac{\partial v}{\partial y}\right)\right) - \frac{\mu}{K_p}v - \frac{\rho_f}{\sqrt{K_p}}v\sqrt{u^2+v^2} \qquad (17.3)$$

$$\text{Energy}: \left(u\frac{\partial \rho_e C_{p,e} T}{\partial x} + v\frac{\partial \rho_e C_{p,e} T}{\partial y}\right) = \frac{\partial}{\partial x}\left(\lambda_e \frac{\partial T}{\partial x}\right) + \frac{\partial}{\partial y}\left(\lambda_e \frac{\partial T}{\partial y}\right) \qquad (17.4)$$

$$\text{Species}: u\frac{\partial \rho_e w_i}{\partial x} + v\frac{\partial \rho_e w_i}{\partial y} = \frac{\partial}{\partial y}\left(D_{i,e}\frac{\partial^2 \rho_e w_i}{\partial x^2}\right) + \frac{\partial}{\partial y}\left(D_{i,e}\frac{\partial^2 \rho_e w_i}{\partial y^2}\right) \qquad (17.5)$$

Table 17.1 Thermophysical properties of the MF

Porosity	Pore per inch	C_f (J kg^{-1} K^{-1})	K_p (m^2)	$C_{p,s}$	Density	Heat conductivity
0.935	33	0.00643	1969 × 10^{-8}	385	8500	400

In the above equations, effective thermophysical properties are used for taking into account of the presence of the solid and gases phases. Where ρ_e, λ_e, $C_{p,e}$ and $D_{i,e}$ are the effective density, thermal conductivity, heat capacity and diffusivity, respectively.

These equations were used for building a code basing on the FORTRAN programming language.

For the case of metal foam insertion, the Darcy-Brinkman-Forchheimer approach is used [11]. Adopting the effective properties of gases and the thermophysical properties displayed in Table 17.1. For the case of the metal foam insertion, the equations are adopted by adding the source terms, and considering the averaged mixture features.

Boundary Layer

$$\text{Inlet}: u = 0.45 \text{ ms}^{-1}; T = 800\,°C; v = w_{co} = w_{h2} = 0, S/C = 3 \quad (17.6)$$

$$\textbf{Outlet}: \frac{\partial u}{\partial x} = \frac{\partial T}{\partial x} = \frac{\partial w}{\partial x} 0 \quad (17.7)$$

$$\textbf{Top and Bottom adiabatic wall}: u = v = 0, \frac{\partial w_i}{\partial y} = 0 \quad (17.8)$$

$$\textbf{Top and Bottom wash coated walls}: u = 0, v = 0 \quad (17.9)$$

$$\frac{\partial T}{\partial y} = -\frac{\sum r_i \Delta H_{R,i}}{\lambda} 0; \frac{\partial w_i}{\partial y} = -\frac{S_i M_i}{\rho D_i} = 0 \quad (17.10)$$

$$S_i = R_I(v_i'' - v_i') \quad (17.11)$$

17.3 Results and Discussions

The average temperature profile is illustrated in Fig. 17.2. The temperature decreases along the reactor length with following the same behavior of reactants evolution. In fact, the temperature has a strength impact on the reaction activation, so that, the depletion of reactant decreases with the same trend of the temperature. The MF insertion decreases the temperature compared to the case where the MF is not applied by 24.58 °C. This due to the more consumed heat by the reaction because the better

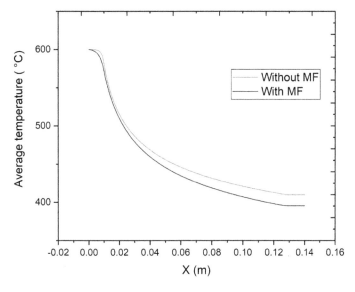

Fig. 17.2 Temperature evolvement over the SMR reactor

heat transfer conductivity of the mixture which is so low in the gas phases. The temperature decay gives more stability for the process and enlarge the choice of used materials in manufacturing the SMR reactor. Figure 17.3 shows the improvement of methane conversion when using the MF as a function of the reactor length. The

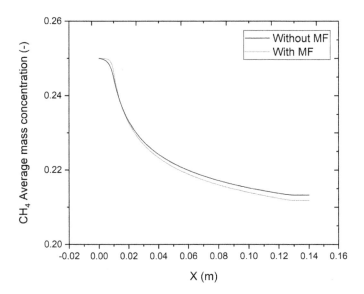

Fig. 17.3 Methane average mass concentration profile as a function of reactor length

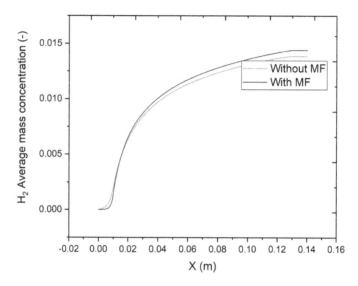

Fig. 17.4 Hydrogen average mass concentration profile as a function of reactor length

methane conversion is an important parameter to evaluate the reactor performances. In the case of using the metal foam, the methane conversion is improved by 5.3%. The metal foam give more time to the process to react inside the reformer.

The hydrogen generation mass fraction over the reactor is displayed on Fig. 17.4. The produced hydrogen is enhanced by 6.42% when inserting the MF. On the other hand, the reactor can be reduced by 9.67% when MF is used with conserving the same efficiency in the case where the MF is not inserted. So that, the catalyst amount is reduced and the cost is decreased. The MF can give the mixture more embarrassing characteristic. Therefore, it can be expected to be more efficient and has more performances

17.4 Conclusion

In this study, SMR was investigated by using a two-dimensional numerical model under steady state conditions. Two catalyst beds take place in opposite position, based on Ni/Al_2O_3 for the SMR reaction in a rectangular channel. The results shown that the performance of the catalytic bed reforming reactor in terms of the methane conversion can be improved when inserting metal foam, making the reactor more efficient, and give it a high level of intensification.

References

1. M.A. Pen, J.P. Gomez, J.G. Fierro, New catalytic routes for syngas and hydrogen production. Appl. Catal. A **144**(1–2), 7–57 (1996)
2. A. Basile, S. Liguori, A. Iulianelli, Membrane reactors for methane steam reforming (MSR), in *Membrane Reactors for Energy Applications and Basic Chemical Production*, Elsevier, pp. 31–59 (2015)
3. M.A. Murmura, M. Diana, R. Spera, M.C. Annesini, Modeling of autothermal methane steam reforming: comparison of reactor configurations. Chem. Eng. Proc. Process Intensification **109**, 125–135 (2016)
4. Z. Chen, J.R. Grace, C.J. Lim, A. Li, Experimental studies of pure hydrogen production in a commercialized fluidized-bed membrane reactor with SMR and ATR catalysts. Int. J. Hydrogen Energy **32**(13), 2359–2366 (2007)
5. A. Karim, J. Bravo, D. Gorm, T. Conant, A. Datye, Comparison of wall-coated and packed-bed reactors for steam reforming of methanol. Catal. Today **110**(1–2), 86–91 (2005)
6. J. Bravo, A. Karim, T. Conant, G.P. Lopez, A. Datye, Wall coating of a $CuO/ZnO/Al_2O_3$ methanol steam reforming catalyst for micro-channel reformers. Chem. Eng. J. **101**(1–3), 113–121 (2004)
7. J. Xu, G.F. Froment, Methane steam reforming, methanation and water-gas shift: I. Intrinsic kinetics. AIChE journal. **35**(1), 88–96 (1989)
8. A. Sciazko, Y. Komatsu, G. Brus, S. Kimijima, J.S. Szmyd, A novel approach to improve the mathematical modelling of the internal reforming process for solid oxide fuel cells using the orthogonal least squares method. Int. J. Hydrogen Energy **39**(29), 16372–16389 (2014)
9. K.K. Lin, M. Saito, Y. Niina, K. Ikemura, H. Iwai, H. Yoshida, Deduction of proper reaction rate of steam methane reforming over catalyst surface validated with a combination of one-and two-dimensional simulations. J. Therm. Sci. Technol. **7**(4), 633–648 (2012)
10. A. Cherif, R. Nebbali, Numerical analysis on autothermal steam methane reforming: effects of catalysts arrangement and metal foam insertion. Int. J. Hydrogen Energy **44**(39), 22455–22466 (2019)
11. K. Vafai, C.L. Tien, Boundary and inertia effects on convective mass transfer in porous media. Int. J. Heat Mass Transf. **25**(8), 1183–1190 (1982)

Chapter 18
Photocatalytic Evolution of Hydrogen on CuFe$_2$O$_4$

H. Lahmar⬤, M. Benamira⬤, L. Messaadia, K. Telmani, A. Bouhala, and Mohamed Trari

Abstract We investigated the technical feasibility of photochemical hydrogen release based on CuFe$_2$O$_4$ powder dispersion in an aqueous electrolyte containing a reducing agent (S$_2$O$_3^{-2}$). The oxide combines an average resistance to corrosion and an optimum inter-referred band Eg of 1.67 eV. The intercalation of a small amount of oxygen should be accompanied by partial oxidation of Cu$^+$ to Cu^{2+} involving a p-type semi-conductivity. The oxidation S^{2-} inhibits the photo-corrosion, and the evolution of H$_2$ increases in parallel with the formation of polysulfides. Most of H$_2$ is produced when p-CuFe$_2$O$_4$ is connected to n-ZnO formed in situ. The release of H$_2$ is mainly on CuFe$_2$O$_4$, whereas, the oxidation of S^{2-} takes place on the surface of ZnO and the hetero-system CuFe$_2$O$_4$/ZnO is optimized with respect to certain physical parameters. The photoelectrochemical production of H$_2$ is a multi-step process whose decisive step is the arrival of electrons at the interface due to their low mobility. Remarkable performance with a rate of 6.74 (µmol g^{-1} min^{-1}) cm^3 h^{-1} hydrogen evolution in 0.1 M S$_2$O$_3^{-2}$ (pH 13) is recorded.

Keywords CuFe$_2$O$_4$ · Hydrogen · Semiconductor

18.1 Introduction

Conversion of solar energy through solar cells and electrochemical converters seems very promising in the future because of the energy crisis observed [1]. The photocatalytic process of our research program aims to study the photo-dissociation of water by the illumination of a suspension of semiconductor oxides (SC). Photocatalysis is booming due to its many applications in the environmental and energy field [2–6]. The development of stable and active photocatalysts under solar irradiation is a major challenge. Much work has been done on this topic in depth over the past two decades because of the considerable interest in the development of low-cost SCs [7–10]. Hydrogen gas is an important chemical that can replace fossil fuel, its production from raw materials such as water has been studied [11–13]. In addition, hydrogen does not cause pollution during combustion; it is renewable and has significant energy capacities [14]. One strategy is to use heterojunctions to improve the high gap efficiency of the SC bands and to shift the spectral photo-response to longer wavelengths. Therefore, the potential level of the H_2O/H_2 redox couple is -0.35 V which can be appropriately positioned relative to CB and hydrogen evolves. Thus, the interest seems to come back to the photo-electrolysis of water, which remains one of the most promising ways to ensure energy storage under mild operating conditions. Energetic photon absorption by a p-type semiconductor (SC) generates electrons/hole (e^-/h^+), the electrons migrate to the interface to reduce the water to gaseous hydrogen as the holes enter the mass to react with the electron donor [15]. Our study proceeds on the reduction of water combined with the oxidation of $S_2O_3^{-2}$ as a hole sensor by increasing the charge separation in the spatial charge region. $CuFe_2O_4$ should be a good sensitizer since its conduction band is positioned at a more negative potential than most oxides. This prediction found experimental support; the $CuFe_2O_4/ZnO$ hetero-system will be very favorable compared to $CuFe_2O_4$.

18.2 Experimental

18.2.1 Preparation of the Photocatalyst

Stoichiometric amounts of $Cu(NO_3)_2$ hydrate (Merck, 99%) and $Fe(NO_3)_3, 9H_2O$ (Fluka > 99%) are dissolved in water, excess nitrates are decomposed at 300 °C under hot plate with magnetic stirring. The amorphous powder is homogenized in an agate mortar and then calcined at 850 °C. ZnO was prepared as reported in our previous work [16] by dissolving $Zn(NO_3)_2 \cdot 6H_2O$ (Merck, 99.5%) in water, the solution was dehydrated and denitrified at 300 °C. The powder was heat-treated at 500 °C, and the oxide shows a pale yellowish color.

18.2.2 Characterization Methods

The X-ray powder diffraction (XRD) patterns have been collected at room temperature, with Cu Kα monochromatic radiation ($\lambda = 1.54056$ Å) of a D8 Advance Bruker diffractometer. The electrical conductivity (σ) was measured with the two-probe technique over the temperature range (300–473 K). The sample (1 cm^2) was prepared using uniaxial pressure of 3 MPa and sintered at 850 °C for 2 h. Then, it was introduced into resin epoxy and polished with SIC paper (1200), only one side is in contact with the solution. The electrical contact was achieved by the painted silver lacquer. The diffuse reflectance spectrum was plotted using a UV-VIS spectrophotometer (Specord 200 Plus). Electrochemical measurements were carried out at room temperature in a standard cell containing (Na$_2$SO$_4$, 0.1 M) solution. The capacitance measurements plots were realized at 10 kHz.

The experiment tests of H$_2$ production by photocatalysis were carried out at 50 °C according to our previous works [13]; above, the water loss by vaporization predominates. Briefly, the photocatalyst powder (100 mg) was introduced in a Pyrex reactor equipped with a cooling system. The powder was dispersed in 200 mL of Na$_2$S$_2$O$_3$ solution and kept under stirring. Before each test, O$_2$ was purged by passing N$_2$ for 30 min. The tungsten lamp (200 W) with a total intensity of 29 mW cm^{-2} was used as a light source. Hydrogen gas was identified by gas chromatography using Clarus® 680 GC PerkinElmer Gas Chromatograph and the volume of H$_2$ was quantified in a water manometer. The photocatalyst was tested at pH = 7 and 13 by the mixture of CuFe$_2$O$_4$ and ZnO in a mass ratio (1:1); KOH was used to control the pH of the solution.

18.3 Results and Discussion

18.3.1 Characterization of Material

The XRD diagram of the CuFe$_2$O$_4$ oxide precipitate (Fig. 18.1a) indicates that the powder is mainly composed of spinel structure (tetragonal, JCPDS card No 34-425).

The Munk-Kubelka equation is used to determinate the optical gap (Eg) graphically:

$$F(R_\infty) = (1 - R_\infty)^2 / 2R_\infty \qquad (18.1)$$

The diffuse reflectance $R_\infty = (I/I_o)_{dif}$ (Fig. 18.1b) is obtained from the converted UV-visible absorption spectrum. Eg was determined by extrapolation of the linear part of the curve, $(\alpha h\nu)^2$ as a function of energy, for $h\nu = 0$ and the transition is directly permitted (Insert, Fig. 18.1b).

Knowledge of the potential of the flat band (V_{fb}, Fig. 18.2) is essential for predicting photocatalytic reactions. It is determined accurately from the capacitance

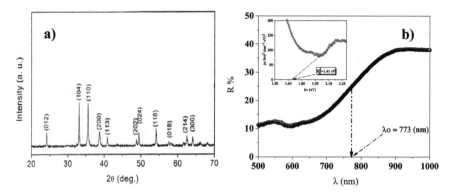

Fig. 18.1 a X-ray diffraction on CuFe$_2$O$_4$ powder; b Diffuse reflectance of CuFe$_2$O$_4$ insert direct optical transition

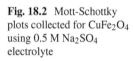

Fig. 18.2 Mott-Schottky plots collected for CuFe$_2$O$_4$ using 0.5 M Na$_2$SO$_4$ electrolyte

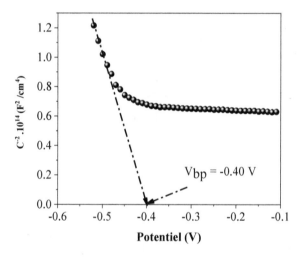

measurements. The capacity of the exhaustion width dominates the overall capacity and obeys the Mott-Schottky relationship:

$$C^{-2} = 2/\varepsilon\varepsilon_o N_A \{V - V_{fb} - kT/e\} \qquad (18.2)$$

C_{sc} Space charge layer capacity (RCS, F/m^2)
ε Dielectric constant of the semiconductor, CuFe$_2$O$_4$ at 293 K
ε_o Permittivity of the vacuum (8.84 × 10^{-12} F m^{-1})
e The charge of the electron.

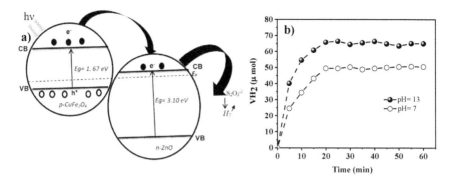

Fig. 18.3 **a** Energy band diagram of the hetero-system CuFe$_2$O$_4$/ ZnO/ electrolyte. **b** Volume of H$_2$ evolved as a function of lighting time

18.3.2 Photocatalytic Activity

The optical gap of CuFe$_2$O$_4$ is available in the literature; clearly, the H$_2$ does not occur in the dark. However, the fact that hydrogen evolves under illumination indicates that the crystallite is biased at the negative potential (V$_{fb}$) and the conduction band is cathodically positioned relative to the H$_2$O/H$_2$ level. This should lead to a deviation, for example, at least 1.5 eV, which is an acceptable value given the black color of our oxide. In these working conditions, the diagram of the energy band can predict thermodynamically if the evolution of the H$_2$ realizable or not. The PEC characterization allows the establishment of our heterojunction energy band diagram (Fig. 18.3a). A sensitizer functions as an electron pump and must have its lowest unoccupied energy level above ZnO–CB and CuFe$_2$O$_4$ fulfills this condition. This property was exploited at pH 7 and 13. CuFe$_2$O$_4$–CB is under ZnO–CB itself less cathodic than the H$_2$O/H$_2$ potential (E$_{red}$). The CB potential is below the potential of HER and this should lead to a spontaneous evolution of H$_2$ even in electrolyte base. When the visible light excites CuFe$_2$O$_4$, this results in the injection of electrons into ZnO–CB, which subsequently releases hydrogen in the following processes:

$$CuFe_2O_4 + h\nu \rightarrow CuFe_2O_4\left(h_{VB}^+ + e_{CB}^-\right) \tag{18.3}$$

$$2e^-(CuFe_2O_4 - CB) + ZnO \rightarrow CuFe_2O_4 + 2e^- + (ZnO - CB) \tag{18.4}$$

$$2H_2O + 2e^-(ZnO - CB) \rightarrow H_2 + 2OH^- + ZnO \tag{18.5}$$

$$S_2O_3^{-2} + 4h^+ + 6OH^- \rightarrow 2SO_3^{-2} + 3H_2O \tag{18.6}$$

Table 18.1 Heterojunctions photocatalysts reported in the literature for hydrogen evolution

Type of hetero system	Production of hydrogen (μmol g^{-1} min^{-1})	References
CuFe$_2$O$_4$/ZnO	6.74	Our results
Pr$_2$NiO$_4$/SnO$_2$	24.3	[13]
CuFe$_2$O$_4$/TiO$_2$	15.0	[17]
ZnFe$_2$O$_4$/SrTiO$_3$	09.2	[18]
Bi$_2$S$_3$/TiO$_2$	02.9	[19]
CuFeO$_2$/SnO$_2$	02.6	[20]

These semi-electrochemical reactions occur simultaneously on the system (CuFe$_2$O$_4$/ZnO), which behaves like a micro PEC compartmentalized space-temperature diode of nanometric dimension. The evolution of H$_2$ proceeds in a non-homogeneous way; it can affect both the type of reaction and the rate of charge transfer on the interface. HER works mainly on ZnO while the oxidation of S$_2$O$_3^{-2}$ occurs on CuFe$_2$O$_4$. The photoactivity is improved by close contact between CuFe$_2$O$_4$ and ZnO, which are coupled in a short-circuit configuration (Fig. 18.3b). The potential of the inter-grain barrier is collapsed, and electrons are channeled to ZnO sites where H$_2$ overvoltage is relatively low.

The good performance is attributed to the charge separation (e$^-$/h$^+$) in comparison with CuFe$_2$O$_4$ alone. The increased activity is also attributed to S$_2$O$_3^{-2}$, which enhances the charge separation and protects CuFe$_2$O$_4$ against photocorrosion. The H$_2$ evolution takes place mainly at the surface of ZnO, while S$_2$O$_3^{-2}$ oxidation occurs over CuFe$_2$O$_4$. The n quantum efficiency (η) is calculated from the following equation:

$$\eta(\%) = \frac{2 \, V \, N \, 100}{10^{-6} \Phi t} \tag{18.7}$$

where V (μmol) is the volume of hydrogen, N the Avogadro number, Φ the photon flux (2.1 × 10^{19} photons s^{-1}) and t(s) the time; factor 2 is involved because HER is two electron-process. A quantum efficiency (η) of 0.37% and production rate of 6.74 μmol g^{-1} min^{-1} are obtained under polychromatic light. The following Table 18.1 gives the hydrogen production values reported in the literature.

18.4 Conclusion

Hydrogen photoproduction was performed on a new hetero-system CuFe$_2$O$_4$/ZnO. The photocurrent indicates a p-type conductivity of the CuFe$_2$O$_4$ oxide which has been prepared by nitrate route. In solution, the oxide is photoelectrochemically stabilized by holes captured via the reducing species. The position of the conduction band predicts a spontaneous evolution of H$_2$ under visible light. This PEC process is limited by the low mobility of the carriers and an improvement in hydrogen evolution has been reported by the addition of ZnO to facilitate the electron transfer of the $CuFe_2O_4$ sensitizer to ZnO–CB. The decrease in the photoactivity was due to the competitive reduction of the final product. The release rate of 6.74 (μmol g^{-1} min^{-1}) obtained is remarkable compared by other papers.

References

1. J. Spitz, A. Aubert, J.M. Behaghel, S. Berthier, J. Lafait, J. Rivory, Matériaux sélectifs pour la conversion photothermique de l'énergie solaire. Rev. Phys. Appl. **14**(1), 67–80 (1979)
2. H. Lahmar, M. Benamira, F.Z. Akika, M. Trari, Reduction of chromium (VI) on the hetero-system $CuBi_2O_4/TiO_2$ under solar light. J. Phys. Chem. Solids **110**, 254–259 (2017)
3. S. Douafer, H. Lahmar, M. Benamira, G. Rekhila, M. Trari, Physical and photoelectrochemical properties of the spinel $LiMn_2O_4$ and its application in photocatalysis. J. Phys. Chem. Solids **118**, 62–67 (2018)
4. H. Lahmar, M. Trari, Photocatalytic generation of hydrogen under visible light on La_2CuO_4. Bull. Mater. Sci. **38**(4), 1043–1048 (2015)
5. S. Douafer, H. Lahmar, M. Benamira, L. Messaadia, D. Mazouzi, M. Trari, Chromate reduction on the novel hetero-system $LiMn_2O_4/SnO_2$ catalyst under solar light irradiation. Surf. Interfaces **17**, 100372 (2019)
6. H. Lahmar, M. Benamira, S. Douafer, L. Messaadia, A. Boudjerda, M. Trari, Photocatalytic degradation of Methyl orange on the novel hetero-system La_2NiO_4/ZnO under solar light. Chem. Phys. Lett. **742**, 137132 (2020)
7. N. Doufar, M. Benamira, H. Lahmar, M. Trari, I. Avramova, M.T. Caldes, Structural and photochemical properties of Fe-doped ZrO_2 and their application as photocatalysts with TiO_2 for chromate reduction. J. Photochem. Photobiol. A **386**, 112105 (2020)
8. H. Lahmar, M. Benamira, Chromate reduction on the novel hetero-system La_2NiO_4/TiO_2 under solar light, in *Proceedings of the 2017 International Renewable and Sustainable Energy Conference. IRSEC 2017* (IEEE, 2018), pp. 1–4
9. K. Telmani, H. Lahmar, M. Benamira, L. Messaadia, M. Trari, Synthesis, optical and photo-electrochemical properties of $NiBi_2O_4$ and its photocatalytic activity under solar light irradiation. Optik **207**, 163762 (2019)
10. A. Doghmane, S. Douafer, Z. Hadjoub, Investigation of pressure effects on elastic parameters of isotropic and anisotropic MgO. J. Optoelectron. Adv. Mater. **16**, 1339–1343 (2014)
11. Y. Bessekhouad, M. Mohammedi, M. Trari, Hydrogen photoproduction from hydrogen sulfide on Bi_2S_3 catalyst. Sol. Energy Mater. Sol. Cells **73**(3), 339–350 (2002)
12. Y. Bessekhouad, M. Trari, Photocatalytic hydrogen production from suspension of spinel powders AMn_2O_4 (A=Cu and Zn). Int. J. Hydrogen Energy **27**(4), 357–362 (2002)
13. M. Benamira, H. Lahmar, L. Messaadia, G. Rekhila, F.Z. Akika, M. Himrane, M. Trari, Hydrogen production on the new hetero-system Pr_2NiO_4/SnO_2 under visible light irradiation. Int. J. Hydrogen Energy **45**(3), 1719–1728 (2020)
14. M. Momirlan, T.N. Veziroglu, Current status of hydrogen energy. Renew. Sustain. Energy Rev. **6**(1–2), 141–179 (2002)
15. F.Z. Akika, M. Benamira, H. Lahmar, A. Tibera, R. Chabi, I. Avramova, M. Trari, Structural and optical properties of Cu-substitution of $NiAl_2O_4$ and their photocatalytic activity towards Congo red under solar light irradiation. J. Photochem. Photobiol., A **364**, 542–550 (2018)
16. H. Lahmar, M. Kebir, N. Nasrallah, M. Trari, Photocatalytic reduction of Cr (VI) on the new hetero-system $CuCr_2O_4/ZnO$. J. Mol. Catal. A: Chem. **353**, 74–79 (2012)
17. A. Kezzim, N. Nasrallah, A. Abdi, M. Trari, Visible light induced hydrogen on the novel hetero-system $CuFe_2O_4/TiO_2$. Energy Convers. Manag. **52**, 2800–2806 (2011)

18. S. Boumaza, A. Boudjemaa, A. Bouguelia, R. Bouarab, M. Trari, Visible light induced hydrogen evolution on new hetero-system $ZnFe_2O_4/SrTiO_3$. Appl. Energy **87**, 2230–2236 (2010)
19. R. Brahimi, Y. Bessekhouad, A. Bouguelia, M. Trari, Visible light induced hydrogen evolution over the heterosystem Bi_2S_3/TiO_2. Catal. Today **122**, 62–65 (2007)
20. A. Derbal, S. Omeiri, A. Bouguelia, M. Trari, Characterization of new heterosystem $CuFeO_2/SnO_2$ application to visible-light induced hydrogen evolution. Int. J. Hydrogen Energy **33**(16), 4274–4282 (2008)

Chapter 19
CFD Study of ATR Reaction Over Dual Pt–Ni Catalytic Bed

Ali Cherif, Rachid Nebbali, and Lyes Nasseri

Abstract In this paper, a numerical study has been conducted for analyzing the autothermal steam methane reforming reactor activated with a dual-bed catalyst (Pt–Ni). The fluid passes through two successive beds in which the first one is devoted for the oxidation reaction and consists of platinum (Pt) catalyst based on the Al_2O_3 support, while the reforming reaction was activated with the following bed that consists of nickel (Ni) catalyst based on the Al_2O_3 support. This configuration was validated, evaluated and the evolution of the different species are illustrated along the reactor length. The reactor was considered as a high thermally performed and well intensified process because the thermal source supplies directly the reforming reaction compared to the externally heated steam methane reforming. The results have shown good hydrogen yield with a generated hydrogen molar fraction of 22.16% as well as the reported conventional autothermal reformer (Wang et al. in Fuel Process Technol 91:723–728, 2010 [1]). Moreover, this dual-bed configuration shows more smoothly temperature profile, and less temperature gradient which can enhance the reactor stability. On the other hand, the temperature peak was about 1438.17 °C which is nearly the same in the conventional reformer (Wang et al. in Fuel Process Technol 91:723–728, 2010 [1]).

Keywords Catalytic combustion · Catalyst · Porous media

19.1 Introduction

The hydrogen is expected to have extensive applications as an energetic vector in future energy strategies. It can be used to supply fuel cells and for ammonia production [1]. Moreover, the increase of the world energy demand and the different environmental issues caused by fossil sources accentuate the demand in clean and renewable

---

A. Cherif (✉) · R. Nebbali · L. Nasseri
Laboratory of Multiphase Flow and Porous Media 'LTPMP', Faculty of Mechanical Engineering and Process Engineering, University of Sciences and Technologies Houari Boumediene, Algiers, Algeria
e-mail: ali.cherif@usthb.dz

energies. Hydrogen can be obtained commonly from hydrocarbons. However, it can be produced often via water electrolysis with high purity rate [2]. Since the twenties of the last century, steam methane reforming (SMR) became the central process for hydrogen production with an efficiency of 65–67% with a simple design. Even though the high performances of SMR process, it needs an external heat source with a high temperature due to the high endothermic reaction [3]. The autothermal reforming (ATR) process in which the heat source and sink are simultaneously occurred in the reactor can override this issue with a high thermal efficiency and intensified reformers. However, the methane conversion is low compared to SMR process with 50% and the temperature evolves with sharp variation. Therefore, much research has been reported aiming to enhance the hydrogen yield of the ATR reactors. Mundhwa and Thurgood [4] studied numerically the SMR reaction coupled thermally with catalytic methane combustion separated by a highly conductive wall for hydrogen production, they found that interspacing the wash-coated catalyst can enhance the methane conversion and avoid the hot and cold spot with decreasing the highest temperature by 33%. The involved catalysts deposition method has high effects on the process. Ismagilov et al. [5] studied experimentally the co-impregnation and the sequential impregnation on ATR methane reactors. The results showed a more influence of catalyst preparation than at low temperature than at higher temperatures. In another study, the authors investigated the addition of a few amount of different precious metals on the Ni (nickel) catalyst activation, the hydrogen generation was promoted as the following order: Pt < Sn < Mo < Re < Pd [6]. A numerical study of conventional ATR and catalytic ATR was conducted by Patcharavorachot et al. [7] They showed that conventional configuration reach a hot spot with less temperature and long distance. While, the hot spot disappears when the catalytic ATR configuration was applied with higher temperature achievement 1418 °C. However, an abrupt variation of temperature was observed. Ayabe et al. investigated the activity sequence of the SMR catalyst over an ATR reactor, the results showed that Rh has the higher activity compared to other precious metals. Moreover, Ni metal can achieve better activity than Rh if more than 5 times of additional weight was used [8]. A new configuration has been designed in our previous paper decreasing the temperature by 45% [9]. In this work, a numerical study was conducted to analyze an adiabatic ATR reactor. In which, the methane combustion (MC) and SMR reactions occur simultaneously in a cylindrical channel aiming to obtain intensified reactor. The reactor's activated area split to two parts, the combustion catalyst takes place at the beginning of the reactor and followed by Ni catalyst for SMR reaction abbreviated as successive dual-bed (SDB).

19.2 Physical Model

The adiabatic ATR reactor configurations studied in this work are illustrated in Fig. 19.1. The cylindrical channel is 1.2 m long and with a 4 cm diameter. In the both

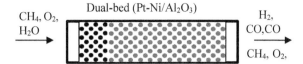

Fig. 19.1 Scheme of the SDB reactor

cases, the catalyst bed is filled with 20% of Pt/Al$_2$O$_3$, while the Ni/Al$_2$O$_3$ catalyst comprises the 80% of the active zone which covers 1 m of the reactor length.

19.3 Mathematical Model

An integrated adiabatic ATR was conducted with a two-dimensional modeling to find the flow parameters of the gaseous fluids inside the reformer. The flow was assumed to be laminar, steady and occurring in the same direction. For the porous media the governing equations derived by the volume-averaging method [10]. The values of the adopted physical method were locally averaged for a representative elementary volume [6], Fig. 19.1 shows the geometry of the internal reformer schematically. The flow field and the heat transfer process inside the porous reformer was described by continuity, momentum, energy and species transfer equations. Consequently, the following transport equations of the averaged physical properties for a laminar flow are applied with considering the Darcy-Brinkman-Forchheimer effects [10]. The finite volume method was adopted to solve numerically the governing equations of continuity, momentum, energy and species. The governing equations are as follows:

$$\text{Continuity:} \quad \frac{\partial \rho_e u}{\partial x} + \frac{\partial \rho_e v}{\partial y} = 0 \quad (19.1)$$

Momentum according to x:
$$\left(u \frac{\partial \rho_e u}{\partial x} + v \frac{\partial \rho_e v}{\partial y} \right)$$
$$= -\frac{\partial p}{\partial x} + \left(\frac{\partial}{\partial x} \left(\mu \frac{\partial u}{\partial x} \right) + \frac{\partial}{\partial y} \left(\mu \frac{\partial u}{\partial y} \right) \right) - \frac{\mu}{K_p} u - \frac{\rho_f}{\sqrt{K_p}} u \sqrt{u^2 + v^2} \quad (19.2)$$

Momentum according to x:
$$\left(u \frac{\partial \rho_e v}{\partial x} + v \frac{\partial \rho_e v}{\partial y} \right)$$
$$= -\frac{\partial p}{\partial y} + \left(\frac{\partial}{\partial x} \left(\mu \frac{\partial v}{\partial x} \right) + \frac{\partial}{\partial y} \left(\mu \frac{\partial v}{\partial y} \right) \right) - \frac{\mu}{K_p} v - \frac{\rho_f}{\sqrt{K_p}} v \sqrt{u^2 + v^2} \quad (19.3)$$

Energy:
$$\left(u \frac{\partial \rho_e C_{p,e} T}{\partial x} + v \frac{\partial \rho_e C_{p,e} T}{\partial y} \right) = \frac{\partial}{\partial x} \left(\lambda_e \frac{\partial T}{\partial x} \right) + \frac{\partial}{\partial y} \left(\lambda_e \frac{\partial T}{\partial y} \right) \quad (19.4)$$

Species:
$$u \frac{\partial \rho_e w_i}{\partial x} + v \frac{\partial \rho_e w_i}{\partial y} = \frac{\partial}{\partial y} \left(D_{i,e} \frac{\partial^2 \rho_e w_i}{\partial x^2} \right) + \frac{\partial}{\partial y} \left(D_{i,e} \frac{\partial^2 \rho_e w_i}{\partial y^2} \right) \quad (19.5)$$

In the above equations, effective thermophysical properties are used for taking into account of the presence of the solid and gases phases. Where ρ_e, λ_e, $C_{p,e}$ and $D_{i,e}$ are the effective density, thermal conductivity, heat capacity and diffusivity, respectively.

These equations were used for building a code basing on the FORTRAN programming language.

Boundary Layer

The non-slip, impermeable wall and the other boundary conditions are illustrated below:

Inlet:

$$u_z = 1 \text{ ms}^{-1}; \quad T = 600\,°C \tag{19.6}$$

$$u_r = n_{co} = n_{h2} = 0 \tag{19.7}$$

$$S/C = 1.4 \text{ (Steam to Carbon molar ratio)} \tag{19.8}$$

$$O/C = 0.6 \text{ (Oxygen to Carbon molar ratio)} \tag{19.9}$$

Outlet:

$$\frac{\partial u_z}{\partial r} = \frac{\partial u_r}{\partial r} = \frac{\partial T}{\partial r} = \frac{\partial n_i}{\partial r} = 0 \tag{19.10}$$

Adiabatic Wall:

$$u = v = 0, \quad \frac{\partial n_i}{\partial z} = \frac{\partial T}{\partial r} = 0 \tag{19.11}$$

19.4 Results and Discussions

For the evaluation and evolvement of the two conducted configurations along the reactor, the temperature and concentrations sketches are illustrated and discussed. Figure 19.2 shows the validation of the SMR reaction used in this study with the work of Zhang et al. [11].

The average temperature profile for the SDB reactor is illustrated in Fig. 19.3. The temperature increases sharply at the inlet of the reactor reaching its highest value about of 1465 °C due to the exponential evolution of the exothermal and complete MC reaction when SDB configuration is applied. This temperature is

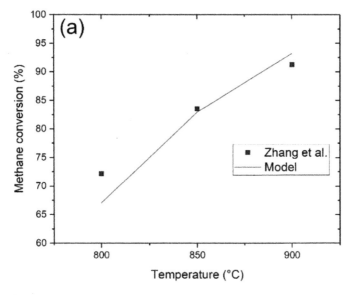

Fig. 19.2 Validation of used SMR kinetics over Ni catalyst experimental results of Zhang et al. [11]; methane conversion as function of temperature **a** (P = 2 MPa), and pressure **b** (T = 900 °C)

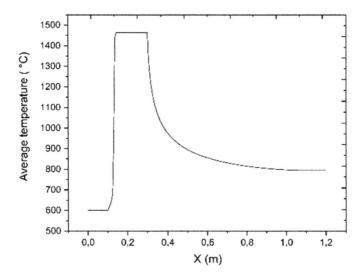

Fig. 19.3 Average temperature profiles in terms of the reactor length

reduced beginning from the leading edge of the Ni until the SMR reaction reaches the equilibrium.

Figure 19.4a, b show the evolvement of methane conversion and hydrogen yield as a function of the reactor length, respectively. The methane conversion is about 60.50%. On the other hand, hydrogen yield achieves 20.21%.

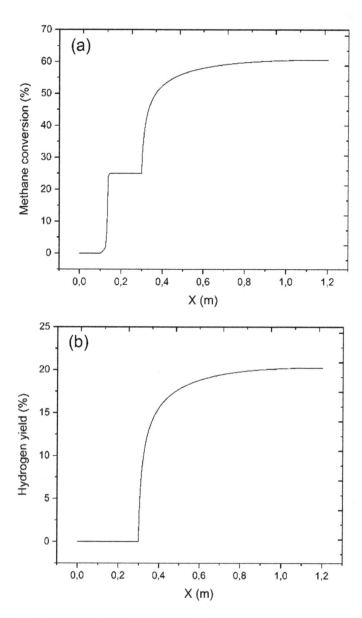

Fig. 19.4 a Methane conversion and b hydrogen yield

19.5 Conclusion

In this work, an investigation of dual-bed catalytic reactor consists of Pt–Ni catalysts is conducted. This type of reactors can afford more decentralization and intensification of the process. However, the found highest temperature is higher than the maximum temperature in traditional SMR reactors. The results showed that the maximum temperature is 1465 °C. While, the hydrogen yield achieved 20%. a hot spot takes place at the trailing edge of the Pt catalyst area.

References

1. L. Wang, L. Yang, Y. Zhang, W. Ding, S. Chen, W. Fang, Y. Yang, Promoting effect of an aluminum emulsion on catalytic performance of Cu-based catalysts for methanol synthesis from syngas. Fuel Process. Technol. **91**(7), 723–728 (2010)
2. S.A. Grigoriev, V.I. Porembsky, V.N. Fateev, Pure hydrogen production by PEM electrolysis for hydrogen energy. Int. J. Hydrogen Energy **31**(2), 171–175 (2006)
3. T.L. LeValley, A.R. Richard, M. Fan, The progress in water gas shift and steam reforming hydrogen production technologies—a review. Int. J. Hydrogen Energy **39**(30), 16983–17000 (2014)
4. M. Mundhwa, C.P. Thurgood, Numerical study of methane steam reforming and methane combustion over the segmented and continuously coated layers of catalysts in a plate reactor. Fuel Process. Technol. **158**, 57–72 (2017)
5. I.Z. Ismagilov, E.V. Matus, V.V. Kuznetsov, M.A. Kerzhentsev, S.A. Yashnik, I.P. Prosvirin, N. Mota, R.M. Navarro, J.L.G. Fierro, Z.R. Ismagilov, Hydrogen production by autothermal reforming of methane over NiPd catalysts: effect of support composition and preparation mode. Int. J. Hydrogen Energy **39**(36), 20992–21006 (2014)
6. I.Z. Ismagilov, E.V. Matus, V.V. Kuznetsov, N. Mota, R.M. Navarro, S.A. Yashnik, I.P. Prosvirin, M.A. Kerzhentsev, Z.R. Ismagilov, J.L.G. Fierro, Hydrogen production by autothermal reforming of methane: effect of promoters (Pt, Pd, Re, Mo, Sn) on the performance of Ni/La$_2$O$_3$ catalysts. Appl. Catal. A **481**, 104–115 (2014)
7. Y. Patcharavorachot, M. Wasuleewan, S. Assabumrungrat, A. Arpornwichanop, Analysis of hydrogen production from methane autothermal reformer with a dual catalyst-bed configuration. Theor. Found. Chem. Eng. **46**(6), 658–665 (2012)
8. M. Luneau, E. Gianotti, N. Guilhaume, E. Landrivon, F.C. Meunier, C. Mirodatos, Y. Schuurman, Experiments and modeling of methane autothermal reforming over structured Ni–Rh-based Si-SiC foam catalysts. Ind. Eng. Chem. Res. **56**(45), 13165–13174 (2017)
9. A. Cherif, R. Nebbali, Numerical analysis on autothermal steam methane reforming: Effects of catalysts arrangement and metal foam insertion. Int. J. Hydrogen Energy **44**(39), 22455–22466 (2019)
10. K. Vafai, C.L. Tien, Boundary and inertia effects on convective mass transfer in porous media. Int. J. Heat Mass Transf. **25**(8), 1183–1190 (1982)
11. N. Zhang, X. Chen, B. Chu, C. Cao, Y. Jin, Y. Cheng, Catalytic performance of Ni catalyst for steam methane reforming in a micro-channel reactor at high pressure. Chem. Eng. Process. **118**, 19–25 (2017)

Chapter 20
Optimization of the Ni/Al$_2$O$_3$ and Pt/Al$_2$O$_3$ Catalysts Load in Autothermal Steam Methane Reforming

Ali Cherif, Rachid Nebbali, and Lyes Nasseri

Abstract In this paper we analyzed an autothermal reformer for the hydrogen production from methane and vapor. The endothermic reforming reaction was heated with the exothermic reaction of methane combustion inside a rectangular reactor. The reforming reaction was activated with nickel (Ni) catalyst layer that coated on one wall of the reactor, while the second wall was coated with platinum (Pt) catalyst in order to activate the combustion reaction. The quantity of the catalysts used for each reaction has an important effect on the process. A low quantity of the catalysts can decrease the hydrogen yield, on the other hand, additional quantity can increase reactor conception costs. So that, the optimization of the process to find out an adequate quantity of catalysts is required to improve the process [1]. The Ni catalyst thickness was varied from 1×10^{-4} m to 4×10^{-4} m while the variation of the Pt catalyst was between 5×10^{-6} m and 4×10^{-5} m. The results showed that the hydrogen yield achieved an optimum value of 36.295% was obtained when the catalyst was operated under a catalyst thickness of 2×10^{-4} m for the Ni catalyst, while, the thickness of the Pt catalyst was of 2×10^{-5} m. the maximum temperature.

Keywords Catalytic combustion · Steam methane reforming · CFD

20.1 Introduction

The environmental problems combined with the depletion of fossil fuels accentuate the demand for alternative energies that preserve the environment. Hydrogen is considered as an interesting solution in more than one way [1] which can gradually cover the fossil fuels that provide energy demand from the eighteenth century to the present, providing over 80% of global energy, followed by coal by (29%), while natural gas by (21%) [2]. Hydrogen is not available in the natural state. So that it must

---

A. Cherif (✉) · R. Nebbali · L. Nasseri
Laboratory of Multiphase Flow and Porous Media 'LTPMP', Faculty of Mechanical Engineering and Process Engineering, University of Sciences and Technologies Houari Boumediene, Algiers, Algeria
e-mail: ali.cherif@usthb.dz

© The Editor(s) (if applicable) and The Author(s), under exclusive license to Springer Nature Singapore Pte Ltd. 2021
A. Khellaf (ed.), *Advances in Renewable Hydrogen and Other Sustainable Energy Carriers*, Springer Proceedings in Energy,
https://doi.org/10.1007/978-981-15-6595-3_20

be produced by the separation of chemical elements of which the hydrogen atom is a component and by the mobilization of a source of energy. The 96% of H_2 production technologies are based on non-renewable sources, the most used process is the natural gas (48%) and oil (30%) reforming, followed by coal gasification (18%). Only 4% of the hydrogen produced is obtained by electrolysis of water [3]. A widely used method for producing hydrogen from natural gas is to use methane and water vapor. The steam reforming of methane [4] is the reaction that converts water vapor and methane to hydrogen and carbon monoxide in the form shown by Eq. (20.1). Associated by a gas-to-water (WGS) conversion reaction shown by Eq. (20.2)

$$CH_4 + H_2O \rightleftharpoons CO + 3H_2 \quad (\Delta H_0 = 206 \, \text{kJ/mol}) \qquad (20.1)$$

$$O_2 + H_2O \rightleftharpoons CO_2 + H_2 \quad (\Delta H_0 = -41 \, \text{kJ/mol}) \qquad (20.2)$$

Since 1920, the steam methane reforming has proved to be an attractive means for the production of hydrogen [5, 6]. It covers nearly half of the hydrogen produced worldwide with an efficiency ranging from 65 to 75% [7]. However, the steam methane reforming is a highly endothermic reaction process where the standard formation enthalpy is 206 kJ/mol as indicated in Eq. (20.1). It needs a sustained supply of heat for activation at a temperature between (700 and 1100 °C). The use of catalysts reduces the amount of energy required as well as the operating temperature of between 200 and 900 °C. Different transient metals can be used as catalysts in the steam methane reforming, the more common used metals are the ruthenium (Ru), palladium (Pb), rhodium (Rh), radium (Rd), nickel (Ni) and platinum (Pt). In our previous research, an achievement of 45% of temperature reduction [8] for the autothermal reformers which can enlarge the used materials and extend the lifetime of the reactor. In this study, the influence of the variation in catalytic loading by changing the size of the two catalysts of an autothermal steam methane reformer was investigated, once the nickel catalyst thickness was determined and changing the size of the platinum catalyst. On the other way by varying the thickness of the platinum catalyst and fixing the size of the nickel catalyst.

20.2 Physical Model

The two-dimensional diagram of the studied configuration of autothermal methane steam reforming is shown in Fig. 20.1. The gaseous mixture passes through an area where the two parallel walls are coated in one side with a platinum catalyst (Pt) on alumina support (Al_2O_3) to activate the reaction of methane combustion. On the other side, the wall is coated with nickel (Ni) as a catalyst on an alumina support (Al_2O_3) to ensure the methane steam reforming reaction.

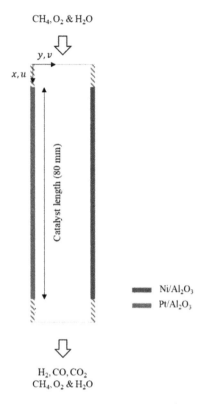

Fig. 20.1 Reactor scheme

20.3 Mathematical Model

An integrated adiabatic ATR was conducted with a two-dimensional modeling to find the flow parameters of the gaseous fluids inside the reformer. The flow was assumed to be laminar, steady and occurring in the same direction. For the porous media the governing equations derived by the volume-averaging method [9]. The values of the adopted physical method were locally averaged for a representative elementary volume, Fig. 20.1 shows the geometry of the internal reformer schematically. The flow field and the heat transfer process inside the porous reformer was described by continuity, momentum, energy and species transfer equations. Consequently, the following transport equations of the averaged physical properties for a laminar flow are applied with considering the Darcy-Brinkman-Forchheimer effects [9]. The finite volume method was adopted to solve numerically the governing equations of continuity, momentum, energy and species.

Boundary Layer
The non-slip, impermeable wall and the other boundary conditions are illustrated below:

Inlet:

$$u_z = 1 \text{ ms}^{-1}; \quad T = 600\,°C \tag{20.3}$$

$$u_r = n_{co} = n_{h2} = 0 \tag{20.4}$$

$$S/C = 1.4 \text{ (Steam to Carbon molar ratio)} \tag{20.5}$$

$$O/C = 0.6 \text{ (Oxygen to Carbon molar ratio)} \tag{20.6}$$

Outlet:

$$\frac{\partial u_z}{\partial r} = \frac{\partial u_r}{\partial r} = \frac{\partial T}{\partial r} = \frac{\partial n_i}{\partial r} = 0 \tag{20.7}$$

Adiabatic wall:

$$u = v = 0, \quad \frac{\partial n_i}{\partial z} = \frac{\partial T}{\partial r} = 0 \tag{20.8}$$

20.4 Results and Discussion

Although the effect of the catalyst is positive of a general point of view because of their influence on the chemical mechanism which leads to a minimal activation energy. However, a quantity of the additional catalyst does not always imply an increase in the hydrogen yield of steam methane reforming system. This fact has been reported by Bravo et al. [10] who compared between the use of the coated catalyst and the fixed bed catalyst, their results demonstrated that the methane conversion for several cases studied is more important for the case of the catalyst coated with hydrogen produced better purity.

In Figs. 20.2 and 20.3, the impact of the variation of the catalytic charge in the reactor is illustrated for the case of a catalytic parallel arrangement. The effect of changing the size of the two catalysts on the hydrogen yield is shown in Fig. 20.2. On the other side, Fig. 20.3 shows the effect on the maximum temperature. It is noted that the operation of the reactor at low catalytic loads of Ni and Pt leads to a less efficient reactor regarding the production of hydrogen. This is attributed to an insufficient amount for the activation of both reactions. It is clear that the increase in the catalytic load of Pt catalyst increases the temperature because of the favoring of the reaction of methane combustion. This change in temperature is reversed for

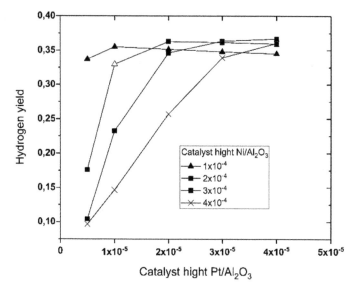

Fig. 20.2 Hydrogen yield as a function of the reactor length

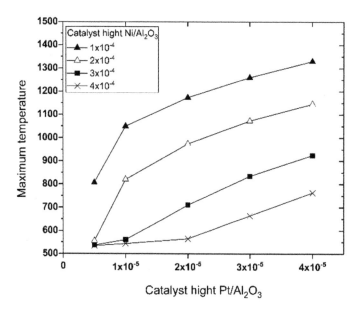

Fig. 20.3 Maximum temperature as a function of the reactor length

the cases where the catalytic load of Ni increases where the reaction of the methane vapor reformer is preferred.

20.5 Conclusion

The catalyst load has a direct impact on the reaction efficiency and the parameter evolvement. Although the increase of the Ni catalyst can enhance directly the hydrogen production, an exceeded amount of the same catalyst can reduce the reaction efficiency. So that, an optimum value of the Ni catalyst load can be determined for each studied case. On the other hand, the Pt catalyst load increment can increase the hydrogen yield linearly for the high amount of Pt catalyst except the case of very low amount of Ni catalyst.

References

1. F. Birol, in *Key World Energy Statistics 2017* (International Energy Agency, 2017)
2. MME e Ministerio Das Minas e Energia. Resenha Energetica Brasileira. Exercício de. 2014. Edição de junho de 2015, http://www.mme.gov.br/documents/1138787/1732840/
3. A. Körner, C. Tam, S. Bennett, J. Gagné, in *Technology Roadmap-Hydrogen and Fuel Cells* (International Energy Agency, 2015)
4. K. Liu, C. Song, V. Subramani, in *Hydrogen and Syngas Production and Purification Technologies* (Wiley, 2010)
5. M.A. Pen, J.P. Gomez, J.G. Fierro, New catalytic routes for syngas and hydrogen production. Appl. Catal. A **144**(1–2), 7–57 (1996)
6. B.C.R. Ewan, R.W.K. Allen, A figure of merit assessment of the routes to hydrogen. Int. J. Hydrogen Energy **30**(8), 809–819 (2005)
7. S.Z. Baykara, Hydrogen: a brief overview on its sources, production and environmental impact. Int. J. Hydrogen Energy **43**(23), 10605–10614 (2018)
8. A. Cherif, R. Nebbali, Numerical analysis on autothermal steam methane reforming: Effects of catalysts arrangement and metal foam insertion. Int. J. Hydrogen Energy **44**(39), 22455–22466 (2019)
9. S.V. Patankar, in *Numerical Heat Transfer and Fluid Flow* (Hemisphere Publications, 1980)
10. A. Karim, J. Bravo, D. Gorm, T. Conant, A. Datye, Comparison of wall-coated and packed-bed reactors for steam reforming of methanol. Catal. Today **110**(1–2), 86–91 (2005)

Chapter 21
Proton Exchange Membrane Fuel Cell Modules for Ship Applications

S. Tamalouzt, N. Benyahia, and A. Bousbaine

Abstract In this article, we proposed a more reliable architecture composed of five fuel cell modules (FC), a storage system composed of battery and supercapacitor was also proposed to support the operation of the fuel cell. The main objective of this work is to study the feasibility of using the global system for small marine applications. In this paper, the global system was modeled and then simulated using Matlab/Simulink. The fuel cell is used as the main power source; each fuel cell is connected with a DC bus via a DC–DC boost converter. The Energy Storage System (HESS) is controlled as a fast-bidirectional auxiliary power source, it contains a battery and supercapacitors and each source is connected to the DC bus via a bidirectional buck-boost DC–DC converter (BBDCC). In order to optimize the HESS, the supercapacitors and the batteries are designed to allow high-efficiency operation and minimal weight. The entire system's energy management algorithm (PMA) is developed to satisfy the energy demand of the boat. Finally, simulation tests are presented in Matlab/Simulink and discussed, where the effectiveness of the proposed system with its control is confirmed.

Keywords Hybrid storage system · Fuel cell system · Ship system

21.1 Introduction

Fuel cells (FCs) are a promising technology, used for electric power generation for autonomous systems especially, in marine transportations (MTs) [1, 2]. The modular design of the FCs enables flexibility in the arrangement of the plant components and

---

S. Tamalouzt (✉)
Laboratoire LTII, University of Bejaia, 06000 Béjaia, Algeria
e-mail: tamalouztsalah@yahoo.fr

N. Benyahia
LATAGE Laboratory, Mouloud Mammeri University, BP 17 RP, 15000 Tizi-Ouzou, Algeria

A. Bousbaine
College of Engineering and Technology, University of Derby, Derby DE22 3AW, UK

© The Editor(s) (if applicable) and The Author(s), under exclusive license to Springer Nature Singapore Pte Ltd. 2021
A. Khellaf (ed.), *Advances in Renewable Hydrogen and Other Sustainable Energy Carriers*, Springer Proceedings in Energy,
https://doi.org/10.1007/978-981-15-6595-3_21

could lead to a more cost-effective and optimized layout of basic ship structures [1]. Also, because of their high efficiency and no emission, their uses for MTs tend to multiply in recent years [1, 3]. However, some disadvantages considered as being inherent shortcomings exist in FCs operations; A slow response in their operations and a low terminal voltage. In addition, no overload capabilities and no acceptance of reverse current. Furthermore, the effectiveness of the fuel cell decreases as the ripples in the output current increase [3–5]. In order to remedy these problems and achieve the normal operating conditions of FC, an ESS with its power conversion systems must be associated in the process. However, a high-power density is necessary in order to make face rapid power changes and to ensure the power supply autonomy of the system. This is why it is necessary to use HESSs [1, 6, 7]. In the present paper, the Bts and the Scs are used as being a HESS, everyone in it is associated with BBDCC.

Several works in the field of transportation systems in general, especially marine transport, have been associated with FC [2, 3, 5, 7–14], but all proposed architectures are based on a single FC module. In these systems, the failure of the FC module induces breakdown of the global system, which implies that such systems are not reliable for marine applications. In the case of FCs with many modules, we can cite some works which are proposed, particularly, the FCs modules connected in parallel [10, 11]. However, in this case, the converter associated with the ESS is used to keep DC voltage constant. Nevertheless, the ESS must supply in addition a constant power to ensure efficient control of DC-bus voltage. This leads to the need for oversizing of the ESS to provide constant power and peak power necessary for ship propulsion.

In this paper, the proposed topology of FC consists of five modules connected to the DC-grid each with a BDCC. This system is associated with HESS based on Bt and Sc, which are connected to the DC-grid through two separate bidirectional BBDCCs. In order to optimize the HESS and make it work in an optimal way, a combination of the Scs and Bts are designed to allow operation at high efficiency and at minimal weight. To keep a constant DC-bus voltage, the voltage regulated converter must provide in addition a minimal power to ensure DC-bus voltage control. In this case, the judicious choice is the BDCC associated with the FC, since it must provide the power needed for propulsion and minimum permanent power to feed its auxiliary. In the present work, an EC and PMA of the whole system are developed, in order to maintain the output DC-voltage from DC-grid equal to its reference value and to ensure power supply in electrical energy of the ship without interruptions. Finally, simulation tests are presented under Matlab/Simulink and discussed.

21.2 Description and Modeling of the Proposed System

The ship's main power system is an autonomous system, where most of the power is used for propulsion and the rest is utilized for auxiliary circuits [5, 15, 16]. The Illustration of the marine propulsion system (MPS) to be studied is shown in Fig. 21.1. This scheme presents principally the proposed architecture of a power system used

21 Proton Exchange Membrane Fuel Cell Modules for Ship Applications

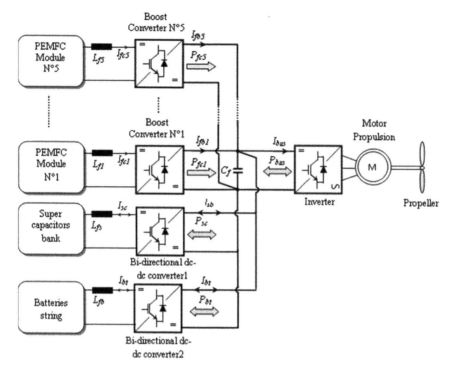

Fig. 21.1 Illustration of the proposed marine propulsion system (MPS)

in marine applications. It is composed of three basic subsystems: the generation subsystem, the storage subsystem, and the propulsion subsystem.

To define the FC model, basing on the Nernst voltage and drops voltage, a single cell output voltage can be defined by Eq. (21.1) [17].

$$\begin{cases} E_{\text{Nernst}} = E_0 - (\Delta S_0/2F)(T - 298.15) + (RT/2F) \ln\left(P_{H_2} \cdot P_{O_2}^{1/2}\right) \\ V_{\text{fc}} = N_s E_{\text{Cell}} \end{cases} \quad (21.1)$$

where the E_{cell} is the Nernst voltage diminished with the drops voltage. The fuel cell polarization curve is shown in Fig. 21.2.

The considered type of the battery has a lower cost than the other batteries type. Lead-acid batteries have high reliability with acceptable efficiency and low maintenance. The open-circuit voltage of one cell [1, 3] can be defined as in Eq. (21.2).

The supercapacitor model consists of a capacitance C_{SC}, an equivalent series resistance ESR representing the charging and discharging resistance, and an equivalent parallel resistance EPR representing the self-discharging losses. This model is suitable for applications where the energy stored in the capacitor is of primary importance. A current pulse was applied to the super-capacitor model, and the output voltage response is shown in Fig. 21.3.

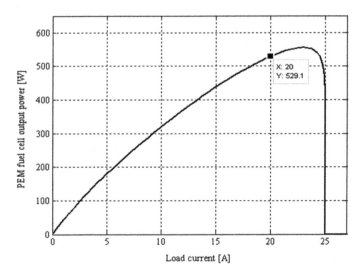

Fig. 21.2 Power versus current polarization curve of PEM fuel cell

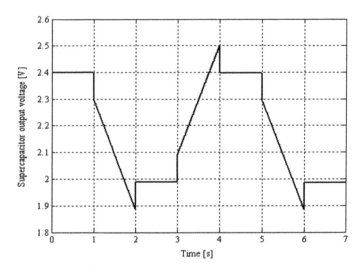

Fig. 21.3 Super capacitor output voltage

The leakage effect, which causes decay in voltage after the initial current pulse, can be observed.

$$\begin{cases} E_m = E_0 - K(273 + \theta)(1 - \text{SOC}) \\ \text{SOC} = 1 - Q_e/C(0, \theta) \\ I_{\text{avg}} = I_m/(1 + \tau_b p) \end{cases} \quad (21.2)$$

The terminal battery stacks output voltage with the total power Bts are defined as:

$$\begin{cases} V_{bt} = N_{bs} E_m \\ P_{bt} = N_{bp} N_{bs} E_m I_{bt} \end{cases} \quad (21.3)$$

21.3 Power Control Strategy and Management System

The EC and PMA of the whole system are described in this part. The BDCC associated with FC is required for transforming the power from the FC to an appropriate form, i.e. the FC voltage and current [1]. A duty ratio around 50% is used for this converter at the FCs stack rated operating point. The reference input voltage for BDCC can be deduced referring to the wave characteristics given in [3]. In steady-state, the FCs stack system operates at the minimum voltage, which means that the current and power of the FC system are operating at rated current and maximum power. The output voltage of the FC is about 26.46 V at a rated current operating point. Therefore, the number of FC stacks needed to be connected in series (N_{fs}) and in parallel (N_{fp}). Knowing that the maximal average power of the ship is 50 kW. Then, five FC modules each interfaced by a boost DC-converter to the DC bus, they all are connected in parallel.

The use of BBDC has the advantage of a faster and more stable response [1, 7, 8]. It is capable of delivering a DC output voltage as required, i.e. higher or lower than the DC input voltage. The BBDC control is based on the hysteresis current controller (HCC), where it has excellent dynamic performances [1, 18]. The current in play by the storage device (Scs or Bts) can be positive or negative.

For medium and high-power applications, Scs and Bts are used in stacks where many cells are connected in series and parallel to obtain acceptable power (current and voltage) [7, 14, 18]. When the load power P_{bus} is high and exceeds the FC modules' average power, the Bts and Scs will operate to ensure the deficiency of this power. The control strategies and the power management of energy sources of the ship, consist of the fact that the FC modules are controlled to provide the propulsion average power, and Scs/Bts based ESS is controlled to provide the peak power. The primary objective of the energy control and power management to preserve the DC bus voltage at its reference and to keep the state of charge (SOC) of the Bts between 40 and 80%, while supplying the required power demand for the ship. If the battery's SOC reaches its lower limit or its upper limit, the difference between the power demand and the average power will be provided by the Scs. Otherwise, the Bts power is limited to be less than maximum power $P_{bM} = 13$ kW in discharge mode and is limited to be less than minimal power $P_{bm} = 3.9$ kW in charge mode, the remainder power is provided by the Scs. The average propulsion power is defined as given in [10].

The auxiliary power provided by the Bts and Scs and the total propulsion power is given by Eq. (21.4), respectively. where ω_{ref} is the reference speed, Q_{ref} is the reference torque of the propulsion motor and P_a is the auxiliary power of the ship.

$$P_{aux} = P_{bus} - P_{av}; \quad P_{bus} = \omega_{ref} Q_{ref} + P_a \tag{21.4}$$

21.4 Simulation Results and Comment

In order to evaluate the EC and management strategy, a simulation study of the proposed system, Fig. 21.1, under Matlab/Simulink is presented. The system performances under random variations in propulsion power have been studied. The maximal bus power is 81 kW and the total average power of the FC modules is 50 kW. The Scs power needed is 18 kW for 30 s, which represents 540 kJ and the Bts power needed is 3.9 kW for 107 s, which represents 417 kJ. The discharge voltage ratio is 0.75.

Figure 21.4 illustrates the waveforms of the different voltages; V_{bus} and V_{FC}. It is clearly seen that the BDCC ensures good regulation of the V_{bus}, it maintains its reference value despite disturbances caused by the variations in the demand power. The voltages at the terminals of the various elements follow the evolution of transit of the powers put into play by the latter.

The interaction and the synchronization of power flow between the different elements of the system affirm the utility, the importance and the efficiency of the HESS proposed in the electric MPS. It is clear that the Scs and Bts are able to

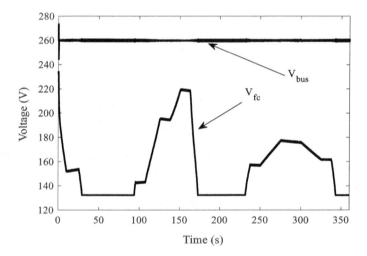

Fig. 21.4 DC bus and FC voltages

provide the instantaneous power demand required in the acceleration and deceleration phases of the propulsion system. When the DC-bus power is higher than the FC modules' total power, the Scs and Bts supply the required power, as illustrated in Figs. 21.5 and 21.6. The Scs and Bts play an important role; they can provide the peak power needed during acceleration and absorb the redundant power released from deceleration. The Scs and Bts current are shown in Fig. 21.6. It can be seen that the BBDCC has a current regulation and the current in the Scs or in the Bts is positive in discharge mode and is negative in charge mode. The SOC of the battery is limited within the range of (40–80%), as presented in Fig. 21.7.

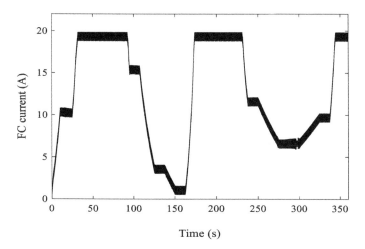

Fig. 21.5 Fuel cell current

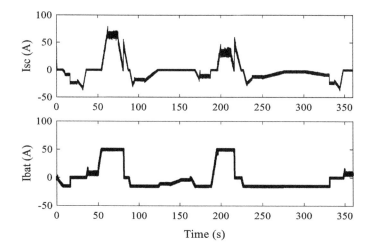

Fig. 21.6 Supercapacitor and battery currents

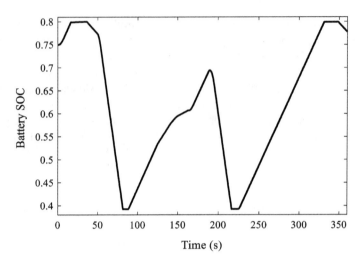

Fig. 21.7 Battery SOC

21.5 Conclusion

In this paper, a more reliable architecture based on five FC modules for small ship applications is proposed. The dynamic models of the PEM fuel cell, battery, and super-capacitor are developed under Matlab/Simulink environment. The double charging effect of the PEM fuel cell and the leakage effect of the super-capacitor are taken into account in modeling. Simulation results show that the behavior of the FC modules voltage, Bts voltage and Scs voltage are within a range of normal operations. The energy control and management showed good performances with respect to sharing power between the FC and HESS, and limit the SOC of the Bts within an optimal range.

References

1. S. Tamalouzt, N. Benyahia, T. Rekioua, D. Rekioua, R. Abdessemed, Performances analysis of WT-DFIG with PV and fuel cell hybrid power sources system associated with hydrogen storage hybrid energy system. Int. J. Hydrogen Energy **41**(45), 21006–21021 (2016)
2. T.J. Leo, T.A. Durango, E. Navarro, Exergy analysis of PEM fuel cells for marine applications. J. Energy **35**, 1164–1171 (2010)
3. N. Benyahia, H. Denoun, M. Zaouia, T. Rekioua, N. Benamrouche, Power system simulation of fuel cell and supercapacitor based electric vehicle using an interleaving technique. Int. J. Hydrogen Energy **40**(45), 15806–15814 (2015)
4. J.P. Trovão, F. Machado, P.G. Pereirinha, Hybrid electric excursion ships power supply system based on a multiple energy storage system. IET Electr. Power Appl. **6**, 190–201 (2016)
5. V. Moreno, M.A. Pigazo, Future trends in electric propulsion systems for commercial vessels. J. Marine research **4**, 81–100 (2007)

6. Y. Bouzelata, N. Altin, R. Chenni, E. Kurt, Exploration of optimal design and performance of a hybrid wind-solar energy system. Int. J. Hydrogen Energy **41**(29), 12497–12511 (2016)
7. S. Tamalouzt, F. Hamoudi, T. Rekioua, D. Rekioua, Variable speed wind generator associated with hybrid energy storage system-application for micro-grids, in *5th International Renewable and Sustainable Energy Conference (IRSEC'2017)* (Tangier, Morocco, 2017), pp. 1–6
8. X. Yu, M.R. Starke, L.M. Tolbert, B. Ozpineci, Fuel cell power conditioning for electric power applications: a summary. IET Electr. Power Appl. **5**, 643–656 (2007)
9. S. Caux, J. Lachaize, M. Fadel, P. Shott, L. Nicod, Modelling and control of fuel system and storage elements in transport applications. J. Process Control **15**, 481–491 (2005)
10. N. Benyahia, H. Denoun, A. Badji, M. Zaouia, T. Rekioua, N. Benamrouche, D. Rekioua, MPPT controller for an interleaved boost DC–DC converter used in fuel cell electric vehicles. Int. J. Hydrogen Energy **27**(39), 15196–15205 (2014)
11. L. Luckose, H.L. Hess, B.K. Johnson, Fuel cell propulsion system for marine applications, in *IEEE Electric Ship Technologies Symposium (ESTS)* (Baltimore, 2009)
12. U.S. Congress, Office of Technology Assessment, in *Marine Applications for Fuel Cell Technology—A Technical Memorandum. 1986 OTA-TM-O-37* (U.S. Government, Washington, DC, 1986). http://www.fas.org/ota/reports/8612.pdf
13. C.N. Maxoulis, D.N. Tsinoglou, G.C. Koltsatis, Modelling of automotive fuel cell operation in driving cycles. J. Energy Convers Manag. **45**, 559–573 (2004)
14. M. Uzunoglu, M.S. Alam, Dynamic, modeling, design and simulation of a PEM fuel cell/ultracapacitor hybrid system for vehicular applications. J. Energy Convers. Manag. **48**, 1544–1553 (2007)
15. IEEE, in *Guide for the Design and Application of Power Electronics in Electrical Power Systems on Ships* (IEEE Industry Applications Society, Std 1662TM, 2008)
16. L. Wang, D.J. Lee, W.J. Lee, Z. Chen, Analysis of a novel autonomous marine hybrid power generation/energy storage system with a high voltage direct current link. J. Power Sources **185**, 1285–1292 (2008)
17. F. Zenith, S. Skogestad, Control of fuel cell power output. J. Process Control **17**, 333–347 (2007)
18. M. Othman, A. Anvari-Moghaddam, J.M. Guerrero, Hybrid shipboard microgrids: system architectures and energy management aspects, in *43rd Annual Conference of the IEEE Industrial Electronics Society (IECON'17)* (Beijing, China, 2017), pp. 6801–6806

Chapter 22
Analysis of off Grid Fuel Cell Cogeneration for a Residential Community

A. Mraoui, B. Abada, and M. Kherrat

Abstract The cogeneration system, studied in this paper, optimize the use of electricity and heat in a one source production system. The electricity and heat load profiles were determined for a subdivision of 50 single-family homes. It was noted that, annually the electricity demand represents 68% of total energy demand and the rest (32%) is a heat demand. To ensure the supply of energy, a photovoltaic generator coupled with a set of fuel cells was considered. Fuel cells are powered by hydrogen supplied by electrolyzers and stored at high pressure in containers. The system can produce electricity that will be converted for domestic use and heat that will be recovered for use as warm water or homes heating when necessary. The whole system is completely autonomous and does not require the grid in any way. The main issue is therefore the intelligent management of the whole through appropriate algorithms whose main objective is to provide energy without any interruption.

Keywords Fuel cells · Cogeneration · Townhouse

22.1 Introduction

With cogeneration systems, it is possible to reduce primary energy consumption. The use of cogeneration with fuel cells reduces greenhouse gas emissions and better integrates renewable energies [1]. A fuel cell cogeneration system can provide 45% of the annual heat requirements and 50% of the annual electricity requirements [10]. For micro CHP systems (use in a single residence), fuel cells are the best option from an economic and environmental point of view [11]. The relatively low operating temperature (about 80 °C) limits the use of the heat generated. However, it is a very positive point if starts and stops are frequent. Also, the heat recovered at this temperature is sufficient for residential use, both for domestic hot water and space heating [4].

A. Mraoui (✉) · B. Abada · M. Kherrat
Division Renewable Hydrogen, Centre de Développement des Energies Renouvelables, BP. 62 Route de l'Observatoire Bouzareah, 16340 Algiers, Algeria
e-mail: a.mraoui@cder.dz

© The Editor(s) (if applicable) and The Author(s), under exclusive license to Springer Nature Singapore Pte Ltd. 2021
A. Khellaf (ed.), *Advances in Renewable Hydrogen and Other Sustainable Energy Carriers*, Springer Proceedings in Energy,
https://doi.org/10.1007/978-981-15-6595-3_22

A hybrid cogeneration system designed to meet the energy demand of 150 houses in the residential community of Sharjah, United Arab Emirates, was studied by Ghenai et al. [3]. The proposed micro-grid system is based on a photovoltaic solar system and hydrogen fuel cell generators to supply alternating current to a load composed of the residential community. Hydrogen, required for the operation of the fuel cell, is produced by an electrolyzer and stored in a high-pressure tank. The electrolyzer operates on the energy produced by the micro-grid. The conversion of direct current (DC) into alternating current is carried out by an inverter. The solar photovoltaic/fuel cell hybrid renewable energy system is directly connected to the DC bus to transfer energy from generation to the inverter. The inverter converts direct current to alternating current to meet the alternating current demand of residential houses.

In a study by Jamshidi and Askarzadeh [5], a hybrid system composed of photovoltaic, diesel generator and fuel cell was considered. A multi-objective optimization of this off-grid application leads to having a cost-effective and reliable energy system. Indeed, by considering fuel cell/electrolyzer/hydrogen tank in the hybrid energy system, the total cost will decrease. The same conclusion was found by Sami et al. [12] using a Smart Home system, considering fuel cell/electrolyzer/hydrogen, that improves electricity production without interruption. The system reacts so flexible with all critical constraints through a precise multi-agent strategy. Several control scenarios were studied, the overall cost is reasonable given equipment maintenance and life cycle. It was found that using both battery and hydrogen for energy storage is the economically more viable solution than using only one form of energy storage [13].

The systems studied use energy balance equations to simulate the production of electricity, heat and hydrogen. These equations do not take into account real constraints. For example, a photovoltaic system with batteries and regulator has an efficiency of about 75–80% using these equations. Taking into account the system performance and the management of the charge/discharge of the batteries, the efficiency is about 50–60%. It is necessary to use detailed modelling to approach practical cases.

In this study, we will consider a hydrogen production system using electrolyzer powered by photovoltaic generators. Hydrogen is used by fuel cells that provide electricity and heat to a group of single-family homes. The profile of the energy demand (electricity and heat) will be modelled for Algerian dwellings load shape. We will carry out a detailed modelling in order to take into consideration all possible types of losses.

22.2 Profile of the Electricity and Heat Demand of a Residential Building with Air Conditioning System

To model the electricity and heat load we combined data from Opérateur du Système Electrique of Algeria [9] and National Energy Balance [8]. To get hourly shapes we adapted the general data using System Advisor Model [2] and Whole Premise Load Shapes [14]. From the obtained data, we observe that the overall energy demand is higher in summer than in winter (Fig. 22.1) due to the demand for air conditioning during this period. The heat load is greater in winter than in summer for heating needs during this period. According to Table 22.1, the percentage of electricity is only 35.4% while heat consumes 64.6% of the total energy produced in January. While in June, the proportion of electricity is 86.1% and heat is 13.9%. The overall annual balance is 68.5% electricity and 31.5% heat. A fuel cell produces about 60% electricity and 40% heat. Globally, a fuel cell could help to provide heat and electricity to the system. However, the problem is that it is very difficult to store heat from one season to another.

The design of the production system is based on the highest daily. The greatest daily load occurs in June (around 71 kWh) and the greatest daily heat load is in December (around 37 kWh), which represents 34% of the total energy demand. From a design point of view, a fuel cell is perfectly suitable for this type of application.

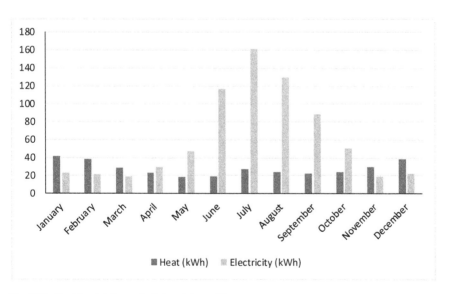

Fig. 22.1 Residential demand of electricity and heat

Table 22.1 Proportion of each energy in the overall balance

Month	Electricity (%)	Heat (%)
January	35.4	64.6
February	35.4	64.6
March	40.3	59.7
April	55.6	44.4
May	71.8	28.2
June	86.1	13.9
July	85.7	14.3
August	84.5	15.5
September	79.7	20.3
October	68.0	32.0
November	38.8	61.2
December	36.5	63.5
Annual	68.5	31.5

22.3 Simulations

The system considered is presented schematically in Fig. 22.2. The system consists of a photovoltaic generator to produce renewable electricity. Hydrogen is produced using an alkaline electrolyzer at moderate temperature and pressure. A 200 bar hydrogen storage container to store the hydrogen that will be used by the cell. The fuel cell provides electricity and heat. We have also included additional elements such as regulators, inverters, heat exchangers etc. to ensure energy transfers.

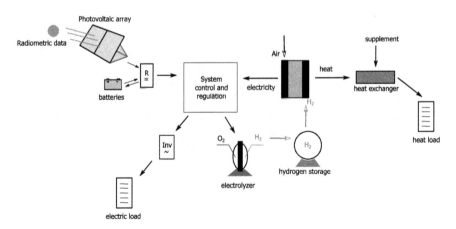

Fig. 22.2 Cogeneration system schematic

22.4 System Modeling

Each element of the system was modeled using empirical equation [6, 7]. The output of the PV generator can be calculated from:

$$P_{PV} = N_{PV} \cdot f_{PV} \cdot P_{STC} \cdot \frac{G_A}{G_{STC}} \cdot (1 + (T_C - T_{STC}) \cdot C_T) \quad (22.1)$$

where N_{PV} is the number of PV panels, f_{PV} the conversion efficiency, P_{STC} the PV array power under standard test conditions (STC), G_A the global solar radiation on the PV array, T_C the temperature of the PV cells, and C_T the PV temperature coefficient.

An electrolyzer can produce hydrogen from electricity, and this hydrogen is then usually stored in tanks. The characteristic of an electrolyzer can be described as:

$$V_{el} = N_{el} \cdot V_{rev} + (r_1 + r_2 \cdot T) \cdot \frac{I_{el}}{A_{el}}$$
$$+ \left(s_1 + s_2 \cdot T + s_3 \cdot T^2\right) \cdot \log\left(1 + \left(t_1 + \frac{t_2}{T} + \frac{t_3}{T^2}\right) \cdot \frac{I_{el}}{A_{el}}\right) \quad (22.2)$$

N_{el} the number of cells, V_{rev} the reversible cell potential, T the temperature and $\frac{I_{el}}{A_{el}}$ the current density. $r_1, r_2, s_1, s_2, s_3, t_1, t_2$ and t_3 are empirical coefficients.

The production rate of hydrogen in a cell is:

$$\eta_F = \frac{(I_{el}/A_{el})^2}{f_1 + (I_{el}/A_{el})^2} f_2 \quad (22.3)$$

where f_1 and f_2 are coefficients.

A fuel cell (FC) can produce electricity from hydrogen (H_2), which can be drawn from hydrogen tanks. We use the voltage electrical model:

$$V_{fc} = \left(E_{OC} - r_{fc} \cdot i_{fc} - a \cdot \ln(i_{fc}) - m \cdot e^{n_0 \cdot i_{fc}}\right) \cdot N_{fc} \quad (22.4)$$

where E_{OC} is the open circuit voltage, i_{fc} is the current density in one cell, N_{fc} is the number of cells, r_{fc}, a, m, n_0 are empirical coefficients.

As an FC generates electricity and heat at the same time, the produced heat can be calculated:

$$Q_{fc} = N_{fc} \cdot \left(1.48 - \frac{V_{fc}}{N_{fc}}\right) \cdot I_{fc} \quad (22.5)$$

Then the hydrogen consumed:

$$\dot{n}_{H2} = \frac{N_{fc} I_{fc}}{2FU} \quad (22.6)$$

where F is the Faraday constant, and U is the utilization efficiency of hydrogen.

The state-of-charge (SOC) represents the state of the battery:

$$\text{SOC}(t) = \text{SOC}(t - \Delta t) + \frac{\eta_{\text{ch}} \cdot P_{\text{ch}}(t) \cdot \Delta t}{\text{CB}} - \frac{P_{\text{dis}}(t) \cdot \Delta t}{\text{CB}} \quad (22.7)$$

where η_{ch} is the charging efficiency, P_{ch} is the charging power, P_{dis} is the discharging power, Δt is the interval time, and CB is the capacity of the battery.

22.5 Results

The produced electricity is more important in summer than in winter season. From Fig. 22.3, we observe that during the first half-year period the fuel cells produce less power than in the second half-year. The main objective is to satisfy the electrical and heat energy loads of the houses. The state of hydrogen tank varies and in some cases the tank is empty, i.e. the fuel cell demand is important. The objective in this case is that the state of charge of the tank at the end-year is at least greater than in the start. The control algorithm should be modified to avoid the state of empty tank to ensure a proper power energy balance. We will modify the design and algorithm so that the state of charge must be greater than 10% in all cases (Fig. 22.4).

The profile of the heat produced is shown in Fig. 22.5. Mainly the fuel cell can satisfy all the demand of hot water. There is a excess during the summer when

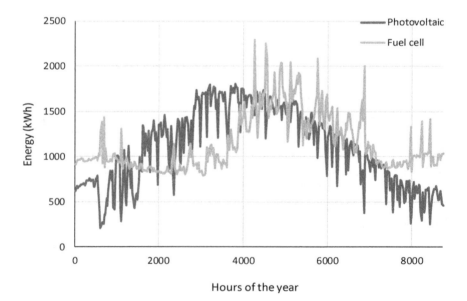

Fig. 22.3 Hourly electricity production

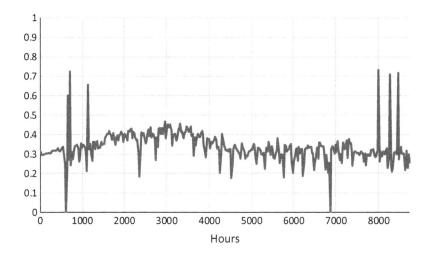

Fig. 22.4 State of hydrogen tank (0 empty, 1 full)

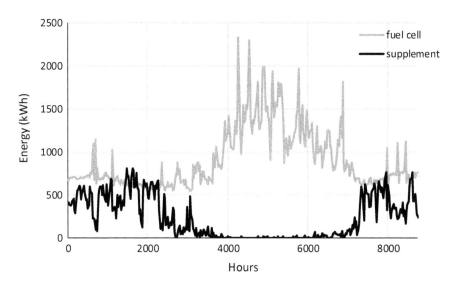

Fig. 22.5 Hourly heat produced by the fuel cell and the necessary supplement

the demand for electricity is high, so the heat produced by the fuel cell is high. In Fig. 22.6, it can be seen that the fluid temperature is increasing. We imposed 60 °C set point for both domestic hot water and space heating. When the temperature exceeds 60 °C there is no need to heat supplement. The fuel cells produce enough thermal energy to the system.

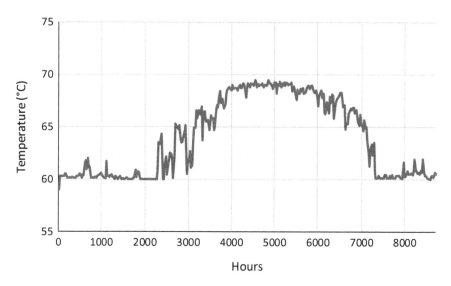

Fig. 22.6 Hot water temperature

The system supplies enough electricity to meet the load, we have included a supplement for thermal energy to compensate for the difference between the electrical demand for and heat demand. Fuel cells implicitly produce heat when they produce electricity. This heat is either recovered or lost. It is difficult to design a seasonal heat storage system unless hydrogen is used in a burner to produce heat. In this case the system must be modified to meet the load. Hydrogen will be an energy carrier for electricity and heat.

22.6 Conclusion

The simulations we have carried out are very promising. However, some elements of system optimization still need to be improved, such as the state of the tank and the management of the heat produced. An optimal control strategy must be determined in order to better manage the system. We did not optimize the management of the system and the sizing was done on the basis of the most unfavorable cases. There is also a need to optimize transfer flows and take into account all real constraints. For example, the fuel cell consumes about 10% of its electrical production in the auxiliary systems necessary for its operation (pumps, compressor, filters, etc.). Heat losses in the pipes must be taken into account, it is also advisable to add hot water tanks in each house (the capacity of each is to be optimized).

References

1. M. De Paepe, P. D'Herdt, D. Mertens, Micro-CHP systems for residential applications. Energy Convers. Manag. **47**(18–19), 3435–3446 (2006)
2. J.M. Freeman, N.A. DiOrio, N.J. Blair, T.W. Neises, M.J. Wagner, P. Gilman, S. Janzou, System advisor model (SAM) general description. National Renewable Energy Lab. (NREL), Golden, CO (United States) (2018)
3. C. Ghenai, T. Salameh, A. Merabet, Technico-economic analysis of off grid solar PV/fuel cell energy system for residential community in desert region. Int. J. Hydrogen Energy. S0360319918316550 (2018)
4. H. Ito, Economic and environmental assessment of residential micro combined heat and power system application in Japan. Int. J. Hydrogen Energy **41**(34), 15111–15123 (2016)
5. M. Jamshidi, A. Askarzadeh, Techno-economic analysis and size optimization of an off-grid hybrid photovoltaic, fuel cell and diesel generator system. Sustain. Cities Soc. **44**, 310–320 (2019)
6. B. Li, R. Roche, D. Paire, A. Miraoui, Sizing of a stand-alone microgrid considering electric power, cooling/heating, hydrogen loads and hydrogen storage degradation. Appl. Energy **205**, 1244–1259 (2017)
7. C.-H. Li, X.-J. Zhu, G.-Y. Cao, S. Sui, M.-R. Hu, Dynamic modeling and sizing optimization of stand-alone photovoltaic power systems using hybrid energy storage technology. Renew. Energy **34**(3), 815–826 (2009)
8. Ministère de l'Energie, Bilan énergétique national 2018. Ministère de l'Energie (2019)
9. Opérateur du Système Electrique, http://www.ose.dz/courbes.php. Accessed: 19 Feb 2020
10. A.D. Peacock, M. Newborough, Impact of micro-CHP systems on domestic sector CO_2 emissions. Appl. Therm. Eng. **25**(17–18), 2653–2676 (2005)
11. H. Ren, W. Gao, Economic and environmental evaluation of micro CHP systems with different operating modes for residential buildings in Japan. Energy Build. **42**(6), 853–861 (2010)
12. B.S. Sami, N. Sihem, Z. Bassam, Design and implementation of an intelligent home energy management system: a realistic autonomous hybrid system using energy storage. Int. J. Hydrogen Energy **43**(42), 19352–19365 (2018)
13. J. Šimunović, F. Barbir, G. Radica, B. Klarin, Techno-economic analysis of PV/wind turbine stand-alone energy system, in *2019 4th International Conference on Smart and Sustainable Technologies (SpliTech)* (2019), pp. 1–5
14. Whole Premise Load Shapes, https://loadshape.epri.com/wholepremise, Accessed: 19 Feb 2020

Chapter 23
Tri-generation Using Fuel Cells for Residential Application

A. Mraoui, B. Abada, and M. Kherrat

Abstract In this paper, we present a bibliographical review of tri-generation systems using renewable fuel. These systems are used to produce cold, heat and power from mainly one fuel source. This mode of operation increases overall efficiency and allows better use of the input fuel. Most tri-generation systems use a fuel cell for its reliability and performance. Furthermore, fuel cells produce heat and electricity at a very high efficiency compared to classical systems. A part of produced heat is recovered to produce cold by an absorption cycle when necessary. The fuel cell can be fed by various fuels, including natural gas, hydrogen produced from renewable or biomass gasification. The overall efficiency depends on the configuration and management of the entire system. The optimal sizing of tri-generation systems is a complicated task. It generally requires two optimization algorithms, one to optimize power flows within the system and the other to optimize the size of the elements. The optimal sizing of tri-generation systems requires two optimization algorithms, one to optimize power flows within the system and the other to optimize the size of each element of the system. Between 75 and 95%, efficiency can be attained depending on the choice of production system and energy management strategy.

Keywords Tri-generation · Fuel cell · Hydrogen

23.1 Introduction

Typically, a tri-generation system produces electricity, heat and cooling from a single source (Fig. 23.1). Products are adjusted to meet the energy demand of a group of residences according to demand. The major advantage of tri-generation is an efficient use of produced energy, lower pollutant emissions and security of supply [11]. Fuel cell technology is integrated into tri-generation systems as it allows better control of

---

A. Mraoui (✉) · B. Abada · M. Kherrat
Division Renewable Hydrogen Energy, Centre de Développement des Energies Renouvelables, BP. 62 Route de l'Observatoire Bouzareah, 16340 Algiers, Algeria
e-mail: a.mraoui@cder.dz

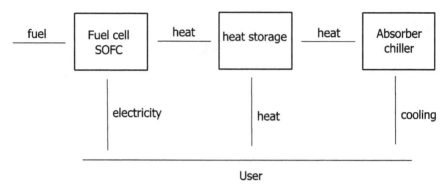

Fig. 23.1 Schematic principle of tri-generation system

the energy produced, high reliability and higher efficiency. For the most applications, solid oxide fuel cells (SOFCs) are the preferred choice for tri-generation systems [15].

Fuel cells produce heat and electricity. Electricity is used via inverters by the end user, heat can be used to heat the home or to provide domestic hot water. The excess hot water produced in summer is used to produce cold for air conditioning needs via absorption systems.

Absorption systems are generally based on the evaporation and condensation of a solution consisting of a binary to produce cold. These systems can use any type of waste heat, steam, hot gas or hot liquid. The binary water/lithium bromide is commonly used for air conditioning or cooling in which the low temperature is between 5 and 10 °C.

Most research uses simulations to study tri-generation. The data are either synthesized or collected from subsystems. However, there are some experimental studies, but there appears to be a real lack of data in this area [15].

23.2 Fuel Supplied Systems

23.2.1 Natural Gas

Natural gas is the most environmentally friendly of all fossil fuels. Its combustion produces water vapour and CO_2 in smaller quantities than petroleum-based fuels. This generates about 30–50% less emissions. Solid oxide fuel cells (SOFCs) generally have an integrated reformer that converts desulphurized natural gas into hydrogen that can be used by the high temperature fuel cell system. The operating temperature of an oxide solid fuel cell is between 600 and 1000 °C [8]. The fuel (NG) enters a mixer for mixing with the recycled fuel from the outlet. It reacts with heated air to produce a gas rich in hydrogen and carbon monoxide. This operating mode is simplified since the reaction takes place at the anode, thus avoiding the use of an external reformer. The nikel contained at the anode is the catalyst of the reaction which is thermodynamically favoured at high temperature.

The same system has been studied by Ramadhani et al. [9], but with the option of supplying hydrogen by electrolysis to a fuel cell car refuelling station. The polygeneration system also allows the sale of excess hydrogen, which reduces operating costs. The reformer can be an external unit combined with a proton exchange membrane fuel cell (PEMFC). This system has been studied by Baniasadi et al. [1] and has demonstrated good reliability and high performance. Elmer et al. [4] used a liquid desiccant system instead of the water absorber/Li-Br. The desiccator is a viable technology in the development of these systems. The authors used data from a commercial fuel cell (BlueGEN SOFC) that operates at a temperature of 750 °C.

Ozcan and Dincer [8] used an organic rankin cycle (ORC) to produce additional electricity by recovering some of the heat produced by the SOFC. Toluene is the working liquid, it allows to obtain a high efficiency at a moderate pressure range. The size of the ORC system can be medium or small, optimizing the cost. The superheated toluene vapour expands in the expander, releases heat into the regenerator, condenses at the condenser by supplying heat to the domestic water heating system. The cold production is ensured by a water/Li-Br absorption system that recovers the secondary heat produced by the SOFC cell.

23.2.2 Renewable Input

Lototskyy et al. [7] used a reversible solid oxide fuel cell (R-SOFC), powered by a hydrogen container. Their system uses heat management by metal hydrides. In electrolyser mode, the R-SOFC uses the electrical energy supplied by a photovoltaic generator field to produce hydrogen during the sunny period. The produced hydrogen is stored in a high temperature metal hydride that generates a significant amount of heat at about 300 °C. The heat produced and the heat released by the R-SOFC is used to produce steam for the electrolyser. In the absence of sunlight, part of the residual heat is used to release hydrogen from the metal hydride container and the rest is sent to the heat management system for heating and cooling purposes.

23.2.3 Input Biomass

Biomass gasification is an essential technology to achieve the objective of sustainable development. The main product of this reaction is syngas, which can be used in various processes. Segurado et al. [13] studied a conventional tri-generation system using syngas to supply SOFCs. The system can thus produce cold, heat and renewable electricity. This technique is an important tool for the flexibility of the energy system. Biomass can be composed of different solid waste or by-products from industry or other sources. Wegener et al. [15] considers that downdraft gasification is the most practical choice because of the high quality of the syngas. Most of these systems have a nominal power rating less than 500 kW. Small gasification systems are generally more cost-effective than large systems due to lower maintenance costs.

23.2.4 Polysources

Sezer and Koç [14] propose an innovative multi-generation system using only renewable energy sources. They use a heliostat field that concentrates the solar energy to a set of photovoltaic collectors, increasing their efficiency to 30%. The heat generated at the photovoltaic cells is recovered by a boiling heat exchanger system that produces water vapor at 110 °C while maintaining the temperature of the photovoltaic cells at 125 °C. The steam produced is sufficient to supply a seawater desalination system to produce fresh water. The brine from desalination is used to produce osmotic energy. Wind turbines have been designed to produce the same amount of electrical energy as the concentrated photovoltaic system. Using the same principle, Inac et al. [6] propose a system that uses four main components: a photovoltaic field, an anaerobic digester for biogas production, a solid oxide fuel cell, and a proton exchange membrane electrolyzer in a system that produces cold, heat and electricity.

23.3 Sizing Tri-generation Systems

The sizing of tri-generation systems depends mainly on two factors: the operating strategy, which determines the distribution of the power flow throughout the system, and the forecasting error of the induced data (charging profile, photovoltaic production, etc.). Conventional approaches to the sizing of hybrid renewable energy systems are the empirical method, the ampere hour method and the software method that uses mathematical optimization techniques [12].

The empirical method is based on the experience acquired with existing systems. It provides guidelines for the sizing and operation of renewable energy sources, a battery bank, DC bus voltage and electronic power equipment according to the required charging energy, peak demand and availability requirements. For renewable sources, a value of 40–60% of the load is suggested. For the sizing of the battery bank, there are many rules of thumb; it is recommended that a system composed only of renewable sources should have a storage capacity of three to five days, while a remote telecommunications station should have five to ten days of storage. The empirical method does not provide an optimal solution.

For the ampere hour method (Ah), the energy consumption of each load is calculated [12]. Losses in power converters, batteries etc. are taken into account. The power requirement of renewable sources is determined from the daily energy requirements of the load. The method is based on the efficiency of the individual components and on the average values of energy consumption from renewable energy sources. The limitations of the empirical rule and Ah dimensioning methods can be overcome by the implementation of mathematical optimization algorithms [5].

There are two main optimization methods used to optimize the operation of renewable energy systems. The genetic algorithm (GA) is used to calculate the optimal size of the system. While another method such as particulate swarms (PSO) is used to optimize the operational strategy. Other methods can be used, but it seems that GA and PSO are the most promising [5].

23.4 Performance Analysis

According to Ozcan and Dincer [8], the electrical efficiency of a commercial fuel cell (Siemens-Westinghouse) is 43.3% and its thermal efficiency is 43.7%. The efficiency for cold production is 52.6% and the efficiency for hot water production is 46.7%. This gives an overall efficiency between 87.95 and 95.9% depending on how the system works. Chahartaghi and Kharkeshi [2] achieve an efficiency of 81.55% for the system they propose. Lototskyy et al. [7] claim that tri-generation leads to a 36% improvement in energy efficiency.

With their system, Ramadhani et al. [9] achieve primary energy savings, cost savings and emission reductions of about 73%, 50% and 70% respectively. Ebrahimi and Derakhshan [3] achieved an overall tri-generation cycle efficiency of around 75% and fuel savings compared to a conventional cycle of around 39%. Their economic analysis concludes that a hydrogen price of $2/kg will be competitive with conventional cycles.

Chahartaghi and Kharkeshi [2] with their system attained 55% exergy efficiency. While Sattari Sadat et al. [10]. reduced it to 54% only. Inac et al. [6] obtained 49% and Sezer and Koç [14] reduced it to as low as 30% with 74% energetic efficiency. Lower the exergy, better is the impact of the system.

23.5 Conclusion

A tri-generation system produces cold, heat and power from an energy source. Fuel cell technology is widely used in this type of system for its reliability, efficiency and better control of the energy produced. The CAP makes it possible to produce heat and electricity at very attractive efficiencies. The excess heat is used in absorption cycles (such as the water/Li-Br cycle) to produce cold for air conditioning.

Solid oxide fuel cells have an integrated reformer. They can be powered by natural gas and convert it into hydrogen to produce heat and electricity. Other systems using renewable sources use a photovoltaic field or several renewable electricity sources to produce electricity that will be used by an electrolyser to produce hydrogen that can be used by the fuel cell. Biomass gasification seems to be a very promising option because it allows waste or by-products to be used to convert them for use by the tri-generation system.

The optimal sizing of tri-generation systems is a complicated task. It generally requires two optimization algorithms, one to optimize power flows within the system and the other to optimize the size of the elements. Efficiencies can vary from 75 to 96% depending on the adopted strategy.

References

1. E. Baniasadi, S. Toghyani, E. Afshari, Exergetic and exergoeconomic evaluation of a trigeneration system based on natural gas-PEM fuel cell. Int. J. Hydrogen Energy **42**(8), 5327–5339 (2017)
2. M. Chahartaghi, B.A. Kharkeshi, Performance analysis of a combined cooling, heating and power system with PEM fuel cell as a prime mover. Appl. Therm. Eng. **128**, 805–817 (2018)
3. M. Ebrahimi, E. Derakhshan, Thermo-environ-economic evaluation of a trigeneration system based on thermoelectric generator, two-bed adsorption chiller, and polymer exchange membrane fuel cell. Energy Convers. Manag. **180**, 269–280 (2019)
4. T. Elmer, M. Worall, S. Wu, S. Riffat, Assessment of a novel solid oxide fuel cell tri-generation system for building applications. Energy Convers. Manag. **124**, 29–41 (2016)
5. G. Human, G. van Schoor, K.R. Uren, Power management and sizing optimisation of renewable energy hydrogen production systems. Sustain. Energy Technol. Assess. **31**, 155–166 (2019)
6. S. Inac, S.O. Unverdi, A. Midilli, A parametric study on thermodynamic performance of a SOFC oriented hybrid energy system. Int. J. Hydrogen Energy **44**(20), 10043–10058 (2019)
7. M. Lototskyy, S. Nyallang Nyamsi, S. Pasupathi, I. Wærnhus, A. Vik, C. Ilea, V. Yartys, A concept of combined cooling, heating and power system utilising solar power and based on reversible solid oxide fuel cell and metal hydrides. Int. J. Hydrogen Energy **43**(40), 18650–18663 (2018)
8. H. Ozcan, I. Dincer, Performance evaluation of an SOFC based trigeneration system using various gaseous fuels from biomass gasification. Int. J. Hydrogen Energy **40**(24), 7798–7807 (2015)
9. F. Ramadhani, M.A. Hussain, H. Mokhlis, M. Fazly, J.M. Ali, Evaluation of solid oxide fuel cell based polygeneration system in residential areas integrating with electric charging and hydrogen fueling stations for vehicles. Appl. Energy. **238**, 1373–1388 (2019)
10. S.M. Sattari Sadat, A. Mirabdolah Lavasani, H. Ghaebi, Economic and thermodynamic evaluation of a new solid oxide fuel cell based polygeneration system. Energy **175**, 515–533 (2019)
11. I. San Martín, A. Berrueta, P. Sanchis, A. Ursúa, Methodology for sizing stand-alone hybrid systems: a case study of a traffic control system. Energy **153**, 870–881 (2018)
12. G. Seeling-Hochmuth, Optimisation of hybrid energy systems sizing and operation control. University of Kassel (1998)
13. R. Segurado, S. Pereira, D. Correia, M. Costa, Techno-economic analysis of a trigeneration system based on biomass gasification. Renew. Sustain. Energy Rev. **103**, 501–514 (2019)
14. N. Sezer, M. Koç, Development and performance assessment of a new integrated solar, wind, and osmotic power system for multigeneration, based on thermodynamic principles. Energy Convers. Manag. **188**, 94–111 (2019)
15. M. Wegener, A. Malmquist, A. Isalgué, A. Martin, Biomass-fired combined cooling, heating and power for small scale applications—a review. Renew. Sustain. Energy Rev. **96**, 392–410 (2018)

Chapter 24
Response Surface Methodology Based Optimization of Transesterification of Waste Cooking Oil

R. Alloune, M. Y. Abdat, A. Saad, F. Danane, R. Bessah, S. Abada, and M. A. Aziza

Abstract According to the directive of the European Union on the incorporation of biodiesel in fuels by 2020, the aim of this study is to contribute to understanding the feasibility of using ester methyl of waste cooking oil (EMWC Oil) in diesel engine. However, biodiesel is a renewable environmental friendly fuel consisting of esters methyl of vegetable oil, generally produced by transesterification reaction of oils seeds and animal fats. In this study, biodiesel synthesis by transesterification of waste cooking oil has been realized. The biodiesel is produced via transesterification reaction using methanol (6:1 molar ration), 0.5% of sodium hydroxide at 55 °C for 60 min of duration and Stirring speed of 200 rpm. The application of design experiment methodology for response surface has allowed us to determine the optimum factors influencing on the transesterification reaction efficiency. The synthesized biodiesel has been subject to several characterizations in order to evaluate its quality by comparing its different physicochemical properties with those described in international norms.

Keywords Biodiesel · Transesterification · Response surface methodology

24.1 Introduction

Depletion of the fossil fuels and environment degradation mainly caused by vehicles emissions, lead to search an alternative fuel that is available, renewable, technically feasible, biodegradable, economically profitable and environmentally friendly [1, 2]. Biodiesel is one of one potential candidate to replace conventional diesel fuel. It

---

R. Alloune (✉) · F. Danane · R. Bessah · S. Abada · M. A. Aziza
Centre de Développement Des Energies Renouvelables, BP. 62, Route de L'Observatoire Bouzaréah, Algiers 16340, Algeria
e-mail: r.alloune@cder.dz

M. Y. Abdat · A. Saad
Département Thermo-Energétique, Faculté de Génie Mécanique et Génie Des Procèdes, Université Des Sciences et de La Technologie Houari Boumediene, Bab Ezzoar 16111, Algiers, Algeria

© The Editor(s) (if applicable) and The Author(s), under exclusive license to Springer Nature Singapore Pte Ltd. 2021
A. Khellaf (ed.), *Advances in Renewable Hydrogen and Other Sustainable Energy Carriers*, Springer Proceedings in Energy,
https://doi.org/10.1007/978-981-15-6595-3_24

consists of fatty acid monoalkyles esters, which are produce from oils, fats, recycled oils and waste greases [3–5].

During the last three decades, biodiesel was produced from edible vegetables oils, mostly sunflowers, rapeseed, safflower, soybean, palm oils, coconut oil and peanut oil [6, 7]. Consequently, food prices have experienced their highest levels since the 1970s and world production of oils seeds are decreased. This may conduct to deficiency on worldwide demand for food; which is dangerous for food security of poor people in the world. For these reasons, there is a need to find other sources of oils that can replace partially or totally edible oils to reduce their prices [8]. Non-edible and waste cooking are becoming promising sources to generate a biodiesel for diesel engine.

Diesel engines with internal combustion fuel a large part of the equipment in several sectors, and therefore have a very important market share in the world. These engines consume a great deal of diesel, a fuel derived from crude oil derivatives. This fact has several consequences, giving rise to many challenges that should not be overlooked. In fact, these challenges need to be address decisively in order to ensure the sustainable and long-term operation of these machines in the future. Biodiesel is a fuel derived from vegetable oils, making it part of the biomass industry, which is one of the most exploited sources of renewable energy.

The vegetable oil chosen is the used cooking oil recovered from the canteens and restaurants, this choice is entirely justified, because it makes it possible, on the one hand, to valorize a waste by turning it into fuel, on the other hand, of avoid this waste treatment both complicated and very expensive [9].

The aim of this work is the optimization of the parameters of the transesterification reaction by mathematical treatment of experimental data, by applying the so-called surface response method on a previously chosen model.

24.2 Methods and Modelisation

24.2.1 System Presentation

The aim of this work is to establish an experimental plan to optimize the parameters influencing the process of the transesterification reaction.

The experimental design is a fairly widespread notion in the field of research and in industry allowing the optimization of experiments by highlighting the relationship between the quantity of interest and the variables of an experiment.

As with each experiment, there are several factors that can influence the response (which is the biodiesel yield in this case). It is therefore necessary to choose the following parameters: the temperature, the molar ratio of alcohol and the percentage of the catalyst (% by weight) that are considered to be the most influential parameters according to the literature. As for the fixed parameters, the duration of the transesterification reaction is fixed at one hour, the agitation speed at 200 rpm, the oil/alcohol ratio is 1:6.

24.2.2 Experimental Procedure

The transesterification reaction is used to produce biodiesel from waste cooking oil. The biodiesel was prepared using 100 g of waste cooking oil (WCOil), methanol proportion was 6:1 (molar ratio of WCOil to methanol) with 0.5% of sodium hydroxide (NaOH) as a catalyst (0.5% of the weight of oil) for 60 min of duration and Stirring speed of 200 rpm.

Firstly, the WCOil was preheated then charged in the reactor. After that, the solution methanol/catalyst was added to the batch reactor and the transesterification reaction started at 55 °C.

Finally, biodiesel produced is heated to 100 °C to remove remaining water particles. The biodiesel yield is estimate after separation and purification of final product using the Eq. (24.1) [5]:

$$\text{Yield of biodiesel} = \frac{\text{weight of biodiesel produced}}{\text{weight of oil used in reaction}} \times 100 \qquad (24.1)$$

24.2.3 Model Development

Presentation of the Box-Behnken plans

The box-Behnken plan for three factors is built on a cube. Experimental points are placed not at the vertices of the cube or hypercube, but in the middle of the edges or at the center of the faces (squares) or at the center of the cubes. This arrangement has the consequence of distributing all the experimental points at equal distance from the center of the field of study, thus on a sphere or on a hypersphere according to the number of dimensions. Points are added to the center of the field of study [10].

The Box-Behnken plan for three factors is illustrated in Fig. 24.1. The cube has 12 edges. We usually add experience points to the center of the field of study, usually three. The 3-factor Box-Behnken plan therefore has 12 + 3 trials, or 15 trials. We can notice that with four points in the center instead of three, we obtain a plane that meets the criterion of almost-orthogonality.

The study of the experimental design focuses on highlighting the influence of the factors chosen on the response, namely the biodiesel yield of the transesterification reaction. A series of experiments was carried out according to the plan of Box-Behnken to maximize product performance. The experimental design consists of a single response (biodiesel yield) and three factors (temperature, catalyst percentage and alcohol ratio).

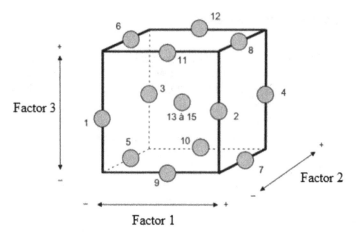

Fig. 24.1 Composite plan for the study of three factors [11]

24.3 Results and Discussions

24.3.1 Field of Study

It should be noted that the choice of the values corresponding to the factors was made with the consideration of the ranges of values found in the literature. Therefore, the temperature working range was defined between 40 and 60, a percentage catalyst between 0.5 and 1.5%.

Finally, the proportions of the alcohol ratio chosen in an arbitrary manner. Each factor has three levels; a low level (−1), a medium level (0) and a high level (+1).

The analysis of the experimental plan was carried out using the Design-Expert© v. 11.0.0 software. A quadratic model is developed from the results obtained from the experiments conducted. This model actually represents a correlation giving the relation between the response and the factors and which allows the optimization of the performance by predicting all the answers of the field of study without carrying out all the required experiments.

The mathematical model corresponding to the adopted plane is a polynomial model of the second order, which takes into account the main effects β_i and the interactions β_{ii} and β_{ij}.

Equation (24.2) y = f (xi) will be written as follows:

$$y = \beta_0 + \sum_{i=1}^{3} \beta_i x_i + \sum_{i=1}^{3} \beta_{ii} x_i^2 + \sum_{i=1}^{2} \sum_{j=i+1}^{3} \beta_{ij} x_i x_j \qquad (24.2)$$

where:

β_0 is the answer in the center of the domain: $y_0 = \beta_0$,

β_i, β_{ii} and β_{ij} are the linear, quadratic and interaction coefficients, respectively.

Yield equation

The data processing, the system modeling and the optimization of the reaction efficiency, were carried out taking into account the parameters identified.

Equation (24.3) gives the quadratic regression model fitted to the experimental data; it expresses the effects of the variables considered on the efficiency of the process.

$$R = 83.83 + 1.81\beta_1 - 5.31\beta_2 - 2.93\beta_3 - 2.91\beta_1\beta_2 - 4.77\beta_1\beta_3 \\ + 1.12\beta_2\beta_3 + 1.65\beta_1^2\beta_2^2 - 1.65\beta_1^2 - 2.20\beta_2^2 - 5.26\beta_3^2 \quad (24.3)$$

where, β_1 is Effect of Temperature, β_2 is Effect of alcohol ratio; β_3 is Effect of percentage of catalyst.

24.3.2 Optimizing Yield Results

The equations of the model make it possible to construct the response surfaces of the experimental plane. The optimization of the transesterification reaction is done by setting one of the factors on an arbitrary value of the working domain and varying the other two along their intervals. In this context, we will start by first fixing the percentage of the catalyst at a value of 0.63% while varying the temperature and the ratio of alcohol. Then set the alcohol ratio to 50% and vary the temperature and the percentage of the catalyst.

Finally, the % of the catalyst and the alcohol ratio vary while the temperature is set at a constant value of 40 °C.

Percentage of catalyst constant at 0.65%:

From Figs. 24.2 and 24.3, it can be seen that high values of the yield correspond to high temperatures and small proportions of alcohol.

24.4 Conclusion

Although transesterification has a proven track record in biodiesel production, further studies are needed to overcome its limitations, which are mostly a high reaction temperature, resulting in considerable production cost and energy loss. To this end, it is imperative to improve the process by performing an optimization study of minimizing the inputs provided while having acceptable yields. The results of the optimization resulted in an optimal solution corresponding to low temperatures with yields exceeding 80%.

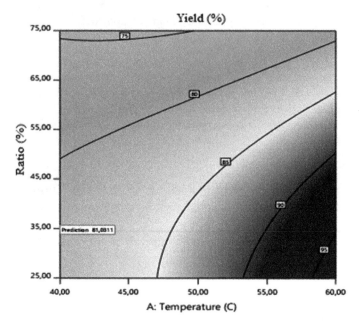

Fig. 24.2 Contour R = f (B, C) for a percentage of catalyst = 0.65%

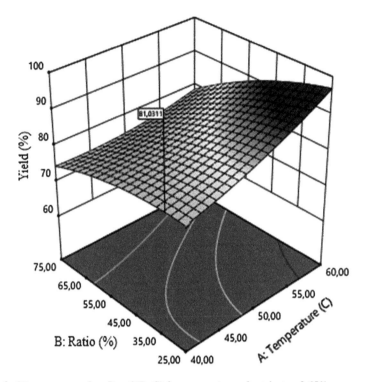

Fig. 24.3 3D response surface R = f (B, C) for a percentage of catalyst = 0.65%

References

1. A.M. Liaquat, M.A. Kalam, H.H. Masjuki, M.H. Jayed, Potential emissions reduction in road transport sector using biofuel in developing countries. Atmosmespheric. Environ. **44**(32), 3869–3877 (2010)
2. M. Mofijur, H.H. Masjuki, M.A. Kalam, A.E. Atabani, I.R. Fattah, H.M. Mobarak, Comparative evaluation of performance and emission characteristics of Moringa oleifera and Palm oil based biodiesel in a diesel engine. Ind. Crops Prod. **53**, 78–84 (2014)
3. A.E. Atabani, I.A. Badruddin, T.M. Mahlia, H.H. Masjuki, M. Mofijur, K.T. Lee, W.T. Chong, Fuel properties of Croton megalocarpus, Calophyllum inophyllum, and Cocos nucifera (coconut) methyl esters and their performance in a multicylinder diesel engine. Energ. Technol. **1**(11), 685–694 (2013)
4. S.K. Hoekma, A. Broch, C. Robbins, E. Ceniceros, M. Natarajan, Review of biodiesel composition, properties, and specifications. Renew. Sustain. Energ. Revews. **16**(1), 143–169 (2012)
5. S.M. Palash, H.H. Masjuki, M.A. Kalam, A.E. Atabani, I.R. Fattah, A. Sanjid, Biodiesel production, characterization, diesel engine performance, and emission characteristics of methyl esters from Aphanamixis polystachya oil of Bangladesh. Energy Convers. Manag. **91**, 149–157 (2015)
6. M.A.H. Altaie, R.B. Janius, U. Rashid, Y.H. Taufiq-Yap, R. Yunus, R. Zakaria, N.M. Adam, Performance and exhaust emission characteristics of direct-injection diesel engine fueled with enriched biodiesel. Energy Convers. Manag. **106**, 365–372 (2015)
7. A. Kumar, S. Sharma, Potential non-edible oil resources as biodiesel feedstock: an Indian perspective. Renew. Sustain. Energ. Revews. **15**(4), 1791–1800 (2011)

8. I. Nehdi, H. Sbihi, C.P. Tan, S.I. Al-Resayes, Evaluation and characterisation of Citrullus colocynthis (L.) Schrad seed oil: Comparison with Helianthus annuus (sunflower) seed oil. Food Chem. **136**(2), 348–353 (2013)
9. A. Mohammadshirazi, A. Akram, S. Rafiee, E.B. Kalhor, Energy and cost analyses of biodiesel production from waste cooking oil. Renew. Sustain. Energy Rev. **33**, 44–49 (2014)
10. J. Goupy, L. Creighton, *Introduction aux plans d'expériences*, 3rd edn. (Dunod, France, 2006)
11. J. Goupy, Modélisation par les plans d'expériences, Techniques de l'ingénieur. Mesures et contrôle, R275. 271–R275. 223 (2000)

Chapter 25
Numerical Investigation on Concentrating Solar Power Plant Based on the Organic Rankine Cycle for Hydrogen Production in Ghardaïa

Halima Derbal-Mokrane, Fethia Amrouche, Mohamed Nazim Omari, Ismael Yahmi, and Ahmed Benzaoui

Abstract Hydrogen is clean energy career that can be produced through renewable energies. Solar plant with parabolic trough concentrators systems is a one of the most promising renewable energies sources for producing energy. This technology is tightly related to the solar insolation levels. Algeria has a potential of Direct Normal Insolation (DNI) that exceeds 2000 kWh/m^2/year with an average daily sunlight duration that exceeds 10 h. Thus are the key features for the development of Solar plant. This paper deals with numerical simulations of a small-scale electrolytic hydrogen production using solar plant with parabolic trough concentrators via an Organic Rankine Cycle. The city of Ghardaïa, located in south of Algeria was selected as a study site. The results have demonstrated among the three organic working fluids (Benzene, Toluene, R-123) selected, benzene is the fluid offering the best efficiency. Moreover, the amounts of hydrogen produced are closely related to the organic fluid used as well as the intensity of the solar radiation.

Keywords Hydrogen · Water electrolysis · Organic rankine cycle · Parabolic trough power plant

25.1 Introduction

Hydrogen, associated with other elements, abounds in nature and because it is an environmental friendly; it is today considered by the scientific and political authorities to be the perfect fuel [1]. Hydrogen as fuel is a particularly adapted solution to the energetic and environmental challenges currently posed. It has unique characteristics

H. Derbal-Mokrane (✉) · M. N. Omari · I. Yahmi · A. Benzaoui
Laboratoire Thermodynamique et Systèmes Energétiques. Faculté de Physique, USTHB, B.P. 32, 16111 El-Alia, Algiers, Algeria
e-mail: hderbal@gmail.com

F. Amrouche
Centre de Développement Des Énergies Renouvelables, CDER, 16340 Algiers, Algeria

that make it an ideal energy candidate [1]. Hydrogen produced by solar means is mainly using thermochemical, electrochemical and electrolytic photo processes [1]. Electrolysis of water is a very common commercial process, used for small scale production where very high purity of hydrogen is required. Whereas, this is one of the most energy-consuming methods, electrolysis has the main advantage of not generating greenhouse gases [2].

Solar plant with parabolic trough concentrators systems is one of the most promising options among available renewable energies sources for producing electricity. This technology is tightly related to the solar insolation levels. Within geographical situation of Algeria, there are a large number of potential sites with high solar insolation levels. Indeed, Algeria is a huge territory that has an average daily sunlight duration that exceeds 2000 h a year and can reach 3900 h in the highlands and Sahara. The Direct Normal Insolation (DNI) is almost 1700 kWh/m^2/year in the north and 2263 kWh/m^2/year in the south of the country, which is not negligible potential. Therefore, this paper will be devoted to the production of hydrogen by electrolysis of water whose energy needs are provided by a solar plant with parabolic trough concentrators, implemented in the south of Algeria, within the city of Ghardaïa.

Organic Rankine Cycles (ORCs) are a technology suitable for the exploitation of different energy sources and are suitable for medium–low temperature heat sources and/or for small available thermal power. In order to valorize small-scale parabolic trough concentrator fields, the harvested thermal energy will be used to feed an ORC Rankine cycle which, unlike the Rankine steam cycle, it uses a fluid having a lower boiling point and a higher efficiency at low temperature. The fluids used can be either refrigerants or hydrocarbons, for this study, three organic working fluids (Benzene, Toluene, R-123) were selected. To achieve this goal, the study of the Rankine cycle plants [3] and the technology of the parabolic concentrators [4, 5] models have been carry out. This was made to do the simulations on the production of heat and electricity by the power station as well as the production of hydrogen applied to Ghardaïa sunlight data.

25.2 Modeling of the Solar Power Station

The parabolic trough concentrator (PT) uses reflective surfaces to concentrate solar radiation onto a linear vacuum absorber tube. A heat transfer fluid (synthetic oil, water or steam, molten salts, etc....) that is heated to medium temperature (from 150 to 400 °C) pass through the system. This fluid is then, sent to conventional exchangers to transform the working fluid into superheated steam that is used directly or exploited in a thermodynamic cycle to produce electricity (Fig. 25.1).

The mathematical model is based on the establishment of an energy balance of the concentrator that includes direct solar radiation, optical and thermal losses. Thus, to determine the useful energy delivered by the heat transfer fluid. This energy is expressed by:

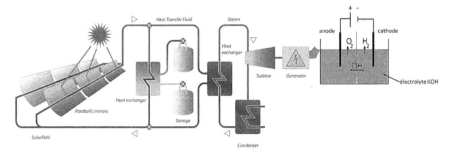

Fig. 25.1 Simplified block diagram of the considered system

$$Q_u = Q_{sra} - Q_{lost} \tag{25.1}$$

Q_u: Useful energy delivered (W/m^2); Q_{sra}: Solar radiation absorbed by the collector (W/m^2); Q_{lost}: Thermal energy lost at the receiver (W/m^2)

The equation for the absorbed solar radiation is [6–8]:

$$Q_{sra} = \text{DNI} \cos\theta \, \text{IAM} \, \eta_{field} \, \eta_{HCE} \, K_{shadow} \, K_{EndLoss} \, K_{oper} \tag{25.2}$$

where:

Q_{sra}	solar radiation absorbed by the receiver tubes (W/m^2)
DNI	direct normal insolation (W/m^2)
θ	angle of incidence (deg)
IAM	incidence angle modifier (–)
η_{field}	field efficiency that accounts for losses due to mirror optics and imperfections (–)]
η_{HCE}	HCE efficiency that accounts for losses due to HCE optics and imperfections (–)
K_{shadow}	performance factor that accounts for mutual shading of parallel collector rows during early morning and late evening (–)
$K_{EndLoss}$	performance factor that accounts for losses from ends of HCEs (–)
K_{oper}	fraction of the solar field that is operable and tracking the sun (–)

$$Q_{lost} = U_L (T_c - T_a) a \tag{25.3}$$

U_L: Overall conductivity of heat losses at the absorber; T_c: Average temperature of the coolant; T_a: Ambient temperature; a : Surface of the absorber.

For this work, SAM software was used to simulate the energy produced by the solar power plant situated in Ghardaïa, and the physical cylindro-parabolic power station model was chosen to model the system. Some important characteristics of the plant are given in Table 25.1:

Table 25.1 Characteristics of the solar power plant

Site chosen: Ghardaïa	Solar field	Factory block	Storage system
Longitude 3.8 °E	Reflector: Solargenix SGX-1	Capacity 8.5 kWe	Molten salt storage fluid: Hitec salt
Latitude 32.4 °N	Receptor: Schott PTR70 2008	Rankine Cycle	Duration 6 h
DNI 2105.01 kWh/m²/y	HTF Therminol VP1	Dry cooling	Type 2 tanks

25.3 Modeling of the Electrolyzer

An electrolyzer is a device used to splits a water molecule into hydrogen and oxygen. It consists of an anode, a cathode, and an electrolyte (ionic conductive medium). Electrolyzers are distinguished by the electrolyte materials used and the operated temperature. The low temperature electrolysis includes; alkaline electrolysis, Proton Exchange Membrane electrolysis (PEM) and Anion Exchange Membrane (AEM) electrolysis [9]. For this work, the alkaline process was adopted, because the alkaline electrolyzer are the most widespread industrially. The decomposition of water by electrolysis is described as:

$$H_2O \rightarrow H_2 + 1/2\,O_2 \quad (25.4)$$

With an enthalpy of water dissociation: $\Delta H = 285$ kJ/mol.

The required power (in W) for the electrolysis of a water flow $m_{\dot{H}_2O}$ is given by [5]:

$$W = U_{\text{thermoneutral}} \cdot 96487000 m_{\dot{H}_2O}/\eta_f \quad (25.5)$$

25.4 Organic Rankine Cycle Modeling

The most commonly used thermodynamic cycle to produce electricity by solar concentrating still the traditional steam Rankine cycle. However while using this system at low or medium temperatures (100–450 °C), as it is the case for a small-scale solar plant (a few kilowatts to three megawatts) with parabolic trough concentrators, the efficiency is greatly degraded. The Organic Rankine Cycle (ORC) using so-called organic fluids (Fig. 25.2) is therefore a good alternative to the steam cycle, and it has numerous advantages [10].

The model is given by the simplest configuration of the Rankine cycle. It includes a single stage turbine without recovery, a condenser, a pump and a boiler (Fig. 25.3). The working fluid went through four different thermodynamic states. The cycle operates between two pressure values, that are the evaporation pressure and the

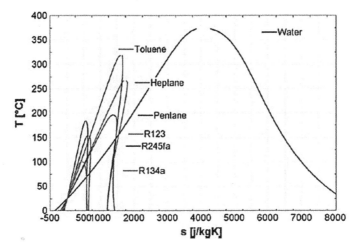

Fig. 25.2 Saturation curves of different organic fluids and water

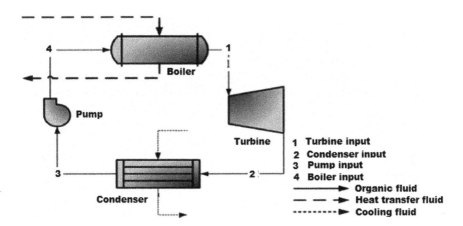

Fig. 25.3 Diagram of the organic rankine cycle model

condensing pressure. The performance of the power cycle is defined as the ratio of net work supplied by the cycle to the amount of heat consumed by the cycle.

$$\eta_{cycle} = \frac{\eta_{concentrator}(\dot{W}_{turbine} - \dot{W}_{pump})}{\dot{Q}_{boiler}} \quad (25.6)$$

The EES (Engineering Equation Solver) software was used to simulate the electricity produced through the Rankine organic cycles.

In order to compare in-between the electricity produced through different fluids in the Rankine cycle, we have set the circulation flow in the cycle to 0.074 kg/s, the monthly thermal powers of heating of the working fluid "Qutile" as well as the same isentropic efficiencies of the pump and the turbine of 85 and 80% respectively.

For each working fluid, the condenser outlet temperature and the pressure at the inlet of the turbine have been set. The outlet pressure at the condenser is the saturation pressure corresponding to the temperature at this point. The solar field has been dimensioned to make the heating power supplied by the latter included in the operating power range of the fluids studied and used in this given cycle.

25.5 Results and Discussions

The establishment of the mathematical equations necessary for the modeling of the solar plant with parabolic trough concentration, the organic Rankine cycle involved in the production of electricity and the water electrolyzis was the crucial step to the numerical simulations. Indeed, based on these equations, it was possible to study the effects of physical, meteorological and geographical parameters on the production of thermal energy, electricity and hydrogen.

At first, the monthly electricity produced by a traditional parabolic concentrator plant using an Andasol 1 storage type, with a power of 8.5 kWe, that is implemented in Ghardaïa site was studied. This was done to size the solar field according to the estimated thermal energy needed to supply the heat to the Organic Rankine cycle.

Figure 25.4 gives the average monthly thermal energy provided by the solar field for the Ghardaïa site. This energy is proportional to the Direct Normal Insolation DNI (Fig. 25.5).

Figure 25.6 shows different efficiencies expressing the production of heat and electricity by the solar field.

It can be noticed that the maximum values are recorded in the summer period.

Afterwards, the monthly electrical production was calculated for the cycle using conventional working fluid, which is the water, and the organic ones: Benzene,

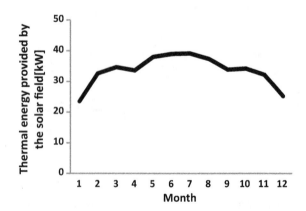

Fig. 25.4 The average monthly thermal energy provided by the solar field

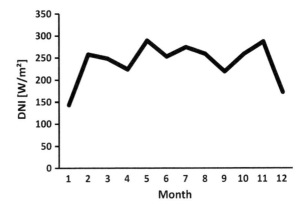

Fig. 25.5 Direct normal insolation DNI monthly in the Ghardaïa site

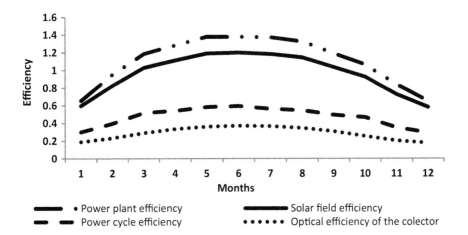

Fig. 25.6 Monthly efficiency of the plant implemented in Ghardaïa

Toluene, R-123. The heat source is a solar field generating a net heating power of 35 kWth. This step was crucial to define which of the working fluids, is the most efficient and most suitable for this low power plant provided by the solar field.

According to Fig. 25.6, it can be seen that among the three organic fluids studied for the production of electricity, the Benzene is the fluid allowing the greatest production of mechanical power at the turbine followed by Toluene, R123 and then by water.

The fluids studied, including water, present a mechanical power produced at the turbine proportional to the thermal output power of the solar field with a lower sensitivity of the refrigerants to the decrease in heating power in winter. Figures 25.7 and 25.8

The pumping power (compression) of the fluid can affect the net electrical power produced by the power cycle and therefore its total efficiency. Indeed, the greater the

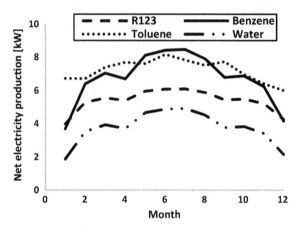

Fig. 25.7 Net electricity production for different working fluids

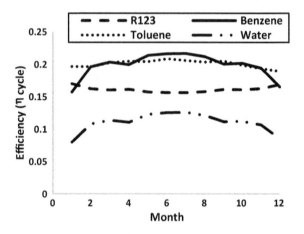

Fig. 25.8 Monthly efficiency of the plant installed in Ghardaïa

power necessary for the compression of the working fluid, the smaller the net power produced by the cycle and therefore the lower the efficiency of the cycle.

The results of the efficiency obtained through this simulation looks like those of [11] and do not exceed 24%; and thus joins what is described in the literature [12, 13].

Figure 25.9 shows the amounts of hydrogen produced by the alkaline electrolyzer for the various organic fluids used in this study. The quantities produced are closely related to the available electrical energy, which is more important for benzene and toluene.

Fig. 25.9 Average daily hydrogen production

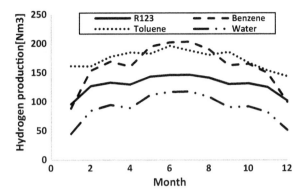

25.6 Conclusion

This study helped to demonstrate the feasibility of hydrogen production by electrolysis using a solar ORC plant in Algeria. Indeed, with a Direct Normal Insolation (DNI) that exceeds 2000kWh/m^2/year, an average daily sunlight duration that exceeds 10 h this country has key elements for the success of this type of project. Moreover, according to the results, the Organic Rankine Cycles are more efficient than the Rankine Steam Cycles while used for low power generation (a few kilowatts). It was also shown that, among the three organic fluids considered for the production of electricity by a small-scale power plant (5–8.5 kWe) through parabolic trough solar power plant installed in Ghardaïa, benzene is the one offering the best efficiency and production capacities throughout the year.

References

1. I. Staffell, D. Scamman, A. Velazquez Abad, P. Balcombe, P.E. Dodds, P. Ekins, N. Shahd, K.R. Warda, The role of hydrogen and fuel cells in the global energy system. Energ. Environ. Sci. **12**(12), 463–491 (2019)
2. N. Bento, La transition vers une économie de l'hydrogène: infrastructures et changement technique. Doctorate thesis. Pierre Mendès-France University—Grenoble II (2010)
3. G. Van Wylen, R. Sonntag, Thermodynamique appliquée. 2eme edn PEARSON (2002)
4. M.S. Shahin, Coupling of an electrolyzer with rankine cycle for sustainable hydrogen production via thermal solar energy. Master thesis. American University of Sharjah (2015)
5. M.J. Blanco, L. Ramirez Santigosa, *Advances in Concentrating Solar Thermal Research and Technology* (Woodhead Publishing Series in Energy, Elsevier, 2017)
6. A. Patnode, Simulation and Performance Evaluation of Parabolic Trough Solar Power Plants. Master Thesis, University of Wisconsin—Madison (2006)
7. R. Forristall, Heat transfer analysis and modeling of a parabolic trough solar receiver implemented in engineering equation solver. Technical report. National Renewable Energy Laboratory, NREL/TP-550 34169. Port (2003)
8. E. Jacobson, N. Ketjoy, S. Nathakaranakule, W. Rakwichian, Solar parabolic trough simulation and application for a hybrid power plant in Thailand. Sci. Asia. **32**(32), 187–199 (2006)

9. H. Mokrane-Derbal, Contribution à l'étude des centrales de puissance à concentration solaire pour la production d'énergie. Etude des perspectives de production d'hydrogène pour les piles à combustible. PhD. thesis. University of science and technology Houari Boumediene, Algiers (2012)
10. D. Quentin, Étude numérique et expérimentale d'un cycle de Rankine-Hirn de faible puissance pour la récupération d'énergie. Thèse de doctorat décembre (2016)
11. M. Ashouri, M. Hossein Ahmadi, M. Feidt, Performance analysis of organic rankine cycle integrated with a parabolic through solar collector. Conference paper. The 4th World Sustainability Forum (2014)
12. S. Quoilin, *Sustainable Energy Conversion Through the Use of Organic Rankine Cycles for Waste Heat Recovery and Solar Applications* (Thèse de doctorat, Liège, 2011)
13. J. Yang, J. Li, Z. Yang, Y. Duan, Thermodynamic analysis and optimization of a solar organic Rankine cycle operating with stable output. Energy Convers. Manag. **187**, 459–471 (2019)

Chapter 26
Optimization Study of the Produced Electric Power by PCFCs

Youcef Sahli, Abdallah Mohammedi, Monsaf Tamerabet, and Hocine Ben-Moussa

Abstract The object of the present work is the thermodynamic study of the proton ceramic fuel cell; particular attention is given to evaluate and maximize the generated power density by a single cell of Protonic Ceramic Fuel Cell (PCFC). In this work, the real potential is given by the difference between the Nernst potential and the reel polarizations generated during the PCFC operation. The activation polarization of the chemical reactions in the anode and the cathode, the losses due to the species concentration in both electrodes (anode and cathode) and the ohmic losses produced by the Joule's effect in the electrolyte and both electrodes (anode and cathode) are considered as the reel polarizations. The obtained results show that the PCFC power density is proportional to the variations of the operating temperature and the oxygen concentration in the oxidizer; conversely, it is inversely proportional to the evolutions of the fuel humidification and the thicknesses of the electrolyte.

Keywords PCFC · Power density · Optimization

26.1 Introduction

PCFC is one of the fuel cells family that is destined for stationary applications. It is characterized by a solid electrolyte, an intermediate operating temperature (300–900 °C) compared to the solid oxide fuel cell and protonic transport in the electrolyte as PEMFC. Current development research of PCFCs are focused on maximizing the electrical energy produced by these cells and reduce the operating temperature that affecting the lifetime of their components in several ways and in various disciplines [1–3]. In addition, the diminution of the operating temperature can be

---

Y. Sahli (✉)
Unité de Recherche En Energies Renouvelables En Milieu Saharien, URERMS, Centre de Développement Des Energies Renouvelables, CDER, 01000 Adrar, Algeria
e-mail: y.sahli@urerms.dz; sahli.sofc@gmail.com

A. Mohammedi · M. Tamerabet · H. Ben-Moussa
Département de Mécanique, Faculté de Technologie, Université Batna 2, Fesdis, Algeria

© The Editor(s) (if applicable) and The Author(s), under exclusive license to Springer Nature Singapore Pte Ltd. 2021
A. Khellaf (ed.), *Advances in Renewable Hydrogen and Other Sustainable Energy Carriers*, Springer Proceedings in Energy,
https://doi.org/10.1007/978-981-15-6595-3_26

made PCFCs usable in mobile applications (transport). Numerous works in literature were conducted to model the physical phenomena produced in the PCFC operating. Kalinci and Dincer [1] have investigated the PCFC performance using a one-dimensional steady-state electrochemical model. Arpornwichanop et al. [2] have presented a performance analysis of a planar PCFC using a one-dimensional steady-state model coupled with a detailed electrochemical model. Zhang et al. [3] have developed a charge transports model to study produced current leakage impact on the PCFC performance.

In the present work, a thermodynamic study of the Protonic Ceramic Fuel Cell (PCFC) is conducted, while paying special attention to the maximizing the power density produced by this fuel cell type.

26.2 Physical Model

PCFC tension (V) is defined by the following equation [4–6]:

$$V = E_{\text{Nernst}} - \text{Losses} \tag{1}$$

E_{Nernst} is the Nernst potential, it represents the PCFC maximum potential, and it is given by the Nernst equation [1]:

$$E_{\text{Nernst}} = E_0 + \frac{R.T}{n.F}\left[\ln\left(\frac{\Pr_{H_2}.\Pr^{0.5}_{O_2}}{\Pr_{H_2O}}\right)\right] \tag{2}$$

R is the perfect gas constant, T is the operating temperature of the PCFC, n is the number of the transfer electrons, F is the Faraday number, $\Pr_j$ is the partial pressure of each specie j (oxygen, hydrogen and water stem) and E_0 is the ideal potential that is defined by the following equation [1, 2, 7]:

$$E_0 = 1.253 - 2.4516 \times 10^{-4}.T \tag{3}$$

The losses defined by Eq. (1) include three types of tension losses: activations, concentrations and ohmic.

26.2.1 Ohmic Losses

These tension losses are due to the resistance encountered by the protons in the electrolyte and the electrons in the electrodes. Ohmic loss in each PCFC constituent element is given by the product of the current density (i) and the electrical resistance of each element of PCFC (R). Equation (4). [2, 8–10].

$$\eta_{ohm} = i.r_e \tag{4}$$

where (r_e) is the ratio of the thickness of each component of the cell heart (e) and its electrical conductivity (σ). Equation (5). [2, 8–11].

$$r_e = \frac{e}{\sigma} \tag{5}$$

26.2.2 Activation Losses

These tension losses are due to the chemical reaction activation in both electrodes. Equation (6). [4–6, 12].

$$\eta_{Act,j} = \frac{R.T}{\alpha.n.F}.\sinh^{-1}\left(\frac{i}{2.i_{0,j}}\right), \quad j = \text{anode, cathode} \tag{6}$$

where (α) is the charge transfer coefficient and (i_0) is the exchange current density of each electrode, it is given by Eq. (7). [4–6, 12].

$$i_{0,j} = k_j.\exp\left(-\frac{E_j}{R.T}\right), \quad j = \text{anode, cathode} \tag{7}$$

26.2.3 Concentration Losses

These losses are due to the inability of the system to maintain the initial concentrations. Equation (8). [13].

$$\eta_{conc,j} = -\frac{R.T}{n.F}.\ln\left(1 - \frac{i}{i_{l,j}}\right), \quad j = \text{anode, cathode} \tag{8}$$

where ($i_{l,j}$) is the limit current of the anode and the cathode, it is given for these two electrodes respectively by 2.99×10^4 and 2.16×10^4 [12].

26.3 Results and Discussion

The obtained results are presented according to three parts, in the first; we present the impact of the operating temperature and current density on the power density. For the second part, we show the influence of the oxygen concentration and the moisture content according to the different current densities on the power density. Finally, the third part exposes the effect of the electrolyte thickness and the current density on the power density delivered by PCFC.

Figure 26.1 shows the effect of the operating temperature and current density on the power density of a supported anode PCFC, the building materials are (Ni-YSZ) for the anode (LSM) for the cathode and (SCY) for the electrolyte. The thicknesses of the cell heart elements are identical to 60 μm for the cathode, 400 μm for the anode and 60 μm for the electrolyte. The fuel is the hydrogen, it is humidified at (X_{H2O} = 0.03), and the oxidizer is the air (X_{O2} = 0.21). The supply pressure is common for both gases (1 bar).

Figure 26.1 shows that the total loss is inversely proportional to the operating temperature; this proves that the PCFC power density is proportional to the operating temperature.

Figure 26.2 shows the effect of the hydrogen water content and the oxygen concentration in oxidizer on the power density of the PCFC, the building materials are (Ni-YSZ) for the anode, (LSM) for the cathode and (SCY) for the electrolyte. The thicknesses of the cell heart elements are identical to 60 μm for the cathode, 400 μm for the anode and 60 μm for the electrolyte. The supply pressure is common for both gases (1 bar). The PCFC operating temperature is 850°C.

Figure 26.2a shows that the power density of the PCFC is proportional to the oxygen concentration in the oxidizer. The PCFC real potential is proportional to the oxygen concentration rate in the oxidizer.

Figure 26.2b shows that the power density of the supported anode PCFC is inversely proportional to the fuel humidification rate. Logically, the PCFC real potential is inversely proportional to the oxygen concentration rate in the oxidizer.

Figure 26.3 shows the effect of the electrolyte thickness and the current density on the power density of a PCFC, the building materials are (Ni-YSZ) for the anode,

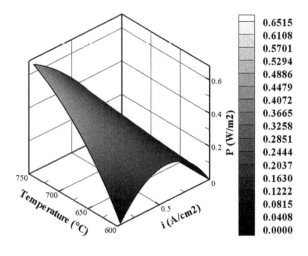

Fig. 26.1 PCFC power density evolution according to the operating temperature and current density

Fig. 26.2 PCFC power density evolution according to the species concentrations and current density. **a** Oxygen concentration in the oxidizer, **b** Water stem concentration in the fuel

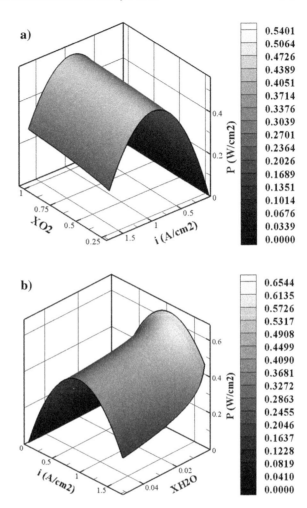

(LSM) for the cathode and (SCY) for the electrolyte. The thicknesses of the anode and cathode are 100 μm. The fuel is humidified hydrogen ($X_{H2O} = 0.03$), and the oxidizer is the air ($X_{O2} = 0.21$) is the oxidizer. The supply pressure is common for both gases (1 bar). The PCFC operating temperature is 850 °C.

The results obtained numerically were showed a good concordance with the experimental results obtained by [1–3].

26.4 Conclusion

The Protonic Ceramic Fuel Cell as all fuel cell types is an electrochemical device that converts the chemical energy of reactions to the thermal and electrical energies.

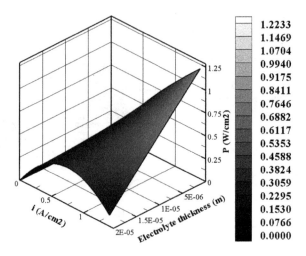

Fig. 26.3 PCFC power density evolution according to the electrolyte thickness and current density

In the present study, an optimization study of the produced electric power by PCFCs is presented. A program in FORTRAN language that is based on the thermodynamic model equations has been developed locally. The results obtained showed a concordance with the experimental results of the literature. According to the analysis of the obtained results, it appears that the realized code can be used as a design tool for the optimization of the produced electric power by PCFCs. The main obtained results in this work are:

- The PCFC power density is proportional to the operating temperature and oxygen concentration in the oxidizer.
- The PCFC power density is inversely proportional to the fuel humidification and electrolyte thickness.

References

1. Y. Kalinci, I. Dincer, Analysis and performance assessment of NH_3 and H_2 fed SOFC with proton-conducting electrolyte. Int. J. Hydrogen Energy **43**(11), 5795–5807 (2018)
2. A. Arpornwichanop, Y. Patcharavorachot, S. Assabumrungrat, Analysis of a proton-conducting SOFC with direct internal reforming. Chem. Eng. Sci. **65**(1), 581–589 (2010)
3. J.H. Zhang, L.B. Lei, D. Liu, F.Y. Zhao, M. Ni, F. Chen, Mathematical modeling of a proton-conducting solid oxide fuel cell with current leakage. J. Power Sources **400**(1), 333–340 (2018)
4. Y. Sahli, B. Zitouni, H. Ben-Moussa, Solid oxide fuel cell thermodynamic study. Çankaya Univ. J. Sci. Eng. **14**(2), 134–151 (2017)
5. Y. Sahli, B. Zitouni, H. Ben-Moussa, Thermodynamic optimization of the solid oxyde fuel cell electric power. University Politehnica of Bucharest Scientific Bulletin Series B-Chemistry and Materials Science **80**(2), 159–170 (2018)
6. Y. Sahli, H. Ben-Moussa, B. Zitouni, Optimization study of the produced electric power by SOFCs. Int. J. Hydrogen Energy **44**(39), 22445–22454 (2019)

7. K.Q. Zheng, M. Ni, Q. Sun, L.Y. Shen, Mathematical analysis of SOFC based on co-ionic conducting electrolyte. Acta. Mech. Sin. **29**, 388–394 (2013)
8. Y. Sahli, B. Zitouni, H. Ben Moussa, H. Abdenebi, Three-dimensional numerical study of the heat transfer on the planar solid oxide fuel cell: joule's effect, in *Progress in Clean Energy Volume I: Analysis and Modeling*, ed. by I. Dincer, C.O. Colpan, O. Kizilkan, M.A. Ezan (Springer, Switzerland, 2015), pp. 449–461
9. A. Arpornwichanop, Y. Patcharavorachot, Investigation of a proton-conducting SOFC with internal autothermal reforming of methane. Chem. Eng. Res. Des. **91**(8), 1508–1516 (2013)
10. H. Abdenebi, B. Zitouni, H. Ben-Moussa, D. Haddad, H. Zitouni, Y. Sahli, Inlet methane temperature effect at a planar sofc thermal field under direct internal reforming condition, in *Progress in clean energy volume II: Novel Systems and Applications*, ed. by I. Dincer, C.O. Colpan, O. Kizilkan, M.A. Ezan (Springer, Switzerland, 2015), pp. 567–581
11. Y. Sahli, B. Zitouni, H. Benmoussa, Etude numérique tridimensionnelle de l'effet de la température d'entrée des gaz sur la production de chaleur dans une pile à combustible SOFC planaire. Revue des Energies Renouvelables **21**(2), 173–180 (2018)
12. V. Menon, A. Banerjee, J. Dailly, O. Deutschmann, Numerical analysis of mass and heat transport in proton-conducting SOFCs with direct internal reforming. Appl. Energy **149**(1), 161–175 (2015)
13. X. Zhang, Y. Wang, J. Guo, T. Shih, J. Chen, A unified model of high-temperature fuel-cell heat-engine hybrid systems and analyses of its optimum performances. Int. J. Hydrogen Energy **39**(4), 1811–1825 (2014)

Chapter 27
Accurate PEM Fuel Cell Parameters Identification Using Whale Optimization Algorithm

Mohammed Bilal Danoune, Ahmed Djafour, and Abdelmoumen Gougui

Abstract Proton Exchange Membrane Fuel cells (PEMFC) are used in many engineering applications as a power source. The device has a mathematical model by which the output voltage-current (V-I) characteristics can be estimated at various operating conditions. However, this model comprises several nonlinear coupled parameters, and that makes the identification problem as challenging task. Owing to that, some advanced techniques been employed to extract the optimal constants of a PEMFC including meta-heuristic optimization algorithms. At this point, this paper deals with a novel swarm-intelligence optimization method named Whale Optimization Algorithm for the purpose of extracting the exact parameters of the PEMFC. This investigation is conducted to examine the effectiveness of the method in providing better results compared to other methods presented in the literature. A commercial PEMFC from Heliocentris with rated power = 40 W is employed to conduct a series of experiments in the laboratory. Moreover, to confirm the effectuality of the method, the results are compared with Particle-Swarm-Optimization (PSO) algorithm. The proposed method showed a significant enhancement in terms of accuracy and convergence speed compared to PSO, where, the error between the predicted and real data was negligibly small (Mean Absolute Error = 0.0726 V).

Keywords PEM fuel cell · Parameters identification · Whale optimization algorithm · Hydrogen energy

27.1 Introduction

For the last decades, the world was (and still) experiencing a difficult global challenges such as limitations in fossil fuels and deterioration in the environment. Due to these reasons, the integration of renewable-energy (RE) has noticeably increased to meet the load demands. One of these sources is called fuel cell. The device has

---

M. B. Danoune (✉) · A. Djafour · A. Gougui
Faculté Des Sciences Appliquées, Laboratoire LAGE, Université Kasdi Merbah Ouargla, Ouargla, Algérie
e-mail: danoune.mohammed.bilal@gmail.com; danoune.m_bilal@univ-ouargla.dz

© The Editor(s) (if applicable) and The Author(s), under exclusive license to Springer Nature Singapore Pte Ltd. 2021
A. Khellaf (ed.), *Advances in Renewable Hydrogen and Other Sustainable Energy Carriers*, Springer Proceedings in Energy,
https://doi.org/10.1007/978-981-15-6595-3_27

paid worldwide attention, and been used at various applications. Without rotating parts, the fuel cells are able to furnish power continuously and efficiently. There exist many kinds of fuel cells that are invented in last decades such as: alkaline fuel cell, Direct methanol, Phosphoric acid, Molten carbonate, Solid oxide, and PEM-fuel cells [1]. Nominated types have the same structure (Anode, Cathode separated by an electrolyte) but it differs in the used type of the fuel and/or electrolyte. However, due to many advantages i.e. low operating temperature (70–85 °C), compact design, lightweight, solid electrolyte, and high efficiency PEMFC is preferred to be used in many applications [2, 7]. Furthermore, the PEMFC has a mathematical model that developed to replicate the output characteristics of the device. This model can help designers to design new systems, and new control techniques. Nonetheless, a precise output characteristics prediction is extremely linked to a precise parameters identification. Unfortunately, the model has a complex and non-linearly coupled parameters, that makes the identification process as challenging task. For these purposes, advanced methods depends on swarm intelligence (so-called Meta-heuristic methods) are extensively employed to extract model parameters. Using V/I measurements, meta-heuristic methods can efficiently and effectively extract the optimal coefficients. A survey in literature showed many algorithms were applied including Adaptive differential evolution [3], improved-multi-strategy-adaptive differential evolution [4], Hybrid-Adaptive-Differential Evolution algorithm [5], Particle Swarm Optimization [6], Grey Wolf Optimizer [7], Multi-Verse Optimizer [8] and Salp Swarm Optimizer [9]. Although, swarm-based methods are showing an intensive progress, and they are being developing day after day. Thus, for further enhancements these new developed methods have to be integrated in this topic. The goal of the current study is to propose a novel effective optimization method titled Whale Optimization Algorithm [11]. To show the superiority, accuracy, and reliability of the suggested method, the results are compared with Particle Swarm Optimization PSO [12]. The rest of paper is written as follows: Section 27.2, gives illustrations about the used model of the PEMFC. Section 27.3, describes both the used objective function and optimization method. Section 27.4, analyses and debates the substantial results that achieved in this study. Finally, Sect. 27.5, summarizes the important points in a general conclusion.

27.2 Mathematical Model of the PEMFC

PEMFC is an equipment that capable of converting chemical energy (in form of hydrogen) into electrical energy through an electro-chemical process. The device is composed of three components: a solid membrane (very often Nafion material) between two electrodes called Anode and cathode. The overall stack voltage can mathematically obtained as follows [2–5]:

$$V_{fc} = N_s \times [E_{nernst} - V_{act} - V_{con} - V_{ohm}] \qquad (27.1)$$

$$E_{nernst} = 1.229 - 0.85 \times 10^{-3} \times (T - 298.15) + 4.3085 \times 10^{-3} \times \ln\left(P_{H2} \times \sqrt{P_{O2}}\right) \quad (27.2)$$

where Ns is cells number connected in series. E_{nernst} is the no load voltage calculated by Nernst equation, V_{act} is the activation voltage, V_{con} is the losses due to conductivity and V_{ohm} is the ohmic voltage of both membrane and electrodes resistances [5–9].

T is the cell temperature, P_{H2} and P_{O2} are the anode-cathode pressures respectively. The activation V_{act} is given in Eq. (27.3) [2–6]:

$$V_{act} = -[\xi_1 + \xi_2 \times T + \xi_3 \times T \times \ln(C_{O2}) + \xi_4 \times T \times \ln(I)] \quad (27.3)$$

Here ξ_1, ξ_2, ξ_3 and ξ_4 are a constant unknown parameters. I is the cell current and C_{O2} is the concentration of the oxygen at the liquid interface as defined by the Henrys law [5–9].

$$C_{O2} = \frac{P_{O2}}{5.08 \times 10^6 \exp^{(-498/T)}} \quad (27.4)$$

V_{con} is the activation voltage which is logarithmically related with current density (J) and maximum current density J_{max}. b is an unknown parameter [5–9].

$$V_{con} = -b \times \ln\left(1 - \frac{J}{J_{max}}\right) \quad (27.5)$$

V_{ohm} represents the losses due to both membrane and electrodes resistances R_m and R_c respectively.

$$V_{ohm} = (R_m + R_c) \times I \quad (27.6)$$

$$R_m = \frac{l \times \rho_M}{A} \quad (27.7)$$

where ρ_M is the membrane resistivity (Ωcm). The symbol of A represents the cell area (cm^2), l is the membrane thickness (cm), and ρ_M is the resistivity [5–9]:

$$\rho_M = \frac{181.6\left[1 + 0.03\left(\frac{I}{A}\right) + 0.062\left(\frac{T}{303}\right)^2\left(\frac{I}{A}\right)^{2.5}\right]}{\left[\lambda - 0.634 - 3\left(\frac{I}{A}\right)\exp\left[4.18\left(\frac{T-303}{T}\right)\right]\right]} \quad (27.8)$$

27.3 Identification Methodology

27.3.1 Objective Function (OF)

In the ideal case, the experimental voltage-currant points should be equal to the estimated data, therefore, the identification problem can be converted into optimization problem, where, the error between the experimental and estimated should be minimum. The Mean Absolute Error (MAE) between the measured and estimated stack voltage is chosen to be the OF in our case. Now, the goal is to adapt the eight (8) unknown parameters (i.e. $\xi_1, \xi_2, \xi_3, \xi_4, \lambda, b, R_C$ and J_{max}) of the model to achieve a good fit by minimizing the OF given below (Table 27.1):

$$f(\xi_1, \xi_2, \xi_3, \xi_4, \lambda, b, R_C, J_{max}) = \frac{\sum_{j=1}^{N} |V_{exp}(j) - V_{est}(j)|}{N} \qquad (27.9)$$

A commonly used boundaries values are reported in Table 27.2 to limit the search space [2–5].

Table 27.1 Heliocentris FC50 PEM fuel cell specifications [10]

Parameter	Value
Number of cells	10
Cell Area (cm^2)	25
Rated power (W)	40
Membrane thickness (μm)	27
Current at rated power (A)	8
Voltage at rated power(V)	5
Maximum operating temperature (°C)	50
Relative humidity in Anode and cathode	1

Table 27.2 Parameter boundaries and fitting results

Parameter	Lower bound	Upper bound	Best	Worst
ξ_1 (-)	−1.2000	−0.8500	−1.1772	−1.1588
ξ_2 (−)	0.0010	0.0500	0.0010	0.0011
ξ_3 (−)	−3.6000E-5	−9.8000E-5	−5.7080E-5	−5.7080E-5
ξ_4 (−)	−2.6000E-4	−9.8000E-4	−2.6000E-4	−2.7380E-4
λ (−)	10	25	22.9	20.5000
b (v)	0.0100	1	0.0411	0.0594
R_c (Ω)	0	10E-3	4.8314E-4	0.0030
J_{max} (A/cm^2)	0.5000	2	1.3103	0.8817
MAE (v)	–	–	0.0726	0.1752

27.3.2 Optimization Process by Whale Optimization Algorithm

The suggested method is called Whale Optimization (WOA) Algorithm. It simulates the social behavior of humpback-whales (HBW). The latter is a probabilistic and population-based search method. It has been proposed by Seyedali [11] in 2016. This method has many advantages like ability to track the global optimum accurately, easy to implement, requires few parameters. The implement of the method is made by three main steps (Encircling the prey, Bubble net attacking, and Bubble net attacking) as cited in [11]. A minimization pseudo-code is developed for the proposed method to determine the 8 unknown parameters of the fuel cell (Fig. 27.1).

```
Initialize the whales population (1, 2, ..., total number)
Calculate the fitness of each search agent by Eq.(9)
x* is the positions corresponds to the minimum fitness
while (t < maximum number of iterations)
for each search agent
    Update a, A, C, l, and p
if1 (p<0.5)
if2 (|A| < 1)
    Update the position of the agent by the Eq $\vec{X}(i+1) = \vec{X^*} i - \vec{A}.\vec{D}$;
else if2 (|A|≥1 )
    Choose a random agent ($X_{rand}$ )
    Update the position of the agent by the Eq $\overrightarrow{X}(i+1) = \overrightarrow{X_{rand}} - \vec{A}.\vec{D}$
end if2
else if1 (p ≥ 0.5)
    Update the position of the agent by the Eq   $\vec{X}(i+1) =$
    $\begin{cases} \overrightarrow{X^*}(i) - \vec{A}.\vec{D} & if p < 0.5 \\ \overrightarrow{X^*}(i) + \vec{D}.e^{b \times l}.\cos(2\pi l) & if p \geq 0.5 \end{cases}$
end if1
    end for
If any agent goes out-of the search space rectify it
Compute the fitness of every agent
Update x* if a better solution exists
t=t+1
end while
Go back to x*
```

Fig. 27.1 Pseudo code of WOA [13]

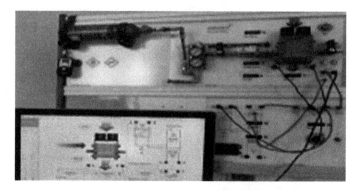

Fig. 27.2 Fuel cell laboratory test bench

27.4 Experimental Characterization

The experimental voltage-current points are obtained from Heliocentris-FC50 testbench, installed at LAGE laboratory. The system is composed of a 40 W fuel cell with the specifications illustrated in Table 27.1, metal-hydride hydrogen bottle, pressure regulator, fan, programmable load and data acquisition system. The load is programmed as a ramp signal to extract the desired voltage-current points. The experiment is conducted at room temperature (24 °C) (Fig. 27.2).

27.5 Results and Discussion

In this part, including WOA, the results of PSO [12] are compared in terms of convergence rate and the exactness in the curves plot. PSO has demonstrated its effectiveness in many research papers, therefore, the method been chosen to be the comparison reference to proof how competitive the method is. Moreover, due-to-the randomness nature of the search method, the results are executed for 100 independent runs with a population of 50 particles in the swarm. The best parameters results (see Table 27.2) are feedback to the model to estimate V/I polarization curves. As seen from Fig. 27.3 that the experimental as fitted data are plotted together. Some differences can be discovered between the exact and the fitted data. From the plotted V/I curves, we can observe that the proposed method is showing less deviation compared to other methods, especially, at no-load points. The calculated total Mean Absolute Error (MAE) of the proposed method is 0.0726 V whereas in PSO = 0.0750 V (see Fig. 27.4). This confirms the effectiveness of the WOA in exploring a high quality of solutions. The main reason of this improvements can be explained by the exploration and exploitation phases in the proposed method. The latter strategy has a good tendency in avoiding the local optimum, and thus extracting high accurate results. In addition to that the Best and Worst results of 100 execution time are

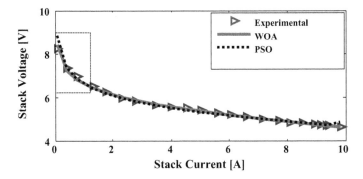

Fig. 27.3 V/I characteristic at 26 °C

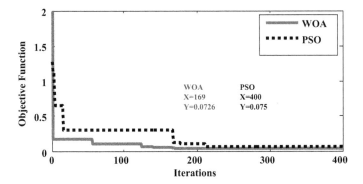

Fig. 27.4 Convergence graph

stored and listed in Table 27.2. From this table we can notice that the error in this 100 runs is ranging in a very narrow band (0.0726–0.1752 V) where 0.0726 V represents the best result (minimum error) and 0.1752 V corresponds to the worst result (maximum error). That means the stochasticity of the method cannot effect on the desired precision, and at every execution time the solution will not be far from the best results.

Furthermore, to diminish the total computation cost, the selected method need to be fast convergence. Therefore, another test is conducted in this second part to exanimate the speed of the method relative to the results of PSO. The results are depicted in Fig. 27.4; we can observe that the suggested method has a rapid convergence, the desired accuracy was reached after low number of iterations (169th). Unlike WOA, PSO has reached target in more few iterations (217th). From these results, we can say that the WOA has a fast convegence rate, and that will positively reflect/effect on the cost of computation of the problem. At this end, it can be stated that the above results supports the superiority of WOA, and the latter can be classified as a powerful method used in PEMFC parameters identification problems.

27.6 Conclusion

In this paper, Whale Optimization Algorithm is suggested to extract the optimal eight (8) parameters of a PEMFC. Interest was paid on tow important factors *i.e.* preciseness and rapidity of the method. Based on these factors, the results of WOA were compared with PSO algorithm to test effectiveness of the proposed method. Unlike PSO, WOA was revealed the satisfyingly results in terms of the precision with an accuracy of WOA = 0.0726 V. In terms of convergence speed, the proposed method was reached the target on less number of iterations (169th), whereas PSO was slow and converged after some more iterations (217th). Finally, by this study it was found that the results of WOA were satisfyingly good. And therefore, we can state that WOA can be efficiently applied to determine the PEMFC parameters.

References

1. J. Larminie, A. Dicks, *Fuel Cell Systems Explained*, 2nd edn. (John Wiley, England, 2003)
2. T. Wilberforce, Z. El-Hassan, F.N. Khatib, A. Al Makky, A. Baroutaji, J.G. Carton, A.G. Olabi, Developments of electric cars and fuel cell hydrogen electric cars. Int. J. Hydrogen Energy **42**(40), 25695–25734 (2017)
3. J. Cheng, G. Zhang, Parameter fitting of PEMFC models based on adaptive differential evolution. Int. J. Electr. Power Energy Syst. **62**, 189–198 (2014)
4. W. Gong, Z. Cai, Parameter optimization of PEMFC model with improved multi-strategy adaptive differential evolution. Eng. Appl. Artif. Intell. **27**, 28–40 (2014)
5. Z. Sun, N. Wang, Y. Bi, D. Srinivasan, Parameter identification of PEMFC model based on hybrid adaptive differential evolution algorithm. Energy **90**, 1334–1341 (2015)
6. R. Salim, M. Nabag, H. Noura, A. Fardoun, The parameter identification of the Nexa 1.2 kW PEMFC's model using particle swarm optimization. Renew Energy **82**, 26–34 (2015)
7. M. Ali, M.A. El-Hameed, M.A. Farahat, Effective parameters' identification for polymer electrolyte membrane fuel cell models using grey wolf optimizer. Renew. Energy **111**, 455–462 (2017)
8. A. Fathy, H. Rezk, Multi-verse optimizer for identifying the optimal parameters of PEMFC model. Energy **143**, 634–644 (2018)
9. A.A. El-Fergany, Extracting optimal parameters of PEM fuel cells using salp swarm optimizer. Renew Energy **119**, 641–648 (2018)
10. Heliocentris FC50 fuel cell technical datasheet. available at: http://heliocentrisacademia.com/portfolio-item/fuel-cell-trainer/. Accessed 07 April 2020
11. S. Mirjalili, A. Lewis, The whale optimization algorithm. Adv. Eng. Softw. **95**, 51–67 (2016)
12. J. Kennedy, R. Eberhart, A new optimizer using particle swarm theory, in *Proceedings, IEEE 6th International Symposium on Micro Machine and Human Science* (1995)

Chapter 28
Hydrogen Versus Alternative Fuels in an HCCI Engine: A Thermodynamic Study

Mohamed Djermouni and Ahmed Ouadha

Abstract The current study assesses, through a thermodynamic analysis, how well six alternative fuels, namely methanol, ethanol, LPG, biodiesel and hydrogen perform compared to LNG fuel in an HCCI engine. In addition to traditional energetic analysis, the second law of thermodynamics, through the concept of exergy, is used to provide detailed information on the thermodynamic processes during an engine cycle. In particular, the sources and magnitudes of the energy wasted in the system are evaluated and methods to reduce them can be afforded. Results indicate that the engine thermodynamic performances with the type of fuel used. The highest engine performance values are obtained using biodiesel as fuel, followed, in order, by ethanol, LPG, methanol, LNG and hydrogen. Furthermore, it is observed that the use of hydrogen in HCCI engine leads to increased exhaust gas temperature which. Thus, its overall performance can be effectively enhanced by using heat recovery techniques.

Keywords Thermodynamic analysis · Hydrogen · Alternative fuels · HCCI engine

28.1 Introduction

With its outstanding advantages such as high efficiency and reliability, Diesel engines are the prime mover of most seagoing vessels. However, Modern marine Diesel engines are faced a new challenge of achieving higher thermal efficiencies with lower pollutant emissions under the IMO Annex VI regulations. During the last years, several technological solutions, such as water addition, internal engine modification, post treatment and alternative fuels have been explored.

M. Djermouni (✉) · A. Ouadha
Laboratoire des Sciences et Ingénierie Maritimes, Faculté de Génie Mécanique, Université des Sciences et de la Technologie Mohamed Boudiaf d'Oran, 31000 Oran, Algeria
e-mail: djermounimohamed@yahoo.fr

A. Ouadha
e-mail: ah_ouadha@yahoo.fr

© The Editor(s) (if applicable) and The Author(s), under exclusive license to Springer Nature Singapore Pte Ltd. 2021
A. Khellaf (ed.), *Advances in Renewable Hydrogen and Other Sustainable Energy Carriers*, Springer Proceedings in Energy,
https://doi.org/10.1007/978-981-15-6595-3_28

In this context, the homogeneous charge compression ignition (HCCI) technology is considered as a promising solution. HCCI engines combine characteristics of both spark-ignited engines and diesel engines. Similar to spark-ignited engines, HCCI uses a pre-mixed fuel-in-air charge, and similar to diesel engines, the mixture is compression ignited. The diluted premixed charge facilitates a relatively uniform auto-ignition event rather than a non-premixed flame found in diesel engines, and thus HCCI engines can achieve fewer emissions of particulate matter. It has been demonstrated that it can achieve fuel economy levels comparable to those of a Diesel engine with low NOx particulate matter emissions.

This technology provides also the ability to use different fuels. A variety of alternative fuels that includes diesel [1–3], hydrogen [4–7], natural gas [8–11], ethanol [12], biogas [13], and dimethyl [14] have been used in previous studies of the literature.

Several studies have been carried out on the suitability and merits of HCCI engines fueled with various fuels. However, only few researches have thermodynamically analyzed HCCI engines are available in the literature [15–20].

The above literature survey shows the satisfactory performance of HCCI engines compared to conventional Diesel engines. Thermodynamic modeling of the operation of an HCCI engine running on various alternative fuels is a reliable, time and cost saving approach to assess their performance under different operating conditions. In this study, a thermodynamic model was developed and used to investigate the effect of various engine parameters on performance characteristics of an HCCI fueled with various alternative fuels, namely hydrogen, LNG, LPG, ethanol, methanol and biodiesel.

28.2 Engine Thermodynamic Modeling

Environmental issues combined to the depletion of conventional fuels have encouraged the use of alternative technologies and fuel sources for marine Diesel engines. The concept of HCCI combustion in which allows smaller amounts of NO, CO, unburned hydrocarbons, and soot emissions compared to conventional IC engines is considered as the most promised technique to comply with international regulations in terms of consumption and pollutant emissions. Alternative fuels such as natural gas, ethanol, methanol, biogas and hydrogen are also considered as an effective method to meet global environmental issues and it is expected that these kinds fuel will find widespread use.

28.2.1 Investigated Engine

The thermodynamic analysis has been performed for an eighteen cylinder, turbocharged Diesel engine to study the effects of some operating parameters such as the compression ratio, the ambient temperature, the equivalence ration and the

Table 28.1 Main engine specifications

Number of cylinders	18
Bore (B) × Stroke (S), mm	510 × 600
Engine compression ratio (r_c)	14
Rotation speed (N), rpm	500
Equivalence ratio (ϕ)	0.55
Compressor pressure ratio (r_p)	4.5

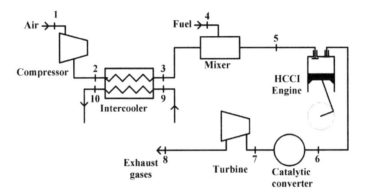

Fig. 28.1 Schematic of the engine system layout

compressor pressure ratio on the performance of the engine. Table 28.1 provides the engine specifications.

The system layout and components are presented in Fig. 28.1. It consists mainly of a turbocharged and intercooled HCCI engine, a catalytic converter and a mixer. The ambient air, at state 1, is compressed using a turbo-compressor to reach T_2 and p_2. The compressed air is then cooled by means of a cooler to the required temperature T_3. The fuel is injected near the intake valve of each cylinder (timed port injection) at a temperature T_4 and pressure p_4. The premixed charge is then introduced to the engine to undergo cyclic processes according to the Otto cycle. The exhaust gases at state 6 passthrough a catalytic converter to get rid of the remained unburned fuel as well as carbon monoxide. The clean flux of gases at temperature T_7 and pressure p_5 expands in the turbine and produces the power required to drive the compressor before being released to the atmosphere at state 8.

28.2.2 Thermodynamic Analysis

The present section deals with the thermodynamic model adopted for the examination of the performances of the aforementioned system. The working fluid differs in composition according to the process under consideration. It could be air, fuel, mixture of reactant (fuel + air), or burned gases of the combustion products. The thermodynamic properties of the working fluid are calculated using polynomials derived by regression methods from thermodynamics tables available in the literature.

The thermodynamic analysis is performed by considering several assumptions which are given as:

- All processes are considered at steady state condition.
- Changes in kinetic and potential energies are neglected.
- Changes in kinetic and potential exergies are neglected.
- Liquid fuels vaporize in the mixer by gaining heat from air delivered by the intercooler.

Mass, energy and exergy balance equations are applied to each component of the system described in Fig. 28.1. Neglecting kinetic and potential energy and exergy changes, these equations are summarized as:

$$\sum \dot{m}_{in} - \sum \dot{m}_{out} = 0 \qquad (28.1)$$

$$\dot{Q}_{in} + \dot{W}_{in} + \sum_{in} \dot{m}_i h_i = \dot{Q}_{out} + \dot{W}_{out} + \sum_{out} \dot{m}_i h_i \qquad (28.2)$$

$$\sum_{in} \left(1 - \frac{T_0}{T_i}\right) \dot{Q}_i + \dot{W}_{in} + \sum_{in} \dot{m}_i ex_i$$
$$= \sum_{out} \left(1 - \frac{T_0}{T_i}\right) \dot{Q}_i + \dot{W}_{out} + \sum_{out} \dot{m}_o ex_o + \dot{E}x_D \qquad (28.3)$$

where $\dot{Q}$ is the rate of heat transfer between the control volume and its surroundings, $\dot{W}$ the rate of work, h the specific enthalpy and $\dot{E}x_D$ is the rate of exergy loss.

The mass, energy and exergy balances in Eqs. (28.1), (28.2) and (28.3) have been applied to each component of the system.

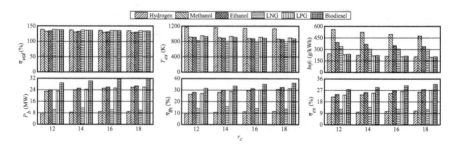

Fig. 28.2 Comparison of thermodynamic performance between alternative fuels in an HCCI engine. (r_p = 4.5, t_{amb} = 25 °C, ϕ = 0.55, N = 500 rpm)

28.3 Results and Discussions

A Fortran code has been developed based on the thermodynamic model described above. The code varies selected operating parameters such as compression ratio (r_c), fuel-air equivalence ratio (ϕ), engine speed (N) and turbo-compressor pressure ratio to calculate the common engine performance parameters: Brake power output (P_b), brake specific fuel consumption ($bsfc$), energy efficiency (η_{en}), volumetric efficiency (η_v) and exhaust gas temperature (T_{ex}). In addition, individual and overall exergy loss rates ($\dot{E}x$) and the exergy efficiency (η_{ex}) have been assessed as second-law parameters.

As expected, all assessed fuels exhibit similar trend, i.e. their performance are improved by increasing the engine compression ratio. Furthermore, the amount of oxygen available in biodiesel, ethanol and methanol also contributes to the enhancement of the combustion process. The highest engine performances (brake power output and energetic and exergetic efficiencies) are associated with biodiesel fuel followed by ethanol, LPG and methanol. LNG performance values are slightly lower than those of the fuels cited above. Although hydrogen has the highest low heating value (120 MJ/kg), it develops the lowest brake power output, which explains its poor performance values compared to the others fuels. This can be explained by its low density (0.0838 kg/m^3) which is too low compared to that of the others fuels. For $r_c = 14$, the average drops in thermal efficiency compared to biodiesel are 11, 14, 16, 54 and 68% for ethanol, LPG, methanol, LNG, and hydrogen, respectively. The same trend is recorded for the exergetic efficiency since the ratio between the lower heating value and the exergy of the fuel is constant. Methanol produces the highest brake specific fuel consumption, followed in order by ethanol, LNG, hydrogen, biodiesel, and LPG by a reduction of 29.8, 42.2, 57.2, 57.5, and 58.6%, respectively. Hydrogen exhaust gas temperature is the largest and it by 19.6, 21.5, 22.5, and 32.3% for LPG, biodiesel, methanol, ethanol, and LNG, respectively. Nevertheless, the volumetric efficiency is insensitive to the variation of the engine compression ratio and the fuel nature. This trend can be explained by the fact that the volumetric efficiency is governed mainly by the temperature and pressure of the reactant mixture induced to the engine cylinders which have been adjusted according to the requirements of HCCI engines operation independently of the fuel considered (Fig. 28.2).

In Fig. 28.3, the engine performances are plotted as function of the equivalence ratio and the compression ratio. The fuel-air equivalence ratio is defined as the actual fuel-air ratio divided by the stoichiometric fuel-air ratio. Results showed that, for all fuels, the brake specific fuel consumption is decreased with the increase of brake power output, energy and exergy efficiencies and volumetric efficiency. This means that the conversion of the thermal energy released by the fuel at higher equivalence ratios is more efficient. Overall, biodiesel fuelled HCCI engine performs better in terms of thermodynamic performance under the operating parameters considered in the current study. This is mainly due to its higher in-cylinder temperature of biodiesel combined with its stoichiometric air-fuel ratio and its oxygen content. It is also observed that the specific fuel consumption decreases with the increase of

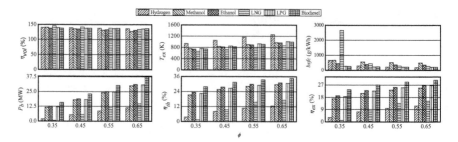

Fig. 28.3 Engine performance variations for alternative fuels at different fuel-air equivalence ratios. $r_c = 14, r_p = 4.5, t_{amb} = 25\ °C, N = 500$ rpm)

fuel-air equivalence ratio. For instance, in the case of hydrogen, an increase of 57% is recorded when shifting from an equivalence ratio of 0.45–0.65. As expected, *bsfc* is higher for hydrogen fuel due to its lower density compared to the others fuels. The engine exhaust gas temperature increases with increasing the equivalence ratio as illustrated in Fig. 28.3. This is can be attributed to high temperatures at the end of combustion which increases with increasing the mass of fuel burned. In addition, the exergetic efficiency increases as consequence of the higher temperatures attained by the combustion for higher equivalence ratios. This results also in lower combustion irreversibilities which contribute to the increase of the exergetic efficiency especially for the oxygenated fuels.

Performance variations with respect to the assessed fuels at various engine speeds are shown in Fig. 28.2. Overall, the engine brake power output, the energy efficiency and the exergy efficiency are increase with the increment of engine speed for all fuels considered. Again, the order of the fuels in respect of their performances is respected for the same reasons. For a given fuel, the brake specific fuel consumption did not show a meaningful variation as the engine speed rises. For all engine speed values, *bsfc* is the highest for methanol, followed, in order by ethanol, LNG, biodiesel, hydrogen and LPG. Hydrogen produces, by far, the highest the exhaust gas temperature. Finally, it is observed that the volumetric efficiency is insensitive to both the engine speed and the fuel nature (Fig. 28.4).

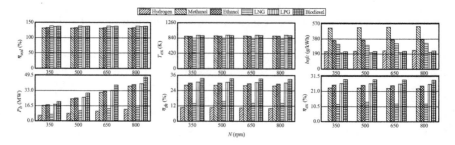

Fig. 28.4 Engine performance variations for alternative fuels at different engine speeds ($r_c = 14, r_p = 4.5, t_{amb} = 25\ °C, \phi = 0.55$)

Performance variations with respect to the assessed fuels at various engine speeds are shown in Fig. 28.2. Overall, the engine brake power output, the energy efficiency and the exergy efficiency are increase with the increment of engine speed for all fuels considered. Again, the order of the fuels in respect of their performances is respected for the same reasons. For a given fuel, the brake specific fuel consumption did not show a meaningful variation as the engine speed rises. For all engine speed values, *bsfc* is the highest for methanol, followed, in order by ethanol, LNG, biodiesel, hydrogen and LPG. Hydrogen produces, by far, the highest the exhaust gas temperature. Finally, it is observed that the volumetric efficiency is insensitive to both the engine speed and the fuel nature.

Figure 28.5 depicts the variation of engine performance parameters with respect to turbo-compressor pressure ratios for the six alternative fuels assessed. As expected, all assessed fuels exhibit similar trend, i.e. their performance are improved by increasing the compressor pressure ratio. Indeed, an increase in the latter enhances the homogeneous mixture preparation and the volumetric. An increase of the volumetric efficiency of 14, 17, 27, 15, and 14% is observed for increasing pressure ratio from 2.4 to 5.5 for hydrogen, methanol, ethanol, LNG, LPG, and biodiesel, respectively. In its turn, the brake power increases by 79, 121, 129, 93, 129, and 130% for the aforementioned fuels.

In order to identify the sources and magnitudes of the energy wasted in the system, a mapping of exergy losses within the whole system for fixed compression ratio, turbo-compressor pressure ratio, ambient temperature, fuel-air equivalence ratio and engine speed ($r_c = 14$, $r_p = 4.5$, $t_{amb} = 25$ °C, $\phi = 0.55$, $N = 500$ rpm) is achieved. It is observed that under the operating parameters selected, exergy losses that occur in the engine dominate. For all fuels assessed, the highest exergy losses are unregistered in the HCCI engine. They range between 73.7 and 92.7% of the total exergy losses in the system. The irreversible losses within the HCCI engine are mainly due to the irreversible nature of the mixing and combustion processes. The remaining components (turbo-compressor, intercooler, mixer, catalytic converter and turbine) produce only 7.3–26.3% of the total exergy loss of the system. Figure 28.6 gives also details about exergy losses within the remaining components of the system. For liquid fuels (LNG, LPG, ethanol, methanol and biodiesel), the highest exergy losses

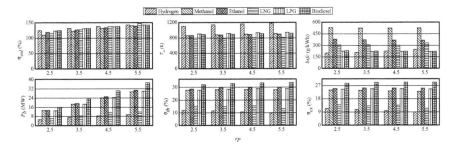

Fig. 28.5 Engine performance variations for alternative fuels at different engine speeds (rc = 14, $t_{amb} = 25$ °C, $\phi = 0.55$, $N = 500$ rpm)

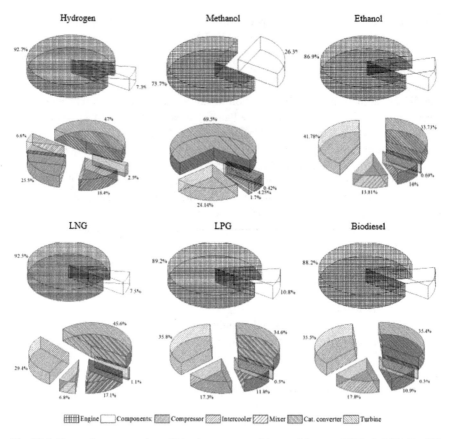

Fig. 28.6 Exergy losses mapping within the system rc=14, rp= 4.5, t_{amb}=25°C, f=0.55, N =500 rpm

originate from the mixer. In this case, heat transfer irreversibilities due to the evaporation of the fuel are also taken into account. However, for hydrogen, a gaseous fuel, the highest exergy losses are produced in the catalytic converter. As a result, from an exergy point of view, the system performance could be enhanced by reducing exergy losses during the combustion process.

28.4 Conclusions

In this study, some alternative fuels, namely LNG, LPG, biodiesel, methanol, ethanol and hydrogen have been assessed, through a thermodynamic analysis, in an HCCI engine. Results showed that the assessed fuels exhibit different trends. Globally biodiesel, ethanol, methanol and LPG perform the best. In addition, the following conclusions are drawn from the current study:

- Overall, the variation in the engine operating parameters (r_c, r_p, N and ϕ) affects considerably the engine performance. The latter are significantly improved at higher values of r_c, r_p, N and ϕ. This is mainly due to the improved combustion process under higher values of these parameters.
- The engine thermodynamic performances are observed to correlate with the type of fuel used. The highest engine performance values are obtained using biodiesel as fuel, followed, in order, by ethanol, LPG, methanol, LNG and hydrogen. Fuels that contain oxygen such as biodiesel, ethanol and methanol allow better combustion and achieve higher in-cylinder temperature.
- The use of hydrogen in HCCI engine leads to reduced energetic and exergetic efficiencies and increased exhaust gas temperature. Thus, its overall performance can be effectively enhanced by using heat recovery techniques.

References

1. J.J. Ma, X.C. Lu, L.B. Ji, Z. Huang, An experimental study of HCCI-DI combustion and emissions in a diesel engine with dual fuel. Int. J. Thermal Sci. **47**, 1235–1242 (2008)
2. A.A. Hairuddin, A.P. Wandel, T. Yusaf, Effect of different heat transfer models on a diesel homogeneous charge compression ignition engine. Int. J. Auto. Mech. Eng. **8**, 1292–1304 (2013)
3. H. Bendu, S. Murugan, Homogeneous charge compression ignition (HCCI) combustion: mixture preparation and control strategies in diesel engines. Renew. Sustain. Energy Rev. **38**, 732–746 (2014)
4. J.M.G. Antunes, R. Mikalsen, A.P. Roskilly, An investigation of hydrogen-fuelled HCCI engine performance and operation. Int. J. Hydrogen Energy **33**, 5823–5828 (2008)
5. M.M. Ibrahim, A. Ramesh, Experimental investigations on a hydrogen diesel homogeneous charge compression ignition engine with exhaust gas recirculation. Int. J. Hydrogen Energy **38**, 10116–10125 (2013)
6. A. Khaliq, Energy and exergy analyses of a hydrogen fuelled HCCI combustion engine combined with organic Rankine cycle. Int. J. Exergy **17**, 240–265 (2015)
7. S.V. Khandal, N.R. Banapurmath, V.N. Gaitonde, Performance studies on homogeneous charge compression ignition (HCCI) engine powered with alternative fuels. Renewable Energy **32**, 683–693 (2019)
8. J.O.O Isson, P. Tunesta, B. Johansson, S.B. Fiveland, R. Agama, M. Willi et al., Compression ratio influence on maximum load of a natural gas fueled HCCI Engine. SAE Paper 2002-01-0111 (2002)
9. D. Jun, K. Ishii, N. Iida, Auto ignition and combustion of natural gas in a 4 stroke HCCI engine. JSME Int. J. B-Fluid Therm. Eng. **46**, 60–67 (2003)
10. D. Yap, S.M. Peucheret, A. Megaritis, M.L. Wyszynski, H. Xu, Natural gas HCCI engine operation with exhaust gas fuel reforming. Int. J. Hydrogen Energy **31**, 587–595 (2006)
11. M. Djermouni, A. Ouadha, Thermodynamic analysis of an HCCI engine based system running on natural gas. Energy Conv. Manage. **88**, 723–731 (2014)
12. J.H. Mack, S.M. Aceves, R.W. Dibble, Demonstrating direct use of wet ethanol in a homogeneous charge compression ignition (HCCI) engine. Energy **34**, 782–787 (2009)
13. S.S. Nathan, J.M. Mallikarjuna, A. Ramesh, An experimental study of the biogas-diesel HCCI mode of engine operation. Energy Conv. Manage. **51**, 1347–1353 (2010)
14. S. Jafarmadar, N. Javani, Exergy analysis of natural gas/DME combustion in homogeneous charge compression ignition engines (HCCI) using zero-dimensional model with detailed chemical kinetics mechanism. Int. J. Exergy **15**, 363–381 (2014)

15. S.K. Trivedi, A. Haleem, Thermodynamic analysis and utilisation of wet ethanol in homogeneous charge compression ignition engine. Int. J. Sustainable Energy **35**, 33–46 (2013)
16. S. Mamalis, D.N. Assanis, Second-law analysis of boosted HCCI engines: modeling study. J. Energy Eng. **141**, C4014014 (2015)
17. Y. Li, M. Jia, Y. Chang, L.K. Kokjohn, R.D. Reitz, Thermodynamic energy and exergy analysis of three different engine combustion regimes. Appl. Energy **180**, 849–858 (2016)
18. M. Djermouni, A. Ouadha, Comparative assessment of LNG and LPG in HCCI engines. Energy Proc. **139**, 254–259 (2017)
19. A. Khaliq, S. Islam, I. Dincer, Energy and exergy analyses of a HCCI engine-based system running on hydrogen enriched wet-ethanol fuel. Int. J. Exergy **28**, 72–95 (2019)
20. M.M. Namar, O. Jahanian, Energy and exergy analysis of a hydrogen-fueled HCCI engine. J. Therm. Anal. Calorim. 137– 205 (2019)

Chapter 29
Thermodynamic Study of a Turbocharged Diesel-Hydrogen Dual Fuel Marine Engine

Fouad Selmane, Mohamed Djermouni, and Ahmed Ouadha

Abstract In this study, a mathematical model is used in order to analyse the influence of some usual engine parameters such as compression ratio, turbocharger compressor pressure ratio, equivalence ratio, and engine speed on the performances of a Diesel-hydrogen dual fuel marine engine. The model takes into account the gas composition resulting from the combustion process and the specific heat temperature dependency of the working fluid. The analysis is based on both the first and second laws of thermodynamics using the concept of exergy analysis. Results showed that, overall, the variation in the engine operating parameters (r_c, r_p, N and ϕ) affects considerably the engine performance. Furthermore, an exergy loss mapping of the system indicates that most of exergy losses (88.2%) occur in the engine due to due to the irreversible nature of the mixing and combustion processes. The remaining components (turbo-compressor, intercooler, mixer, catalytic converter and turbine) are responsible of only 11.2% of the total exergy loss of the system.

Keywords Thermodynamic analysis · Marine engine · Dual fuel · Hydrogen fuel

29.1 Introduction

The strengthen emissions legislation of the International Maritime Organization (IMO) has emerged as a new issue facing the marine diesel engines due to their high sulfur oxides (SOx), nitrogen oxides (NOx) and particulate matter (PM) emissions. Furthermore, the 2018 IMO targets for low- and zero-emission shipping aspire to reduce the total annual greenhouse gases (GHG) emissions by at least 50% by 2050 over 2008 levels and phase them out, as soon as possible in this century.

F. Selmane · M. Djermouni · A. Ouadha (✉)
Laboratoire des Sciences et Ingénierie Maritimes, Faculté de Génie Mécanique, Université des Sciences et de la Technologie Mohamed Boudiaf d'Oran, Oran El-M'nouar, 31000 Oran, Algeria
e-mail: ah_ouadha@yahoo.fr

F. Selmane
e-mail: foselmane@gmail.com

Addressing the above concerns, diesel engines manufactures have proposed dual-fuelling of diesel engines as an attractive solution to mitigate the harmful effects of marine diesel engines emissions. Dual-fuel engines provide high fuel flexibility as they can operate in either conventional mode (diesel) or dual-fuel mode (diesel–gas) depending on the gas fuel availability. This technology is now available and several studies have been performed on gas dual-fuel marine engines [1–11]. In addition, engines companies such as MAN and Wartsila have already met on the market a wide range of gas engines using different kinds of gases such as natural gas, ammonia, methanol and hydrogen.

While many ship owners have adopted LNG as a marine fuel, a serious debate on the role of methane in meeting the IMO GHG reduction targets has been arisen during the last years. It seems that to meet the ambitious IMO targets, shipping needs to switch to the so-called carbon-neutral fuels such as ammonia and hydrogen in the short to mid-term. Whereas, the carbon footprints of this kind of fuels should be considerably lower than those of conventional fuels if they are produced from renewable energy or biomass sources.

Hydrogen, a renewable, high-efficient and clean fuel, can potentially comply with the stringent IMO regulation and save the future of diesel-type engines. Hydrogen is a colorless, odorless and zero (harmful) emission fuel when burned with oxygen. The combustion of hydrogen leads to energy release and water formation only. Compared to conventional and alternative fuels, hydrogen has the higher combustion enthalpy per unit mass (lower heating value). Indeed, 1 kg of hydrogen can provide almost three times more energy than diesel and gasoline fuels. It has also the highest octane rating (>130), meaning that the engine can operate at elevated compression ratio values without experiencing the "knock" phenomenon. Due to its wide flammability range, hydrogen can be used in engines at extremely low equivalence ratios which improve the fuel economy. However, the significantly low density of hydrogen results in reduced energy density of the hydrogen-air mixture inside the cylinder chambers of an engine and hence may lead to low power output.

Several studies have been carried out on the suitability and merits of hydrogen as a fuel for diesel engines [12–26]. A careful examination of the above studies indicates that thermodynamic analysis studies of hydrogen-based dual fuel engines are scanty in the literature. For the design of an engine running on various fuels, accurate thermodynamic models are needed to determine their performances as function of key operating parameters. The present study focuses on a thermodynamic analysis of a marine dual-fuel engine, working both in diesel and hydrogen modes. The developed engine thermodynamic model has been implemented as a computer code and used to assess the performance of the engine as function of various operating parameters.

29.2 Engine Thermodynamic Modeling

Marine Diesel engines operated fuel can be converted to operate with up to 99% per cycle heat from a gaseous fuel such as natural gas, methanol, hydrogen, propane, LPG etc. under dual fuel mode. In dual fuel mode, most of the engine power is provided by the gaseous fuel, while a pilot amount of the liquid Diesel fuel, less than 1% of the total fuel supplied to the engine at full load operation (energy basis), is injected at the end of the compression stroke to act as an ignition source of the gaseous fuel–air mixture. The engine power output is controlled by changing the amount of the primary gaseous fuel, while the pilot diesel fuel quantity is kept constant [27].

Compared to diesel fuel, hydrogen fuel is neutral-carbon fuel with many desirable attributes which make it excellent internal combustion engine. These include wide range of flammability limits, high heating value and stoichiometric air to fuel ratio and low specific gravity. Furthermore, hydrogen is not a mixture which the properties can change depending on the source as is the case with LNG. Although that hydrogen fuel has been used in SI engines, its relatively high auto-ignition temperature makes it a serious option to operate with diesel fuel in dual fuel mode.

29.2.1 Investigated Engine

The engine considered in the current study is a turbocharged and intercooled dual-fuel engine. It is an in-line eighteen cylinders which is widely used in marine applications. The main engine characteristics are illustrated in Table 29.1.

The proposed system is a basic turbocharged main propulsion engine as illustrated in Fig. 29.1. The turbo-compressor compresses air from ambient conditions (state 1) to temperature T_2 and pressure p_2. It is then cooled in the cooler to reach the required temperature T_3. The fuel is injected near the intake valve of each cylinder (timed port injection) at ambient temperature and pressure p_4. The premixed charge is then introduced in the engine to undergo cyclic processes according the Otto cycle. Before released to the atmosphere, the exhaust gases at state 6 passthrough a catalytic converter to get rid of the remained unburned fuel as well as carbon monoxide. The clean flux of gases at 7 expands in the turbine to produce the power require to drive the compressor.

Table 29.1 Main engine specifications

Number of Cylinders	18
Bore (B) × Stroke (S), mm	510 × 600
Engine compression ratio (r_c)	14
Rotation speed (N), rpm	500
Equivalence ratio (ϕ)	0.55
Compressor pressure ratio (r_p)	4.5

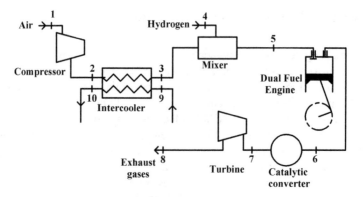

Fig. 29.1 Schematic of the engine system layout

29.2.2 Thermodynamic Analysis

In order to carry out the thermodynamic analysis, some parameters of the working fluid such as enthalpy, internal energy, entropy and exergy are required. In an internal combustion engine, the working fluid changes in composition according to the process undergone. It could be air, fuel, a mixture of reactant (fuel + air), or a mixture of the combustion products. In this study, the thermodynamic properties of the working fluid are calculated using polynomials derived by regression methods from thermodynamics tables available in the literature.

The thermodynamic analysis is performed by considering several assumptions which are given as:

- All processes are considered at steady state condition.
- Changes in kinetic and potential energies are neglected.
- Changes in kinetic and potential exergies are neglected.

Mass, energy and exergy balance equations are applied to each component of the system described in Fig. 29.1. Neglecting kinetic and potential energy and exergy changes, these equations are summarized as:

$$\sum \dot{m}_{in} - \sum \dot{m}_{out} = 0 \qquad (29.1)$$

$$\dot{Q}_{in} + \dot{W}_{in} + \sum_{in} \dot{m}_i h_i = \dot{Q}_{out} + \dot{W}_{out} + \sum_{out} \dot{m}_i h_i \qquad (29.2)$$

$$\sum_{in} \left(1 - \frac{T_0}{T_i}\right) \dot{Q}_i + \dot{W}_{in} + \sum_{in} \dot{m}_i ex_i$$

$$= \sum_{out} \left(1 - \frac{T_0}{T_i}\right) \dot{Q}_i + \dot{W}_{out} + \sum_{out} \dot{m}_i ex_i + \dot{E}x_D \qquad (29.3)$$

where $\dot{Q}$ is the rate of heat transfer between the control volume and its surroundings, $\dot{W}$ the rate of work, h the specific enthalpy and $\dot{E}x_D$ is the rate of exergy loss.

The mass, energy and exergy balances in Eqs. (29.1), (29.2) and (29.3) have been applied to each component of the system.

29.3 Results and Discussions

The thermodynamic analysis has been performed for a turbocharged Diesel engine to study the effects of some operating parameters such as the engine compression ratio, the equivalence ratio, the engine speed and the turbo-compressor pressure ratio on the engine performance. Common engine performance parameters such as the brake power output, the brake specific fuel consumption, the brake thermal efficiency and the engine volumetric efficiency have been considered in the present study. In addition, individual and overall exergy losses and the exergy efficiency have been assessed as second-law parameters.

Figure 29.2 illustrates the effect of varying the equivalence ratio and the engine compression ratio on the performance of the system. The compressor pressure ratio, engine speed and ambient temperature were kept constant ($r_p = 4.5$, $N = 500$ rpm, and $t_{amb} = 25\ °C$) while the fuel-air equivalence ratio, defined as the actual fuel-air ratio divided by the stoichiometric fuel-air ratio, was varied from 0.25 to 0.7 for four engine compression ratio values: 12, 14, 16 and 18. Obviously, an increase in the equivalence ratio yields to an improvement of the brake power output, the energy and exergy efficiencies and the exhaust gas temperature for a fixed engine compression ratio. This is mainly due to the large amount of fuel being burnt at higher equivalence ratio values. Nevertheless, it produces a slight drop in the volumetric efficiency due to the increase of the gaseous fuel that displaces a portion of air mass flow rate. This stands only for 2.3% by shifting from an equivalence ratio of 0.25–0.7. While the brake power increases by 406%. The brake specific fuel consumption is also found decreasing by 51% with the increase of equivalence ratio form 0.25 to 0.7.

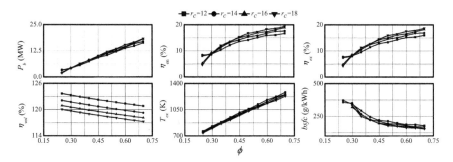

Fig. 29.2 Engine-out performance for different engine compression ratio values and equivalence ratios

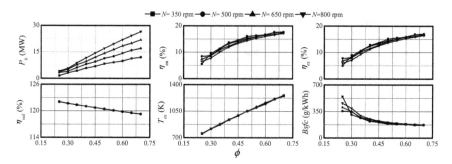

Fig. 29.3 Engine-out performance for different engine speed values and equivalence ratios

This is caused by the increase of heat released by the combustion that is converted effectively to work during expansion with higher equivalence ratios. Furthermore, higher compression ratio values improve the engine performance due to the high temperatures of the reactants achieved at the end of compression which leads to a higher maximal temperatures and pressures. Taking the example of the bake power and the thermal efficiency, they increase by 11 and 15%, respectively, by altering the compression ratio from 12 to 18. The exhaust temperature increases also at higher compression ratios, this could be explained by the fact that the expansion process does take full advantage of the total potential of the hot burned gas produced by the combustion, and portion of the thermal energy is wasted to the exhaust gases. However, this trend is not respected for fuel-air equivalence ratios lower than 0.3 (very lean mixtures) where the combustion performances does not respond to the increase in the compression ratio which reduces the engine performance.

Figure 29.3 shows the combined effect of the fuel-air equivalence ratio and the engine speed on engine performance. Results have been obtained for fixed values of the compressor pressure ratio, engine compression ratio and ambient temperature were kept constant ($r_p = 4.5$, $r_c = 14$. and $t_{amb} = 25\ °C$) for four engine speed values: 350, 500, 650 and 800. The brake power output, the brake efficiency, the exergy efficiency and the exhaust gas temperature are increased as the equivalence ratio is increased, while the volumetric efficiency and the brake specific fuel consumption are decreasing. It is interesting to note that the engine speed has slight effect on the brake thermal efficiency, the exergy efficiency and the brake specific fuel consumption. It is worth mentioning that the enhanced performances is related to the reason that with increasing the engine speed, the reduction of heat loss through cylinder walls along the higher flow rate of air and fuel (higher end-combustion temperatures and pressures) overcome the increase of work loss due to friction. This situation is inversed for the thermal and exergetic efficiency with a further increase of engine speed beyond 500 rpm, where the heat loss through cylinder walls and friction overwhelms the scene. For example, the brake power output increases by 114% by increasing the engine speed from 350 to 800 rpm. Furthermore, both the volumetric efficiency and the exhaust gas temperature remain insensitive to the variation of the engine speed. This is due to the fact that the volumetric airflow is directly

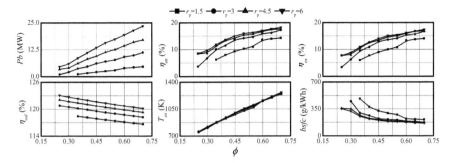

Fig. 29.4 Engine-out performance for different turbo-compressor pressure ratio values and equivalence ratios

controlled by the engine speed. The engine speed affects the engine power output mainly at higher values of the fuel-air equivalence ratio. This can be explained by the reduction of heat losses and the energy of fuel supplied to the engine.

In Fig. 29.4 are plotted the engine brake power, energy efficiency, energy efficiency, brake specific fuel consumption, volumetric efficiency and exhaust gas temperature as function of the equivalence ratio and the turbo-compressor pressure ratio. Results have been obtained for fixed engine compression ratio, engine speed and ambient temperature ($r_c = 14$, $N = 500$ and $t_{amb} = 25\ °C$). According to the graph, all performance parameters increase as the fuel-air equivalence ratio is increased except the volumetric efficiency which decreases for the reasons mentioned above. In addition, the increase of the turbo-compressor pressure ratio has a significant effect on the improvement the charge (reactants) mixture preparation and the volumetric efficiency, a key factor to higher air mass flow rates and therefore higher fuel mass flow rates, which increases the in-cylinder pressure and temperature. For instance, the brake power, thermal efficiency, exergetic efficiency, and the volumetric efficiency increase by 495, 49, 44 and 3% when the pressure is increased from 1.5 to 6. However, it has an insignificant effect on the exhaust gas temperature.

A mapping of exergy losses within the whole system for fixed compression ratio, turbo-compressor pressure ratio, ambient temperature, fuel-air equivalence ratio and engine speed ($r_c = 14$, $r_p = 4.5$, $t_{amb} = 25\ °C$, $\phi = 0.55$, $N = 500$ rpm) has been also performed. It is observed that under the operating parameters selected, exergy losses that occur in the engine dominate. They account for 88.23% of total exergy losses in the system. The remaining components (turbo-compressor, intercooler, mixer, catalytic converter and turbine) are responsible of only 11.77% of the total exergy loss of the system. Exergy losses within the engine are mainly due to the irreversible nature of the mixing and combustion processes.

29.4 Conclusions

Engine brake power output, volumetric efficiency, thermal efficiency, brake specific fuel consumption, exhaust gas temperature, total exergy loss and exergy efficiency data have been evaluated as function of some operating parameters such as compression ratio, equivalence ratio, engine speed and compressor pressure ratio for a diesel-hydrogen dual fuel engine.

It is observed that, overall, the variation in the engine operating parameters (r_c, r_p, N and ϕ) affects considerably the engine performance. The latter are significantly improved at higher values of r_c, r_p, N and ϕ. This is mainly due to the improved combustion process under higher values of these parameters. Furthermore, an exergy loss mapping of the system indicates that most of exergy losses (88.2%) occur in the engine due to the irreversible nature of the mixing and combustion processes. The remaining components (turbo-compressor, intercooler, mixer, catalytic converter and turbine) are responsible of only 11.2% of the total exergy loss of the system.

References

1. A.P. Roskilly, S.K. Nanda, Y.D. Wang, J. Chirkowski, The performance and the gaseous emissions of two small marine craft diesel engines fuelled with biodiesel. Appl. Thermal Eng. **28**, 872–880 (2008)
2. F.J. Espadafor, M.T. García, J.B. Villanueva, J.M. Gutiérrez, The viability of pure vegetable oil as an alternative fuel for large ships. Transp. Res. Part D Transp. Environ. **14**, 461–469 (2009)
3. A.A. Banawan, M.M. El Gohary, I.S. Sadek, Environmental and economical benefits of changing from marine diesel oil to natural-gas fuel for short-voyage high power passenger ships. J. Eng. Marit. Environ. **224**(2), 103–113 (2009)
4. A. Petzold, P. Lauer, U. Fritsche, J. Hasselbach, M. Lichtenstern, H. Schlager, F. Fleischer, Operation of marine diesel engines on biogenic fuels: modification of emissions and resulting climate effects. Env. Sci. Technology **45**, 10394–10400 (2011)
5. C.-Y. Lin, T.-H. Huang, Cost–benefit evaluation of using biodiesel as an alternative fuel for fishing boats in Taiwan. Marine Policy **36**, 103–107 (2012)
6. P. Westling, Stena RoRo, methanol—a good alternative for ferries and short sea shipping, in *Interferry Conference* (Malta, 2013)
7. S. Brynolf, E. Fridell, K. Andersson, Environmental assessment of marine fuels: liquefied natural gas, liquefied biogas, methanol and bio-methanol. J. Clean. Prod. **74**, 86–95 (2014)
8. M.M. El Gohary, Y.M.A. Welaya, The use of hydrogen as a fuel for inland waterway units. J. Marine Sci. Application **13**, 212–217 (2014)
9. H. Pan, S. Pournazeri, M. Princevac, J.W. Miller, S. Mahalingam, M.Y. Khan, V. Jayaram, W.A. Welch, Effect of hydrogen addition on criteria and greenhouse gas emissions for a marine diesel engine. Int. J. Hydrogen Energy **39**, 11336–11345 (2014)
10. MAN B&W Technical Papers: Using Methanol Fuel in the MAN B&W ME-LGI Series (2015)
11. S. Stoumpos, G. Theotokatos, E. Boulougouris, D. Vassalos, I. Lazakis, G. Livanos, Marine dual fuel engine modelling and parametric investigation of engine settings effect on performance-emissions trade-offs. Ocean Eng. **157**, 376–386 (2018)
12. G. Gopal, P.S. Rao, K.V. Gopalakrishnan, B.S. Murthy, Use of hydrogen in dual-fuel engines. Int. J. Hydrogen Energy **7**(3), 267–272 (1982)
13. E. Tomita, N. Kawahara, Z. Piao, S. Fujita et al., Hydrogen combustion and exhaust emissions ignited with diesel oil in a dual fuel engine. SAE Technical Paper 2001-01-3503 (2001)

14. M. Masood, S.N. Mehdi, P.R. Reddy, Experimental investigations on a hydrogen-diesel dual fuel engine at different compression ratios. J. Eng. Gas Turbines Power **129**(2), 572–578 (2007)
15. I.J.S. Veldhuis, R.N. Richardson, H.B.J. Stone, Hydrogen fuel in a marine environment. Int. J. Hydrogen Energy **32**, 2553–2566 (2007)
16. N. Castro, M. Toledo, G. Amador, An experimental investigation of the performance and emissions of a hydrogen-diesel dual fuel compression ignition internal combustion engine. Appl. Thermal Eng. **156**, 660–667 (2009)
17. M.M. Roy, E. Tomita, N. Kawahara, Y. Harada, A. Sakane, An experimental investigation on engine performance and emissions of a supercharged H_2-diesel dual-fuel engine. Int. J. Hydrogen Energy **35**, 844–853 (2010)
18. D.B. Lata, A. Misra, S. Medhekar, Effect of hydrogen and LPG addition on the efficiency and emissions of a dual fuel diesel engine. Int. J. Hydrogen Energy **37**(7), 6084–6096 (2012)
19. A.E. Dhole, R.B. Yarasu, D.B. Lata, A. Priyam, Effect on performance and emissions of a dual fuel diesel engine using hydrogen and producer gas as secondary fuels. Int. J. Hydrogen Energy **39**(15), 8087–8097 (2014)
20. Y. Suzuki, T. Tsujimura, T. Mita, The performance of multi-cylinder Hydrogen/ Diesel dual fuel engine. SAE Int. J. Engines **8**(5), 2240–2252 (2015)
21. Y. Karagoz, T. Sandalci, L. Yuksek, A.S. Dalkilic, S. Wongwises, Effect of hydrogen–diesel dual-fuel usage on performance, emissions and diesel combustion in diesel engines. Adv. Mech. Eng. **8**, 1–13 (2016)
22. A. Tsujimura, Y. Suzuki, The utilization of hydrogen in hydrogen/diesel dual fuel engine. Int. J. Hydrogen Energy **42**, 14019–14029 (2017)
23. P. Dimitriou, M. Kumar, T. Tsujimura, Y. Suzuki, Combustion and emission characteristics of a hydrogen-diesel dual-fuel engine. Int. J. Hydrogen Energy **43**, 13605–13617 (2018)
24. N.R. Ammar, Energy efficiency and environmental analysis of the green-hydrogen fueled slow speed marine diesel engine. Int. J. Multi. Curr. Res. **6**, 1–10 (2018)
25. Y. Sohret, H. Gurbuz, I.H. Akçay, Energy and exergy analyses of a hydrogen fueled SI engine: Effect of ignition timing and compression ratio. Energy **175**, 410–422 (2019)
26. J. Vavra, I. Bortel, M. Takats, A dual fuel hydrogen—diesel compression ignition engine and its potential application in road transport, SAE Technical Paper 2019-01-0564 (2019)
27. D. Woodyard, *Pounder's Marine Diesel Engines and Gas Turbines*, (Butterworth-Heinemann, 2009)

Chapter 30
Effect of Bluff-Body Shape on Stability of Hydrogen-Air Flame in Narrow Channel

Mounir Alliche, Redha Rebhi, and Fatma Zohra Khelladi

Abstract In this work, we study the effect of bluff body shape on the proprieties of Hydrogen-air flame propagating in a narrow channel, specially its stabilization. The study is conducted with a 3D CFD numerical simulation with detailed reactional mechanism. In this work, the Finite Volume Method is applied. In other hand, the k-ε turbulence model was adopted and no artificial flame anchoring boundary conditions was used. The results show dependence between flame structure and bluff body shape. This impact is particularly clear on its thermal properties. Indeed, it appears that the flame has a stretched appearance. This stretch increases particularly at the flame front. However, the heat losses determine the flame anchoring location. In other hand, results analyze shows a dependence between flame location and vortices dilatation of the turbulent flow in the narrow channel with bluff-bodies. The stability of the flame is increased when a triangular or semicircular bluff-body is applied.

Keywords CFD · Turbulent flame · Bluff-Body

30.1 Introduction

Recently, the micro-combustion technique has been widely developed due to the growing demand for portable power devices, the development of micro-electro-mechanical systems (MEMS) technology and the disadvantages of traditional batteries (low density consumption, high weight, long charging time and short charging time). The high energy densities of various fuels used in micro-combustion chambers could be an excellent opportunity to introduce combustion-based micro-generation [1, 2]. However, the power required in micro-generation is not compromised if the chemical energy of the fuel is used efficiently. Low efficiency of micro-combustion systems has been observed by Fernandez-Pello [3] (generating a few

---

M. Alliche (✉) · R. Rebhi · F. Z. Khelladi
Department of Mechanical Engineering, Faculty of Technology, University of MEDEA, Medea, Algeria
e-mail: alliche.mounir@univ-medea.dz

© The Editor(s) (if applicable) and The Author(s), under exclusive license to Springer Nature Singapore Pte Ltd. 2021
A. Khellaf (ed.), *Advances in Renewable Hydrogen and Other Sustainable Energy Carriers*, Springer Proceedings in Energy,
https://doi.org/10.1007/978-981-15-6595-3_30

watts in an extremely low volume). In fact, understanding the combustion characteristics of small-scale combustion chambers play a crucial role in improving the efficiency of the system, especially in the presence of heat losses through the walls of the micro-combustion chamber. Therefore, improving the flame stability and thermal efficiency of micro-combustors has become a new challenge in combustion investigations [4, 5]. Various experimental and numerical investigations have been carried out [5, 6]. This research made it possible to understand the fundamental principles of micro-combustion in terms of flammability limits, flame stability and thermal efficiency.

The application of micro-combustion systems in MEMS has been successfully tested by Waitz et al. [7]. The stability of the flame in combustion chambers smaller than one millimeter has become a new challenge [8]. In fact, the stability of the flame in small-scale combustion chambers is influenced by the absorption and destruction of the combustion radicals [9]. Consequently, the possibility of using catalyzed combustion [10] as well as the application of external heat [11] and heat recirculation [12] has been investigated. It has been found that the stability of the flame in small-scale combustion chambers is strangely dependent on the recirculation of heat through the walls of the combustion chamber [13].To reduce heat loss to the atmosphere and prevent heat recirculation in the unburned area, a specific material should be selected. Therefore, the selection of materials in the micro-scale combustion chambers is essential due to the strong thermal coupling between the reactive mixture and the walls of the combustion micro-system. In general, since the surface/volume ratio of the micro combustion chamber is higher than that of conventional combustion chambers, the possibility of thermal cooling in micro combustion chambers is very high due to the extremely high heat loss of the flame [14]. Maruta et al. [15] pointed out that the instability of the flame in non-premixed small-scale combustion could be attributed to the interaction between the structure of the flame and the flow of flame, heat loss and mass transfer limitations. Sánchez-Sanz et al. [16] studied the source of flame instabilities during premixed combustion from the combustion of narrow channels. The effects of channel width, mixture velocity and Lewis number on the flame propagation were studied to investigate the impact of heat loss on the flame stability in channels [17]. Numerical studies in the micro-combustor process have been developed to simulate different parameters in these devices. As a result, in so-called super adiabatic combustion, the peak temperature of the flame at the reaction zone is higher than the maximum adiabatic flame temperature of combustion without excess enthalpy. When the maximum flame temperature is augmented, quenching cannot occur due to heat loss from walls, so the flame becomes stronger [14, 17]. Although the effects of micro-combustor dimensions and operating conditions on the flame stability and combustion characteristics of micro-combustion have been noted, the impacts of different shapes of bluff body on the flame stability, wall temperature and exhaust temperature have not been properly developed.

In this numerical study, the effects of the different shapes of bluff body applied at the micro-combustor entrance and different inlet velocity magnitude with respect to the fixed equivalence ratio of premixed hydrogen-air on the flame stability and efficiency of the small-combustor is investigated.

30.2 Numerical Modeling

The geometry of the micro-combustor has been selected based on experimental works proposed by [8]. The length of the micro-combustor is 50 mm and its diameter is 5 mm. The cross-sectional view of the studied bluff bodies are located at the entrance of micro-combustor (10 mm from the entrance). Indeed, viscous forces, pressure work and gas radiation were not taken into consideration. In this simulation, the hydrogen was assumed as the fuel due to extremely high burning velocity of hydrogen in comparison with hydrocarbon fuels. In order to protect the inlet geometry, bluff bodies and combustor walls against high temperatures, lean premixed mixture with equivalence ratios of 0.5 is applied to control the maximal temperature. In all of the studied cases, the combustion characteristics were investigated based on hydrogen-air mixture velocity inlet V = 40 m/s. All the governing equations for the micro-combustor have been considered to ensure the convergence of CFD simulated models. In this simulation, we adopt k-ε model. At the internal surface of micro-combustor, non-slip and no species flux normal to the surfaces were considered. Moreover, heat loss from walls to the surroundings is calculated with considering both thermal radiation and natural convection heat transfer. The inlet temperature of the hydrogen-air mixture and the pressure outlet of the emissions were set at 500 K and 1 atm respectively. The Partial Premixed Model was chosen to solve the interaction between turbulence flow and combustion. The summary of the boundary conditions is illustrated in Table 30.1.

The wall material was steel and the surface reaction effects were neglected. In computational domain, the atmosphere inside the combustor is considered as a fluid and the bluff body shapes are subtracted from the mentioned domain. It means in the solution process the bluff body is assumed as a shape which fluid cannot flow through it. Based on this assumption, the material of bluff body and heat conduction model in this section is not considered and the main discussion is about the whole of micro-combustor. Ansys CFX 12 [18] was used to solve the momentum equations, energy, species, mass and heat transfer in different states. The first-order upwind scheme was applied for discretization and for the pressure-velocity coupling, the

Table 30.1 Boundary conditions of the simulation

Inlet	Velocity	Uniform = 40 m/s
	Temperature	Uniform = 500 K
	Pressure reference	1 atm
	Hydraulic diameter	1 mm
	Mole fraction	Regard to the equivalence ratio(0.5)
Outlet	Hydraulic diameter	5 mm
	Pressure gage	0 Pa
Wall	Wall slip	Non-slip
	Material	Steel
	Thermal condition	Mixed
	Heat transfer coefficient	40 W/m^2 K

Fig. 30.1 Example of mesh

SIMPLE Algorithm was used. Indeed, gradient is set least squares cell based, pressure is standard, momentum and energy are second order upwind and turbulent kinetic energy and turbulent dissipation rate are set first order upwind. The unstructured square grid system, refined at the walls, with around 20,000 cells and 40,000 nodes for all cases was set in the final simulation (Fig. 30.1).

30.3 Results and Discussion

30.3.1 Flame Structure

The concepts of chemical kinetics indicate that the chemical reaction time decreases when the reaction speed increases. In the micro-combustion process, increasing the reaction time can be achieved by increasing the reaction temperature. Thus, increasing the flame temperature and reducing the heat loss in a micro-combustion chamber can have a significant effect on the stabilization of the flame.

Figure 30.2 shows the structure of the flame in all micro-combustion chambers having different forms of bluff body when the equivalence ratio is 0.5 and an entry speed of the hydrogen-air mixture is 40 m/s.

The maximum temperature (T = 1603.52 K) occurs in the case of a square section bluff body. A higher value (1682.36 K) is obtained for a lower speed (V = 10 m/s). In general, observations confirm that when the inlet velocity increases from V = 10 m/s to 50 m/s, the flame temperature of the micro-combustion intensifies in all domain except near the bluff body. When the inlet speed of the mixture increases from 20 m/s to 40 m/s, the maximal flame temperature decreases in all the cases studied. Consequently, it can be concluded that for the specific input flow rate of the hydrogen-air mixture, the flame of the micro-combustion chamber with a semi-circular bluff body is more stable than in the other cases. The flame is more attached, presenting a recirculation zone behind the bluff body.

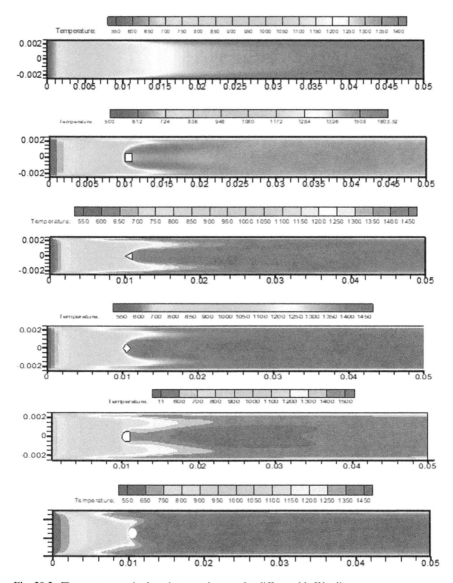

Fig. 30.2 Flame structure in the micro-combustors for different bluff bodies

30.3.2 Flame Stabilization

In order to analyze the various aspects of flame stability in micro-combustors with different bluff bodies, we present the flow fields contours in Fig. 30.3. This figure shows that a small recirculation zone is constituted behind the bluff body. In partial premixed combustion, the recirculation zone retains a significant supply of hot

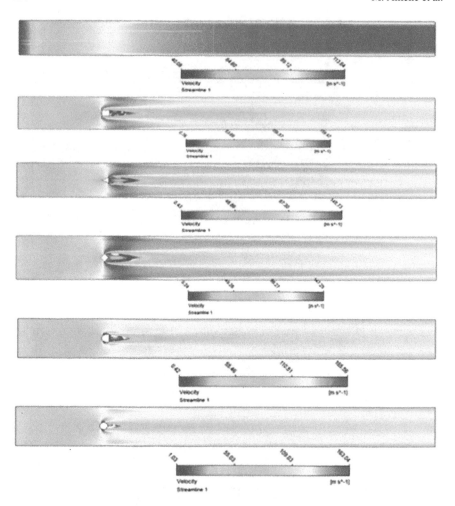

Fig. 30.3 Velocity contours in the micro-combustors for different bluff bodies

combustion products to ignite the fresh reactants continuously. Blow off occurs when the time allowed by the flow is not sufficient for the chemical reactions to proceed to ignition. We note that the blow off limit for micro-combustors without bluff body is very low and it has been found that bluff body extends the stable flame of hydrogen-air mixture gradually. In fact, it has been stipulated that blow-off limit is a function of bluff body shape, fuel type, flow velocity, stoichiometry, pressure, temperature, blockage ratio, and dilution [10]. In all of the studied cases, when the inlet velocity is intensified, the recirculation zone is prolonged. Although the recirculation zone is enlarged with a further inlet velocity augmentation, the middle of the reaction zone tends to be split into two parts. In very large velocities, the flame is no longer able to be sustained due to the very small reaction zone behind the bluff body and blow-off occurs.

Table 30.2 Variation of micro-combustion characteristics with bluff body shape ($V_{in} = 40$ m/s and $\Phi = 0.5$)

BB shape	T_{max} (K)	T_{wall} (K)	T_{outlet} (K)	V_{outlet} (m/s)
Without bb	1447.24	1064.48	1439.79	113.60
Circular	1469.75	1135.97	1468.78	104.79
Triangular	1487.60	1079.19	1466.81	109.07
Diamond	1468.50	1142.76	1468.38	109.68
Semi circular	1547.25	1142.55	1495.25	111.65
Square	1603.52	1148.27	1475.86	109.85

30.3.3 Thermal Efficiency of Micro-Combustion

Thermal radiation from the micro-combustor wall is applied for power generation in some cases, thus the micro-combustor is considered as an emitter and the radiation heat loss from the combustor wall plays an important role [15]. The ratio of the total radiation through the combustor wall to the total energy input is defined as emitter efficiency. From Table 30.2, it can be seen that the mean wall temperature in most of the cases is similar; however, the mean wall temperature in bluff body wall is very high. Moreover, the mean temperature of the wall decreases when the inlet mixture velocity increases. On the other hand, in some applications (for instance micro-gas turbines), the temperature of exhaust gases is important. Maximum exhaust gas temperature increases when inlet velocity of mixture rises. Table 30.2 illustrates the exhaust gas temperature in different states. Compared to the experimental work done by Ref. [12], it can be seen that in triangle bluff body case, the experimental and numerical results have the same trend. Experimental results show that at $\Phi = 0.5$, when the inlet velocity of the fuel is high (about 40 m/s) the blow off occurs. Table 30.2 depicts also the mean outlet velocity of the exhaust gases. When inlet velocity is set **40** m/s, maximum outlet velocity was recorded at 111.65 m/s and minimum outlet velocity, which is related to bluff body wall, is about **42** m/s (Fig. 30.3).

30.4 Conclusion

Flame stability should be ensured in micro-combustors by taking some appropriate strategies. In this work, the combustion characteristics of a micro-combustor with different shapes of bluff bodies under various velocities were investigated numerically to obtain the desirable conditions. In a lean hydrogen air mixture, the maximum temperature (T = 1682.32 K) was recorded in the micro-combustor with semicircular bluff body in lower velocity (V = 10 m/s). If the inlet velocity rises from V = 10 m/s to 50 m/s, the flame temperature of the micro-combustion intensifies in all cases. When inlet velocity of mixture increases, the flame temperature reduces in all studied cases. Therefore, the flame of the micro-combustor with semicircular bluff body is more stable and lifted. Emitter efficiency is very high in a micro-combustor with this shape of bluff body, the temperature of exhaust gases is the lowest in this case and this bluff body shape is not suitable for micro-turbine systems. Due to the very small size of micro-combustors, residence time reduces dramatically and flame stability encounters problems in these devices. Most investigations became necessaries.

References

1. B. Khandelwal, S. Kumar, Experimental investigations on flame stabilization behavior in a diverging micro channel with premixed methane-air mixtures. Appl. Therm. Eng. **30**, 2718–2723 (2010)
2. S.K. Chou, W.M. Yang, K.J. Chua, J. Li, K.L. Zhang, Development of micro power generators—a review. Appl. Energy **88**, 1–16 (2011)
3. A.C. Fernandez-Pello, Micropower generation using combustion: issues and approaches. Proc. Combust. Inst. **29**, 883–899 (2002)
4. J. Zhou, Y. Wang, W. Yang, J. Liu, Z. Wang, K. Cen, Combustion of hydrogen-air in catalytic micro-combustors made of different material. Int. J. Hydrogen Energy **34**, 3535–3545 (2009)
5. C.M. Miesse, R.I. Masel, C.D. Jensen, M.A. Shannon, M. Short, Submillimeter scale combustion. AIChE J. **50**, 3206–3214 (2004)
6. N.S. Kaisare, D.G. Vlachos, Optimal reactor dimensions for homogeneous combustion in small channels. Catal. Today **120**, 96–106 (2007)
7. I.A. Waitz, G. Gauba, Y.S. Tzeng, Combustors for micro-gas turbine engines. J. Fluids Eng. **120**, 109–117 (1998)
8. A. Fan, J. Wan, K. Maruta, H. Yao, W. Liu, Interactions between heat transfer, flow field and flame stabilization in a micro-combustor with a bluff body. Int. J. Heat Mass Transf. **66**, 72–79 (2013)
9. Y. Ju, C. Choi, An analysis of sub-limit flame dynamics using opposite propagating flames in mesoscale channels. Combust. Flame **133**, 483–493 (2003)
10. S.J. Shanbhogue, S. Husain, T. Lieuwen, Lean blowoff of bluff body stabilized flames: scaling and dynamics. Prog. Energy Combust. Sci. **35**, 98–120 (2009)
11. W.M. Yang, S.K. Chou, K.J. Chua, J. Li, X. Zhao, Research on modular micro combustor-radiator with and without porous media. Chem. Eng. J. **168**, 799–802 (2011)
12. J. Wan, A. Fan, K. Maruta, H. Yao, W. Liu, Experimental and numerical investigation on combustion characteristics of premixed hydrogen/air flame in a micro-combustor with a bluff body. Int. J. Hydrogen Energy **37**, 19190–19197 (2012)
13. D.G. Norton, D.G. Vlachos, A CFD study of propane/air microflame stability. Combust. Flame **138**, 97–107 (2004)
14. J. Daou, M. Matalon, Influence of conductive heat-losses on the propagation of premixed flames in channels. Combust. Flame **128**, 321–339 (2002)
15. K. Maruta, T. Kataoka, N.I. Kim, S. Minaev, R. Fursenko, Characteristics of combustion in a narrow channel with a temperature gradient. Proc. Combust. Inst. **30**, 2429–2436 (2005)
16. M. Sánchez-Sanz, Premixed flame extinction in narrow channels with and without heat recirculation. Combust. Flame **159**, 3158–3167 (2012)
17. M. Alliche, P. Haldenwang, S. Chikh, Extinction conditions of a premixed flame in a channel. Combust. Flame **157**, 1060–1070 (2010)
18. Ansys CFX, *12.0 Theory Guide* (Ansys Inc., 2009)

Chapter 31
Experimental Validation of Fuel Cell, Battery and Supercapacitor Energy Conversion System for Electric Vehicle Applications

R. Moualek, N. Benyahia, A. Bousbaine, and N. Benamrouche

Abstract Due to the increasing air pollution and growing demand for green energy, the most of research is focused on renewable and sustainable energy. In this work, the PEM fuel cell is proposed as a solution to reduce the impact of the internal combustion engines on air pollution. In this paper a PEM fuel cell, battery and supercapacitor energy conversion system is proposed to ensure the energy demand for an electric vehicle is achived. The storage system consisting of a battery and supercapacitor offers good performance in terms of autonomy and power availability. In this paper, an energy management of the PEM fuel cell electric vehicle has been first simulated in Matlab/Simulink environment and the results are discussed. Second, a Realtime experimental set up is used to test the performance of the proposed PEM fuel cell electric vehicle system. Experimental results have shown that the proposed system is able to satisfy the energy demand of the electric vehicle.

Keywords Fuel cell · Supercapacitor · Batteries · Electric vehicle

31.1 Introduction

The combustion engine using fossil fuels is increasingly subject to controversies, due to its undesirable and negative impact on ecology. Moreover, it is not the only source of pollution but it is considered as the last link in a long chain of massive and irreversible destruction of the environment. Feeling an imminent danger, the modern society has introduced the concept of eco-responsibility in all industrial processes to reduce its carbon impact. The electric vehicle (EV) and the hybrid electric vehicle (HEV) constitute some of the solutions to this new eco-friendly philosophy.

R. Moualek (✉) · N. Benyahia · N. Benamrouche
LATAGE Laboratory, Mouloud Mammeri University, Tizi-Ouzou BP 17 RP, 15000, Algeria
e-mail: rmoualekfr@yahoo.fr

A. Bousbaine
College of Engineering and Technology, University of Derby, DE22 3AW Derby, UK

© The Editor(s) (if applicable) and The Author(s), under exclusive license to Springer Nature Singapore Pte Ltd. 2021
A. Khellaf (ed.), *Advances in Renewable Hydrogen and Other Sustainable Energy Carriers*, Springer Proceedings in Energy,
https://doi.org/10.1007/978-981-15-6595-3_31

However, the HEV still uses a combustion engine although in a reduced or exceptional way it does however contribute quite significantly to the carbon impact footprint. The key solution is probably to use a pure electric vehicle (PEV). For this kind of vehicles, we can find two very different concepts. The first is powered by batteries, which are, themselves, charged by an external source using fixed and known recharging terminals spread through a given territory. This method is primarily disputed because often the energy available on these charging stations emanate from fossil fuels and non-renewable energies. The second type is relatively less harmful with a reduced carbon impact and offering more autonomy to vehicles by the virtue of using a fuel cell as a primary source [1, 2].

It is this type of the system that this paper focuses on. In this paper, a PEM fuel cell electric vehicle power conversion system has been proposed. The proposed electric vehicle system prototype consists of fuel cell, battery and supercapacitor. The PEM fuel cell electric vehicle is simulated first under Matlab/Simulink and energy management strategy is validated. Real-time setup is used to test the performance of the proposed PEM fuel cell electric vehicle system using Real Time Interface using dspace 1103.

This work is organized as follows: In Sect. 31.2, the system's configuration and the modelling of the PEM fuel cell has been presented. In Sect. 31.3, a numerical simulation of the PEM fuel cell and supercapacitor conversion system has been presented under Matlab/Simulink environment. To validate the proposed system, experimental tests have been obtained and discussed in Sect. 31.4. Finally, the conclusions of this work have been provided in Sect. 31.5.

31.2 System Description

The proposed topology of the PEM fuel cell electric vehicle is shown in Fig. 31.1. It includes a PEM fuel cell controlled throughout a boost dc-dc converter, battery that

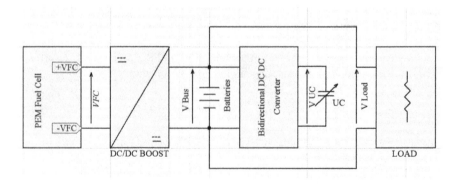

Fig. 31.1 PEM fuel cell based EV topology

is connected to the dc bus and a supercapacitor (U_c) controlled by a buck-boost dc-dc converter. The hybrid storage system offers good performance in terms of autonomy and power availability. In this topology, an energy management of the PEM fuel cell electric vehicle has been considered.

31.3 PEM Fuel Cell Modelling

The nature of the fuel cell and the inherent laws that govern it are such that its net voltage is lower than the voltage generated inside. Voltage drops characterize the operation of the PEM fuel cell. These drops are due to the chemical reactions that take place in the PEM fuel cell [3]. The PEM fuel cell model is given by Eqs. (31.1) and (31.2):

$$V_{fc} = n_{fc} V_{cell} = n_{fc}(E_{nernst} - V_{act} - V_{ohm} - V_{conc}) \quad (31.1)$$

$$E_{Nernst} = 1.229 - 8.5\,e^{-4}(T_{fc} - 298.5) + 4.308\,e^{-5}(Ln(P_{H2}) + (1/2)Ln(P_{O2})) \quad (31.2)$$

where, V_{act} is due to delayed activation of Hydrogen molecules; V_{ohm}, which in turn is due to the nature of the materials making up the fuel cell, V_{conc} due to the concentration of the hydrogen during crucial current requirements, n_{fc} and V_{cell} are the number of cells and the cell voltage respectively. E_{Nernst} is the thermodynamic potential of the cell. The P_{Oi}, P_{H2} and T_{fc} are oxygen and hydrogen partial pressures and cell temperature. The PEM fuel cell characteristic shown in Fig. 31.2 depends on the gaseous flow at the level of these electrodes and hence depends on the duration of their depletion in gas. In the case of a sudden change in the load current, the electrical transient that manifests itself in a more or less significant voltage drop that will depend not only on the availability of gas and air at the electrodes, but also the speed of the chemical reaction at the origin of electron production, as well as the nature of the materials from which the cell is made up. This transient will end as soon as the throttle pressure is readjusted [4, 5].

The PEM fuel cell V-I and P-I polarization curves are shown in Fig. 31.2; the fuel cell is supplied with oxidant (oxygen present in the air) at a pressure of 1 bar. In a practical case, a forced ventilation duct is sufficient. For fuel systems without hydrogen regulation, the pressure is mechanically set at 1.5 bars. Elimination of the water and heat produced is carried out through an air flow via the cathode [6, 7].

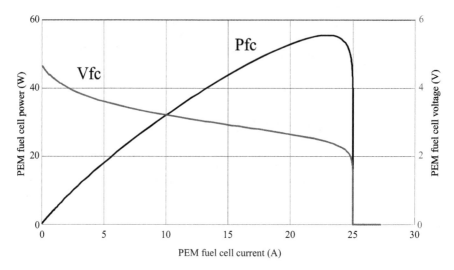

Fig. 31.2 PEM fuel cell V-I and P-I polarization curves

31.4 Simulation Tests

The proposed PEM fuel cell electric vehicle is tested under Matlab/Simulink. The Simulink bloc diagram developed in this work is shown in Fig. 31.3. The PEM fuel cell, battery and supercapcitor models are realized under Powersyst Toolbox of Matlab. The PEM fuel cell is controlled as a main power source and the battery/Supercapacitor as a backup power source. The tests are performed in a variable electric vehicle power load. In addition, the energy management of the whole system is designed to share the PEM fuel cell electric vehicle power demand. The

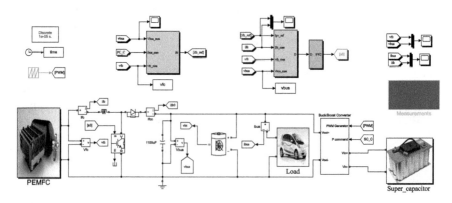

Fig. 31.3 Simulink bloc of the proposed system

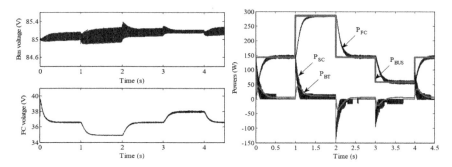

Fig. 31.4 PEM fuel cell model simulation results

load bloc represents the electric vehicle and the storage batter model. The simulation results shown in Fig. 31.4, illustrate that the proposed conversion system is able to satisfy the electric vehicle power load.

31.5 Experimentation and Validation

Real-time experimental part is based on the emulation of the PEM fuel cell. This emulation is realized by a buck dc-dc converter; this converter is controlled to produce the behaviour of the PEM fuel cell. The benefit of this emulation is the possibility to perform the experimental tests without need the real fuel cell, in addition, the input gases pressures and temperature can be changed in this emulator. This PEM fuel cell emulator is associated with a Boost dc-dc converter (see Fig. 31.5). Figure 31.6 show the electronic circuit of the PEM fuel cell emulator. The fuel cell emulator system consists of mains power supply of 80 V associated with a dc-dc buck converter controlled by a simple PI regulator and mathematical model of the fuel cell (see Fig. 31.5). The results given in Fig. 31.7 are obtained under a purely resistive variable

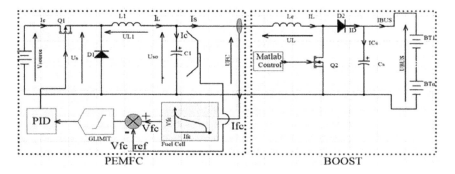

Fig. 31.5 Schematic diagram of the PEM fuel cell emulator

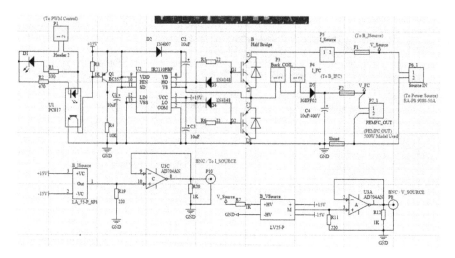

Fig. 31.6 Electronic circuit of PEMFC emulator

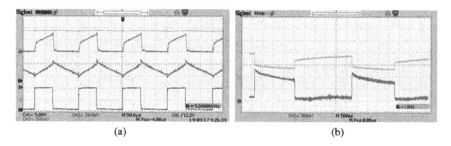

Fig. 31.7 PEM fuel cell emulator results

load. Figure 31.7a shown the output voltage of the battery in cyan, the buck dc-dc converter input current in pink, the load current in red and the control signal of the transistor Q1. The operation of the PEM fuel cell emulator under sudden variation of the load current is illustrated by Fig. 31.7b. This variation is typical and is in conformance with the PEM fuel cell model. The output voltage of the PEM fuel cell emulator is used to supply a boost dc-dc converter used to controller the power from PEM fuel cell emulator. The resulting voltage is used to power the battery rack directly.

The energy management algorithm fixes three phases of operations: When V_{bus} < 84 V. The PEM fuel cell must supply the battery and the load power. When 84 < V_{bus} < 88 V, the batteries are kept in the slow charging mode and the PEM fuel cell must charge the battery and supply load. When V_{bus} > 88 V: The Fuel cell switches off and the batteries ensure the supply of the load alone. The developed real-time setup is illustrated in Fig. 31.8.

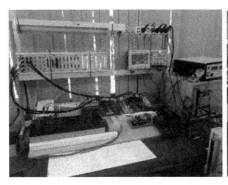 

Fig. 31.8 Complete operational system

31.6 Conclusion

The topology and the energy management proposed in this work can satisfy the power demand of the electric vehicle and ensure the respect of the constraints imposed on the energy sources (dynamics of the PEM fuel cell and provide or absorb the power peaks by the hybrid storage system). The simulation and experimental results show that this topology remains efficient, safe and simple although the presence of an additional converter. In addition, it offers good performance in terms of autonomy and power availability.

References

1. N. Benyahia, H. Denoun, M. Zaouia, T. Rekioua, N. Benamrouche, Power system simulation of fuel cell and supercapacitor based electric vehicle using an interleaving technique. Int. J. Hydrogen Energy **40**(45), 15806–15814 (2015)
2. P. Ching-Tsai, L. Ching-Ming, High efficiency high step up converter with low switch voltage stress for fuel cell system applications. IEEE Trans. Indus. Electron. **6**(57), 1998–2006 (2010)
3. C.C. Chan, The state of the art of electric, hybrid and fuel cell vehicles. Proc. od IEEE **2**(90), 247–275 (2002)
4. T. Phatiphat, D. Bernard, R. Stéphane, S. Panarit, Fuel cell high power applications. IEEE Ind. Electron. Mag. **1**(3), 32–46 (2009)
5. A. Khaligh, L. Zhihao, Battery, Ultracapacitor, fuel cell and hybrid energy storage system for electric, HEV, fuel cell and plug in HEV, State of the art. IEEE Trans. Veh. Technol. **6**(59), 2806–2814 (2010)
6. A. Khosroshahi, M. Adapour, M. Sabahi, Reliability evaluation of conventional and interleaved DC-DC boost converters. IEEE Trans. Power Electron. **10**(30), 5821–5828 (2013)
7. S.S. Williamson, A.K. Rathore, K. Akshay, F. Musavi, Industrial electronics for electric transportation, current state of the art and future challenges. IEEE Trans. Indus. Electron. **5**(62), 3021–3032 (2015)

Chapter 32
Compromise Between Power Density and Durability of a PEM Fuel Cell Operating Under Flood Conditions

H. Abdi, N. Ait Messaoudene, and M. W. Naceur

Abstract It is well known that an adequate choice of operating parameters can result in a high power density and an extension life of a proton exchange membrane fuel cell PEMFC. The aim of this study is to optimize the operating parameters at the inlet of a PEMFC. At first, objective functions are defined to study the problem, according to the desired objective. We have proposed three case studies of single-objective problems. These latter have been solved by the Particle Swarm Optimization algorithm PSO. In a second step, the objective is to simultaneously obtain the highest possible power density and to minimize the dispersion of the current density along the flow channel from its average value. The Non-dominated Sorting Genetic Algorithm NSGA-II is adopted for solving the multi-objective problem. The multi-objective approach allows finding several compromise solutions that allow the user to make the appropriate choice depending on the intended application. These solutions are presented by the Pareto front, containing the set of optimal solutions.

Keywords PEMFC · Optimization · PSO · NSGA-II

32.1 Introduction

Proton exchange membrane fuel cells PEMFC are promising technological solutions in many areas of applications, because of their multiple advantages such as high power density, fast start-up and ease of operation [1, 2]. However, there are still many challenges that need to be considered, such as improving the durability and power output of PEMFC system. These are related to the appropriate operating conditions.

The problem considered in this work is to determine optimum operating parameters at the inlet of the PEMFC. Two optimization approaches are applied to a PEMFC

H. Abdi (✉)
Département de Mécanique, University of Blida 1, PO BOX 270, Blida, Algeria
e-mail: abdih1@yahoo.fr

H. Abdi · N. Ait Messaoudene · M. W. Naceur
Laboratoire Des Applications Energétiques de L'Hydrogène (LApEH), University of Blida 1, Ouled Yaïch, Algeria

operating under partial flooding conditions at the cathode. Particle Swarm Optimization algorithm (PSO) and Non-dominated Sorting Genetic Algorithm (NSGA-II) are used. PSO algorithm is used for solving the single-objective problems. On the other side, NSGA-II is used to find compromise solutions that optimize the dual objective of maximum power density and system durability.

This paper is arranged as follows: Section 32.2 is devoted to the description of the mathematical model of PEMFC; Sect. 32.3 is concerned by the optimization of operating parameters; the simulation results are presented in Sect. 32.4; the conclusion is stated in Sect. 32.5.

32.2 Steady-State Model of PEM Fuel Cell

The mathematical model used in this work has been widely used in the literature [3, 4]. Figure 32.1 shows the elements of a PEM unit cell and the modeled regions.

Under quasi-stationary conditions, the molar flow rates of reactants are [3, 4]:

$$\frac{dN_i(x)}{dx} = \xi_i \frac{w\,I(x)}{4\,F}, \qquad \xi_{H_2} = -2, \quad \xi_{O_2} = -1 \tag{32.1}$$

The variation of the molar flow rate of the liquid water in each flow channel is influenced by the rate of condensation/evaporation.

$$\frac{dN^l_{w,k}(x)}{dx} = \left(\frac{k_c\,h\,w}{R(T_k(x)+273)}\right) \times \left(P^v_{w,k}(x) - P^{sat}_{w,k}(x)\right), \qquad k=a,\,c \tag{32.2}$$

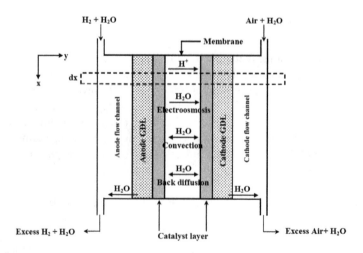

Fig. 32.1 Elements and modeled regions of PEMFC

where k_c is the homogeneous rate constant for the condensation and evaporation of water reaction. h and w are respectively the height and the width of the channel.

The variations of the molar flow rate of water vapor in the flow channel of the anode and the cathode are respectively given by [4]:

$$\frac{dN^v_{w,a}(x)}{dx} = -\left(\frac{dN^l_{w,a}(x)}{dx}\right) - \frac{w\,\alpha\,I(x)}{F} \quad (32.3)$$

$$\frac{dN^v_{w,c}(x)}{dx} = -\left(\frac{dN^l_{w,c}(x)}{dx}\right) + \frac{w\,I(x)}{2F} + \frac{w\,I(x)\,\alpha}{F} \quad (32.4)$$

α is the net flux of water molecules carried by protons across the membrane.

The energy balance at the flow channels takes into account the latent heat due to the condensation/evaporation of the water and the heat transfer between the solid surface and gases [4].

$$\sum_j (N_j(x)\,C_{p,j}\frac{dT_k(x)}{dx} = (H^v_{w,k} - H^l_{w,k})\frac{dN^l_{w,k}(x)}{dx} + U\,a\,(T_s - T_k(x)) \quad (32.5)$$

U is the overall heat exchange coefficient; (a) is the exchange area per unit length.

The cell potential can be calculated from the open circuit voltage, activation overpotential and ohmic drop voltage as follows [3]:

$$E_{cell} = E_{oc} - \frac{R\,(273 + T_s)}{0.5\,F} \ln\left(\frac{I(x)}{I^\circ P^{cat}_{O_2}(x)}\right) - \frac{I(x)t_m}{\sigma_m(x)} \quad (32.6)$$

where F, R, I^0, $\sigma_m(x)$ and $P^{cat}_{O_2}(x)$ are respectively, the Faraday's constant, gas constant, exchange current density for the oxygen reaction, membrane conductivity and oxygen partial pressure at the catalyst interface of the GDL.

The average current density is given by:

$$I_{avg} = \frac{1}{L}\int_0^L I(x)\,dx \quad (32.7)$$

The standard deviation of the local current density can be calculated as:

$$Sd = \sqrt{\frac{\sum_{i=1}^n (I(x_i) - I_{avg})^2}{n+1}} \quad (32.8)$$

The method used to solve the model is detailed in [3, 4]. The input parameters and properties used in computations are taken from [4].

The main parameters that affect the performance of the PEMFC are the operating parameters at the inlet of the PEMFC. Determining the optimal values of these latter requires a combination of the mathematical model and optimization algorithms.

32.3 Objective Functions and Optimization Algorithms

In this work, four optimization cases are considered, depending on the desired output target of the system. Figure 32.2 shows these cases with their objective functions. It is obvious that the average current density is one of the parameters characterizing the electrical power delivered by the PEMFC, while the standard deviation of the local current density is also a key indicator of the durability of the PEMFC system. In addition, nine operating parameters of the PEMFC are selected as decision variables, their upper and lower limit values are given in Table 32.1. The selected algorithms parameters considered in this work are listed in Table 32.2.

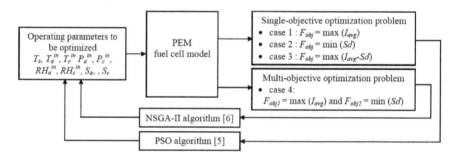

Fig. 32.2 Case studies of optimization parameters of PFMFC

Table 32.1 Upper and lower limits values of decision parameters

Parameter	Lower value	Upper value
Cell temperature, T_s, (°C)	60	80
Inlet temperature of anode and cathode gas, T_a^{in}, T_c^{in}, (°C)	60	80
Inlet pressure of anode and cathode, P_a^{in}, P_c^{in}, (atm)	1.5	3
Inlet anode and cathode humidity, RH_a^{in}, RH_c^{in}, (%)	50	100
Anode stoichiometry, S_a, (–)	1.2	1.8
Cathode stoichiometry, S_c, (–)	1.8	2.5

Table 32.2 Parameter setting for PSO and NSGA-II

Parameter	PSO algorithm	NSGA-II algorithm
Population size	400	400
Maximum number of iterations	200	1000
Crossover probability	–	80%
Mutation probability	–	10%

32.4 Optimization Result and Discussion

The optimal solutions obtained by PSO algorithm for single-objective optimization problem of the three cases are summarized in Table 32.3. Figures 32.3a, b show, respectively, the variation of liquid water saturation at the cathode and the distribution of local current density along the flow channel. It is found that the greater the liquid water saturation, the non uniform the distribution of the locale current. This explains the increase in the dispersion of the current density along the channel from its mean value. By Comparing the compromise solution of the case 3 and the case 1; we shows that the average current density of the cell for case 3 decreases by 12.7% and the standard deviation decreases by 53.55%. A slight decrease in delivered power density is obtained, but a considerable improvement in the durability of the PEMFC.

The result of the optimal Pareto front is shown in Fig. 32.4. The point (A1) is an optimal result corresponding to the case where the average current density is supposed to be the only objective function. Similarly, the point (A2) is an optimal result where the standard deviation of the current density is considered as the only

Table 32.3 Optimal values of decision variables and objectives functions

Parameters	Case 1	Case 2	Case 3
T_s (°C)	80	80	80
T_a^{in} (°C)	80	64.45	80
T_c^{in} (°C)	60	80	67.89
P_a^{in} (atm)	1.5	1.5	1.5
P_c^{in} (atm)	3	1.5	3
RH_a^{in} (%)	100	50	68.87
RH_c^{in} (%)	62.16	100	100
S_a	1.8	1.2	1.8
S_c	2.5	2.5	2.5
Fobj.			
I_{avg} (A/cm²)	1.2983	0.7468	1.1333
Sd (A/cm²)	0.4476	0.0197	0.2079
Power (W/cm²)	778.98 × 10^{-3}	448.08 × 10^{-3}	679.98 × 10^{-3}

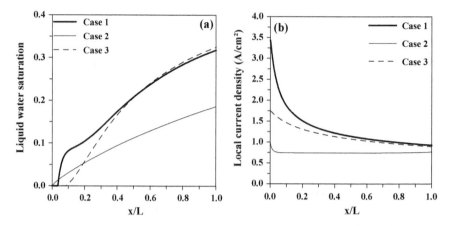

Fig. 32.3 Variation along the flow channel of: **a** liquid water saturation; **b** current density

Fig. 32.4 Distribution of Pareto-optimal solutions using NSGA-II

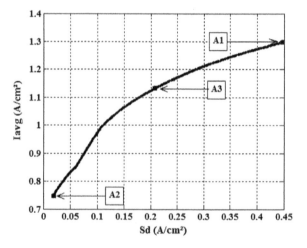

objective function. While the point (A3), is an optimal result of compromise between current density and standard deviation. These solutions correspond exactly to those found by the PSO algorithm. Also, we note that for the multi-objective optimization, the final result is not represented by a single solution, but by a set of compromise solutions that allow the user to make the appropriate choice for its use.

32.5 Conclusion

Simulation results suggest that the two approaches proposed for the optimization of operating parameters of the PEMFC are useful to solve the single and multi-objective problems. A good distribution of solutions on the Pareto front can be observed. They indeed make it possible to maximize the power by increasing the average values of current density while at the same time making it possible to improve the life of the cell by minimizing the variations of this same current density along the cell. Several compromises solutions are proposed in order to allow the user to make the appropriate choice to its application.

References

1. J. Antonio Salva, A. Iranzo, F. Rosa, E. Tapia, E. Lopez, F. Isorna, Optimization of a PEM fuel cell operating conditions: obtaining the maximum performance polarization curve. Int. J. Hyd. Energy **41**(43), 19713–19723 (2016)
2. X. Chen, W. Li, G. Gong, Z. Wan, Z. Tu, Parametric analysis and optimization of PEMFC system for maximum power and efficiency using MOEA/D. Appl. Therm. Eng. **121**, 400–409 (2017)
3. G. Karimi, A. Jamekhorshid, Z. Azimifar, X. Li, Along-channel flooding prediction of polymer electrolyte membrane fuel cells. Int. J. Energy Res. **35**, 883–896 (2011)
4. H. Abdi, N. Ait Messaoudene, L. Kolsi, M. Belazzoug, Multi-objective optimization of operating parameters of a PEM fuel cell under flooding conditions using the non-dominated sorting genetic algorithm (NSGA-II). Therm. Sci. (2018). https://doi.org/10.2298/TSCI180211144A
5. M. Clerc, J. Kennedy, The particle swarm-explosion, stability, and convergence in a multidimensional complex space. IEEE Trans. Evolution. Comput. **6**(1), 58–73 (2002)
6. K. Deb, A fast and elitist multi-objective genetic algorithm: NSGA-II. IEEE Trans. Evol. Comput. **6**(2), 182–197 (2002)

Chapter 33
Optimal Design of Energy Storage System Using Different Battery Technologies for FCEV Applications

B. Bendjedia, N. Rizoug, and M. Boukhnifer

Abstract The aim of this paper focuses on optimal sizing of Energy Storage System (ESS) based on Fuel cell using different battery technologies for Fuel Cell Electric Vehicles (FCEV) applications. The main objective consists to optimize ESS sizes, cost with a longer lifecycle for a Fuel Cell Electric Vehicle with 650 km driving cycle range. In this way, we study the influence of battery technologies which is considered as a secondary source on the storage system characteristics. The obtained results show that considerable gains have been achieved in terms of weight, volume and cost using Ultra High Power (UHP) technology comparatively with high power and high energy technologies. Otherwise, an experimental validation is carried out to complement the simulation results via a 1 kW test bench with Hardware In the Loop approach. As expected, UHP technology confirms its capabilities of decreasing the applied stress on the fuel cell and consequently reduces the hydrogen consumption.

Keywords Batteries · Fuel cell · FCEV · Sizing

33.1 Introduction

A recommended structure composed of batteries and fuel cell is used in electric vehicles. Batteries ensure the power peaks and the energetic component (fuel cell) is used as an energy reservoir to insure the vehicle autonomy [1, 2]. Actually, the optimization of the ESS sizes remains a real challenge for electric vehicle applications. It is needed to increase as possible the storage system lifetime with minimal sizes. An optimal sizing of ESS deals with a multi-objective optimization considering the

---

B. Bendjedia (✉)
LACoSERE Laboratory, University of Laghouat, Laghouat, Algeria
e-mail: b.bendjedia@lagh-univ.dz

N. Rizoug
ESTACA'LAB, ESTACA, Laval, France

M. Boukhnifer
LCOMS, Université de Lorraine, Metz, France

© The Editor(s) (if applicable) and The Author(s), under exclusive license to Springer Nature Singapore Pte Ltd. 2021
A. Khellaf (ed.), *Advances in Renewable Hydrogen and Other Sustainable Energy Carriers*, Springer Proceedings in Energy,
https://doi.org/10.1007/978-981-15-6595-3_33

cost, weight and volume with taking into account the fuel consumption. Subsequent paragraphs, however, are indented.

In this paper, we interest to the sizing of fuel cell system and batteries pack for a Fuel Cell Electric Vehicle with 650 km driving cycle range. The main contribution is studying the influence of the battery technologies on ESS sizes, costs and hydrogen economy by considering four different technologies. Moreover, an experimental validation under 1 kW test bench is carried out. An evaluation of the impact of the used technology on the fuel consumption and life cycle of the whole system is investigated to show the obtained gains using Ultra High Power technology.

33.2 Hybrid Fuel Cell/Battery Description

The hybrid power sources fuel cell/battery combine high energy density of fuel cells with the high power density of batteries. The considered structure in this study is shown in Fig. 33.1. Fuel cell is primary source and batteries pack is the secondary one.

33.2.1 Technical Specifications

Driving Cycle (mission)
The ARTEMIS cycle is a profile selected to size the ESS. Two cycles (urban and road) as shown in Fig. 33.2, were established to describe the course and reproduce

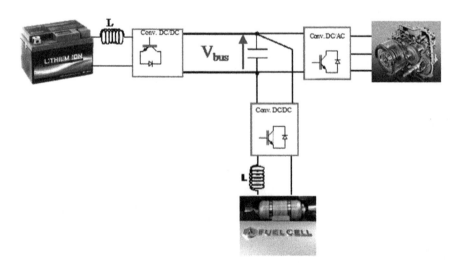

Fig. 33.1 Hybrid source scheme

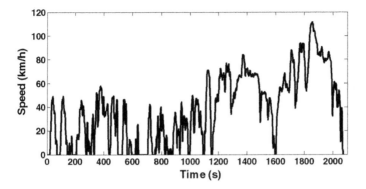

Fig. 33.2 Artemis cycle profile

Table 33.1 FCEV characteristics [4]

Parameters	Values
Vehicle mass (M)	860 kg
Frontal area (S)	2.75 m^2
FC cell power	300 W
Volumetric, wight power density	3.1 kW/L, 2 kW/kg

the real conditions of use of the vehicle. The combination of urban and road cycles can simulate a rolling distance of 22 km over a period of about 34 min [3]. Then, in our case, the Artemis cycle with an average slope of 2.5% is used. In the objective to achieve a distance of 650 km, the Artemis driving cycle is repeated 30 times.

Vehicle and Fuel cell characteristics
To estimate the power and energy needed for the proposed profile, a dynamic model of the vehicle is used under Matlab-Simulink software. The parameters of the vehicle and fuel cell system are shown in the Table 33.1.

33.2.2 Energy Management Strategy

It is needed to split the power between two or more sources, for this reason an energy management strategy can provides the proper power reference of each source. One of the simple and efficient strategies is the filtering management strategy based on a filter to separate high-frequency components to be sent to the storage system and the residuals are dedicated to the main source. This strategy is widely used to manage the hybrid sources [2, 5], for sizing [6], and other strategies performances evaluation.

33.3 Sizing Results of the Hybrid Source

In this section, we present sizing results already published in [7]. They are obtained using different battery technologies (see Table 33.2).

From the obtained results shown in Fig. 33.3, it is confirmed that UHP technology brings best performances in terms of cost, weight and hydrogen consumption.

Otherwise, another interesting part concerns the study of the hybrid source life time. It is needed to evaluate the used technologies in terms of applied stress on the hybrid source components. The next section is dedicated to test the applied stress on the two sources via an experimental validation.

Table 33.2 Characteristics of used battery technologies [8, 9]

Parameters	MHI	7.2HP	13HP	UHP
Nominal capacity Ccel_bat (Ah)	20	7.2	13	0.45
Voltage U_cel_bat (V)	3.7	3.7	3.7	3.7
Max charge/discharge current I_cel_Disc (A)	100/100	14.4/21.6	39/104	22.5/22.5
Weight Wcel_bat (kg)	0.85	0.22	0.325	0.014
Volume Vcel_bat (L)	0.42	0.12	0.174	0.0073
Cost (Euros)	/	/	30.36	4.6

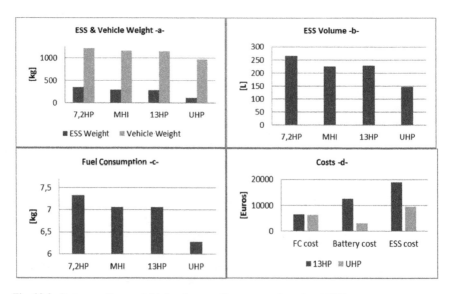

Fig. 33.3 Sizing results **a** weight; **b** volume; **c** H_2 consumption; **d** cost [10]

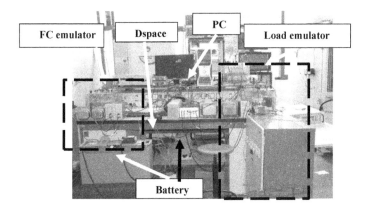

Fig. 33.4 1 kW experimental test bench

33.4 Experimental Validation

An experimental test bench (Fig. 33.4) is used to evaluate the performance of the hybrid source using different battery technologies and to confirm the simulation results.

The maximum output power of the FC emulator is adjusted to 500 W. For technical and commercial reasons related to the purchase of the used battery cells, a hardware in the loop approach as close to the real operating condition is used to emulate the technology behaviour.

33.5 Experimental Results

The evaluation of the used technologies is to show their impact on the fuel consumption and lifespan of the whole system. It was concluded that the current (or power) dynamics have the major influence. Then the Root Mean Square (RMS) power (image of dynamic current) and absolute energy are used in this work to know how often battery pack and fuel cell are solicited [11].

Figure 33.5 shows the different needed evaluation parameters as RMS powers, absolute energy and the fuel consumption respectively.

It is shown in Fig. 33.5a, b that UHP technology possesses the lowest applied stress on the batteries which is expected, regarding its high charging/discharging power rates.

From these two figures, it is noted that a considerable gains are obtained in term of lifetime improvement compared to other technologies. From the Fig. 33.5c, it is observed that MHI and 13 HP provide also the same applied stress on the fuel cell and practically the same fuel consumption (see Fig. 33.5d). However, the 7.2 HP has a higher fuel cell RMS power which leads to the highest fuel consumption. As

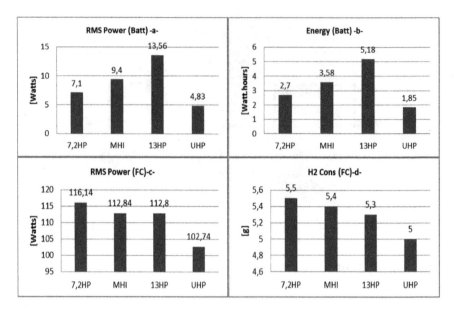

Fig. 33.5 Experimental results with different battery cell's technologies **a** RMS battery power; **b** absolute battery energy; **c** RMS fuel cell power; **d** H_2 consumption of fuel cell

Fig. 33.6 Synthesis of the performances and the applied stresses of the hybrid source using the four technologies

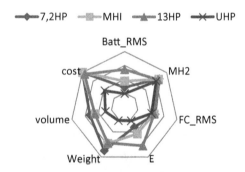

expected, the UHP technology offers the lowest applied stress on the fuel cell which leads to the lowest fuel consumption. The obtained gain can be evaluated to 10% in terms of fuel consumption. As shown in Fig. 33.6, UHP offer the best performances and characteristics compared with other technologies.

33.6 Conclusion

This paper focuses on optimal sizing of an Energy Storage System composed by fuel cell system and batteries. The main objective is to improve the weight, cost, lifespan, and fuel consumption of the hybrid source. From the simulation results, UHP technology guarantees the best performances compared to other technologies.

A complementary experimental part is presented to complete the evaluation of the different ESS performances in term of applied stress on the hybrid source components. As expected, the UHP technology cells contribute to decreasing the applied stress on the fuel cell system and consequently reduce the hydrogen consumption. The obtained results confirm that this technology is the most suitable to use for such ap-plication regarding the improved lifetime and ESS characteristics.

References

1. M.A. Hannan, F.A. Azidin, A. Mohamed, Hybrid electric vehicles and their challenges: A review. Renew. Sustain. Energy Rev. **29**, 135–150 (2014)
2. T. Azib, C. Larouci, Chaibet, M. Boukhnifer, Online energy management strate-gy of a hybrid fuel cell/battery/ultracapacitor vehicular power system. IEEJ Trans. Electr. Electron. Eng. **9**(5), 548–554 (2014)
3. T. Mesbahi, N. Rizoug, P. Bartholomeus, P. Le Moigne, A new energy management strategy of a battery/supercapacitor hybrid energy storage system for electric vehicular applications. Power Electron. *7th IET International Conference on Power Electronics, Machines and Drives (PEMD 2014)*, vol. 1, (7), pp. 8–10 (2014)
4. F.C. Stack information, *Mirai Product Information 2016 Mirai Product Information*, pp. 1–3 (2016)
5. H. Alloui, K. Marouani, M. Becherif, M.N. Sid, M.E.H. Benbouzid, A control strategy scheme for fuel cell-vehicle based on frequency separation, in *2014 1st International Conference on Green Energy*, (ICGE 2014), pp. 170–175 (2014)
6. A.. Hammani, R. Sadoun, N. Rizoug, P. Bartholomeüs, B. Barbedette, P. Le Moigne, Influence of the management strategies on the sizing of hybrid supply composed with battery and supercapacitor, in *2012 1st International Conference on Renewable En-ergies and Vehicular Technology*, REVET 2012, pp. 1–7 (2012)
7. S. Caux, W. Hankache, M. Fadel, D. Hissel, On-line fuzzy energy management for hybrid fuel cell systems. Int. J. Hydrogen Energy **35**(5), 2134–2143 (2010)
8. Kokam Cell Types, 2016, www.kokam.com/cell/. Last Accessed 21 Jan 2020
9. D. Mukai, K. Kobayashi, T. Kurahashi, N. Matsueda, Development of Large High-performance Lithium-ion Batteries for Power Storage and Industrial Use. Mitsubishi Heavy Ind. Tech. Rev. **49**(1), 6–11 (2012)
10. Bendjedia, B., Rizug, N., Bboukhnifer, M.: Influence of Battery Technologies on Sizing of Hybrid Fuel Cell Sources for Automotive Applications, in *2018 International Conference on Applied Smart Systems,*ICASS 2018, pp. 1–7 (2019)
11. F. Herb, P. Akula, R. Trivedi, K.L. Jandhyala, A. Narayana, M. Wöhr, Theoretical analysis of energy management strategies for fuel cell electric vehicle with respect to fuel cell and battery aging. pp. 1–9 (2013)

Chapter 34
Simultaneous Removal of Organic Load and Hydrogen Gas Production Using Electrodeposits Cathodes in MEC

Amit Kumar Chaurasia and Prasenjit Mondal

Abstract In this study, Microbial electrolysis cells (MECs) are used for the treatment of actual paper and pulp industry wastewater and production of biohydrogen, for the evaluation of economical and low-cost cathodes. Nickel, Nickel-Cobalt and Nickel-Cobalt-Phosphate electroplating on SS and Cu rod are explored as-fabricated cathode catalysts in MECs for the estimation of their electrocatalytic activity in terms of energy recovery and treatment efficiency using paper-pulp industry real wastewater. Developed cathodes are explore in MECs in a controlled temperature of 30 ± 2 °C, under applied voltage of 0.6 V at neutral pH with paper-pulp industry wastewater and activated sludge as inoculum. Simultaneously treatment of paper pulp industry wastewater with hydrogen production rates (1.1–3.82 m3/m3-d) and 54–65% of initial COD removal. Obtained results suggest that fabricated cathodes have potential to become alternative to Pt. for the treatment of real industrial wastewater using MECs. It also indicates that the hydrogen production and MECs performance greatly depends on the composition of wastewater and inoculum. MECs performances ware evaluated in terms of hydrogen production rate, columbic efficiencies, overall energy efficiency, and volumetric current density.

Keywords Electrodeposited cathode · Cathodic catalysts · Paper-pulp industry wastewater · Electrocatalytic activity · MECs energy recovery

34.1 Introduction

Microbial electrolysis cells (MECs) are extended versions of microbial fuel cells technologies that recover the energy from organics present in waste material using electrochemically active bacteria. MFCs use the action of exoelectrogenic microbes to

---

A. K. Chaurasia · P. Mondal (✉)
Department of Chemical Engineering, Indian Institute of Technology, Roorkee 247667, India
e-mail: pmondfch@iitr.ac.in

A. K. Chaurasia
e-mail: akcnitj@gmail.com

© The Editor(s) (if applicable) and The Authors(s), under exclusive license to Springer Nature Singapore Pte Ltd. 2021
A. Khellaf (ed.), *Advances in Renewable Hydrogen and Other Sustainable Energy Carriers*, Springer Proceedings in Energy,
https://doi.org/10.1007/978-981-15-6595-3_34

oxidize organic compounds present in the actual wastewater and generate electricity over the transfer of electrons to the anode, while in MECs the electrons produced by exoelectrogens are diffuse at cathode for the production of H_2 from H^+ [1]. MECs is getting much attention in last decade due to it is a promising approach for the generation of renewable energy and valuable products from almost all types of waste materials. MECs combined the removal of organic material with higher hydrogen production (>60% energy recovery) at ambient atmospheric conditions but recovery of hydrogen/energy needs to be improved [2]. In MECs, cathode catalysts for proton reduction to hydrogen at cathode has a significant role in decreasing energy input and increasing hydrogen production or energy recovery [3]. Most of the literature reported that Pt. as standard cathode material utilized in MECs due to low hydrogen-over potential but noble metal platinum cannot be used for extensive practical application of MECs due poisoning by sulphides and high cost [4]. Economical application of MECs required development of alternative cathodes materials, which have low hydrogen-over potential, low cost, more active surface area, good electrocatalytic activity, and reuse capability. Many research group reported that first-row transition metals, stainless steel, nickel alloys, Fe, Zr, Mo, TiO_2 nanocrystals, titanium–nickel and electrodeposits of graphite felt-nickel, Ni–Ti, Ni–Fe–Co, Ni–Fe–Co–P, Ni–Mo, Ni–Fe and Ni–W are promising cathodes catalyst due to their low overpotential and improved electro-catalytic activity [5]. Recent literature reported that the electrode-plating/addition or grouping of two or more transitions metal improved the cathode catalyzed reaction in MECs, it reduce the overpotential losses of electrode, mechanical stability with high catalytic activity by combined influence of their catalytic properties.

Most of the literature using these cathodes are investigate in synthetic wastewater using a single substrate. There are very few literature reported that the effectiveness of these alternative cathode catalyst using actual or real industrial wastewater. Among these studies, some tests were done using industrial and domestic wastewaters such as winery, potato, dairy wastewater. The most probable reason for these fabricated cathodes has not been employed in actual wastewater is due to the complex nature of actual wastewater, inorganic constituents of actual wastewater can greatly affects the cathodes electrocatalytic activity but also leads to the catalyst poisoning due to irreversible absorbance of cofactor/co-enzyme or metals ions at catalyst active site.

In this study, electrodeposits of Ni, Ni–Co, and Ni–Co–P on SS and Cu were selected to evaluate hydrogen production and energy recovery of the paper-pulp industry actual wastewater in MECs. The aim of current study was to evaluate the developed cathodes in actual wastewater and viability of practical MECs operation. The paper-pulp industry wastewater was chosen because it generates large quantity of wastewaters during processing of agricultural raw materials and its contained high COD (100 g/L), BOD (35–40 g/L) and dyes (color) with biodegradability index <0.4. The fabricated cathodes have been explored as a cathode in MEC with anode (graphite rod) under controlled temperature of 30 ± 2 °C at input voltage of 0.6 V.

Table 34.1 Physicochemical characteristics of paper and pulp industry actual wastewater

Parameters	Values (actual wastewater)
COD (g/L)	5.88
Soluble COD (g/L)	5.18
BOD5 (g/L)	1.8
pH	6.72
TS (g/L)	1.44
TDS	15.84
TSS	0.08

34.2 Materials and Methods

Paper and pulp industry wastewater samples were collected from Star paper Mills Saharanpur, India and wastewater characteristics are shown in Table 34.1. Samples were collected at on-site from the primary clarifier of the industry and placed in ice after that stores at 4 °C overnight at processing laboratory. Bacterial source used in this study as mixed culture were taken from Municipal sewerage treatment plant Haridwar, India. The eight different electrodeposited cathodes used in this study are the same cathodes, which are described, in our previous published study as Chaurasia et al. [5]. Fabricated cathodes represented in this manuscript as SS2 (Ni deposit), SS3 (Ni–Co deposit) and SS4 (Ni–Co–P deposit) and similarly Cu_2, Cu_3 and Cu_4. The hydrogen gas produced in MECs was measured on a daily basis using water displacement techniques, and collected in Tedlar gas sampling bags via silicon tube. It qualitatively analysed by Gas chromatography (GC NEWCHROME 6800) and more than 96% hydrogen gas fraction was detects in GC. Each MECs batch cycles consist of 7 days and all the calculation with analysis were done as described in our previous study [5].

34.3 Reactor Configuration and Operation

Eight double chamber MEC (10 cm × 10 cm × 10 cm) with 800 ml working volume were made by using Plexiglas material consisting of graphite rod as an anode and electrodeposited electrode as the cathode at optimum electrode spacing (4 cm). All the experimentation in this study were carried out as in similar pattern with same working procedure of our previous study [5]. All the MECs experimental setup fed with 700 ml actual wastewater with 100 ml concentrated activated sludge as incolumn in controlled temperature 30 ± 2 °C at the applied voltage of 0.6 V. To inhibits the methanogens and maximize the hydrogen production by adding 10 ml of 100 μM of BES (2-Bromo ethane sulfonate) in each MECs reactor. Two times, in a day sample was analyse for COD and further chemical analysis and volume of produced gas was quantify by water displacement techniques on daily basis.

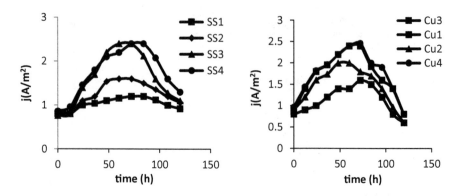

Fig. 34.1 Current density profile of developed cathode of paper and pulp industry actual wastewater in MECs **a** electrodeposits on SS and **b** electrodeposits on copper

34.4 Results and Discussion

34.4.1 Current Generation

The current generation from different electrodeposited cathodes across the MEC reactors are shown in Fig. 34.1. Current density behaviour indicate that the electrodeposited cathodes give higher current rates than without electrodeposited cathodes such as SS1 and Cu_1 and it's primarily due to the effects of electrodeposits and activity of exoelectrogens. The SS4 cathode achieve highest current density of 2.45 A/m^2 and Cu_4 has 2.39 A/m^2 in comparison of bare cathodes using actual wastewater. The rate of hydrogen production predominantly depends on the availability of electrons in the cathode (current density) and the current flows in the circuit trigger by microbial activity. As the growth of microorganism, increase simultaneously increases the current density. The current density profile on actual wastewater has slight variation to our previous study [5] on synthetic wastewater (single substrate acetate). The development of the current density profile also suggest that fast oxidation of simple organics then complex organic compound present in actual wastewater. This clearly revel that MECs performance has less impacts of substrate/nutrient medium. No hydrogen gas production has were detected when the current tends to zero, which demonstrates that, the current and gas produced in the MEC reactor is due to the microbial activity and their metabolism.

34.4.2 COD Removal

The percentage COD removals for different cathodes are ranging from 54 to 65% as shown in Fig. 34.2b with respect to initial COD of wastewater (5.88 g/L) and it

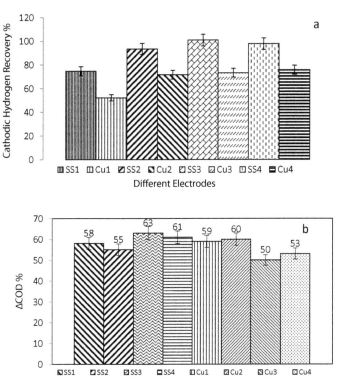

Fig. 34.2 Cathodic Hydrogen recovery of different cathodes on actual wastewater (**a**), Percentage COD removal of actual wastewater with different electrodeposited cathodes (**b**)

indicates COD removal performances depends on cathode catalyst as well as nature of the industrial wastewater. The SS3 cathodes containing MECs shows 64 ± 2% COD removal, other hand SS4 has 60 ± 2% and (61 ± 2%) in case of Cu_2. Lowest COD removal detected in the case of Cu_4 (54 ± 2%) cathode containing MEC reactor. The high COD removals achieved with SS3 and Cu_2 cathodes indicated the paper and pulp industry real wastewater are highly biodegradable.

34.4.3 Energy Analysis

The columbic efficiency is used to express the energy yield compared with the energy input of the MEC reactor. Columbic efficiency of the electrodeposited cathode reflects the trends observed in hydrogen production are shown in Fig. 34.2a, and estimated columbic efficiency is given in Table 34.2. The electrodeposited on SS cathodes

Table 34.2 Comparisons of MECs performance on actual wastewater

Developed electrode/cathode	Q (m^3 H$_2$ m^{-3}d^{-1})	V_{H2} (ml)	C_E (%)	η_E(%)	Y_{H2} (n$_{H2}$/ns)
SS1	1.4 ± 0.1	96 ± 2	77.5 ± 2	77.3 ± 4	1.12 ± 0.1
SS2	2.1 ± 0.4	126 ± 2	88.3 ± 2	92.5 ± 4	1.42 ± 0.1
SS3	2.8 ± 0.5	164 ± 2	93.1 ± 2	97.6 ± 4	1.82 ± 2
SS4	3.8 ± 1	212 ± 2	94.9 ± 2	81.21 ± 4	2.88 ± 2
Cu$_1$	1.1 ± 0.1	84 ± 2	53.2 ± 2	52.36 ± 4	1.06 ± 0.1
Cu$_2$	1.98 ± 0.4	121 ± 2	69.6 ± 2	61.45 ± 4	1.32 ± 0.1
Cu$_3$	2.6 ± 0.4	153 ± 2	72.2 ± 2	71.25 ± 4	1.47 ± 0.1
Cu$_4$	3.87 ± 1	257 ± 2	76.7 ± 2	74.52 ± 4	2.43 ± 0.2

improved the cathodic efficiency but it is insignificant in the case of Cu electrodeposited cathode. It is occur due to the fixed value of the applied voltage of 0.6 V for all the cathodes.

Obtained results show that COD removal and energy recoveries highly depend on cathodes catalyst as well as the nature of constituents present in actual wastewater; it also indicates that the electrodeposition has positive effects on COD removal and energy recoveries. The energy efficiency is the ratio of hydrogen produced to electrical energy input. The energy efficiency (η_E) of the MEC with different fabricated cathodes are summarized in Table 34.2. Estimated results suggest that a significant amount of energy was harvested from the substrate, not from electricity in the case of electrodeposited cathodes.

All the calculations are done in this study as described in our previous study [5] and represented in Table 34.2. Obtained results suggest that electrodeposited cathodes show stable and enhanced performance on actual wastewater. Among all the developed cathodes SS4 cathodes show the maximum hydrogen production and energy recovery as compared to bare cathodes (SS1 and Cu$_4$).

34.5 Conclusion

MECs parameter determined from the experimental results are suggested that the produced electrodeposited cathodes have increased the hydrogen production performances from two to four-fold than palladium and Pt. catalyst using actual paper-pulp industry wastewater. It is also observed that the electrodeposited cathodes enhanced the cathode performance towards HER on actual wastewater. It reveals that a combination of two or three non-noble metal cathode catalyst have potential to replace the Pt. for the commercial application of MEC as well as in real wastewater.

References

1. A. Tenca, R.D. Cusick, A. Schievano, R. Oberti, B.E. Logan, Evaluation of low cost cathode materials for treatment of industrial and food processing wastewater using microbial electrolysis cells. Int. J. Hydrogen Energy **38**(4), 1859–1865 (2013)
2. A. Kadier, P. Abdeshahian, Y. Simayi, M. Ismail, A.A. Hamid, M.S. Kalil, Grey relational analysis for comparative assessment of different cathode materials in microbial electrolysis cells. Energy **90**(2), 1556–1562 (2015)
3. P. Mondal, P. Kumari, J. Singh, S. Verma, A.K. Chaurasia, R.P. Singh, Oil from algae, in *Sustainable Utilization of Natural Resources* (2017)
4. M. Rani, U. Shanker, A.K. Chaurasia, Catalytic potential of laccase immobilized on transition metal oxides nanomaterials: degradation of alizarin red S dye. J. Environ. Chem. Eng. **5**(3), 2730–2739 (2017)
5. A.K. Chaurasia, H. Goyal, P. Mondal, Hydrogen gas production with Ni, Ni–Co and Ni–Co–P electrodeposits as potential cathode catalyst by microbial electrolysis cells. Int. J. Hydrogen Energy (2019). (In Press)

Chapter 35
An Improved Model for Fault Tolerant Control of a Flooding and Drying Phenomena in the Proton Exchange Membrane Fuel Cell

A. A. Smadi, F. Khoucha, A. Benrabah, and M. Benbouzid

Abstract The proton exchange membrane fuel cell (PEMFC), an electrochemical device for converting the chemical energy into electrical energy and heat, is currently considered one of the most promising systems for development of renewable and non-polluting energies. Water management in the PEMFC remains one of the major obstacles to be solved for the commercialization of this technique on a large scale. Flooding and drying are the two main degradation mechanisms that occur when water management is inadequate, and have a direct impact on the resistance and the fuel cell (FC) active area. Therefore, since the impact of these faults is known, the output fuel cell voltage is also predictable. This paper presents an improved fuel cell model that reproduce the output fuel cell voltage behavior in event of water management fault. A future solution is also proposed for fault tolerant system by using a new DC-DC converter with a high gain. Finally, model results are verified using MATLAB Simulink.

Keywords Water management · FT control · PEMFC model · Flooding and drying

35.1 Introduction

For many years, fossil fuels such as coal, oil and natural gas have been used as the main energy sources. Due to the growth of the world's population, energy demand is increasing every year. On the other hand, the use of fossil fuels has a strong impact on the environment due to their production and CO_2 emissions. These findings lead us to seek new energy resources that are renewable, non-polluting and can replace

---

A. A. Smadi (✉) · F. Khoucha · A. Benrabah
Ecole Militaire Polytechnique, UER ELT, Algiers, Algeria
e-mail: smadi.ahmed.abdelhak@gmail.com

F. Khoucha
e-mail: fkhoucha04@yahoo.fr

F. Khoucha · M. Benbouzid
University of Brest, UMR CNRS 6027 IRDL, 29238 Brest, France

© The Editor(s) (if applicable) and The Author(s), under exclusive license to Springer Nature Singapore Pte Ltd. 2021
A. Khellaf (ed.), *Advances in Renewable Hydrogen and Other Sustainable Energy Carriers*, Springer Proceedings in Energy,
https://doi.org/10.1007/978-981-15-6595-3_35

fossil fuels. Fuel cell technologies therefore appear to be among the best solutions and have received a lot of attention in recent years. However, many problems are facing development and commercialization of this technology on a large scale.

Degradations due to poor water management are important. Mismanagement of water leads to FC premature aging by physical degradation. Flooding and drying are the two main degradation mechanisms that occur when water management is inadequate [1, 2].

Flooding is a serious problem for PEMFC operation. It is defined as an accumulation of liquid water that can occur at the anode and/or the cathode of the FC. However, flooding occurs preferentially at the cathode because it is the place where water is produced [3]. Anode flooding is less common but can occur at low hydrogen flow rates, low current densities, and low temperatures [4, 5]. At both electrodes, flooding can be accelerated by saturated or over-humidified reactive gases [3]. This phenomenon can amount to a complete blockage of the gases passage. The transition from a "healthy" FC to a "flooded" FC can be done in minutes [6]. Flooding can be effectively avoided by adjusting gas flow or cell temperature [7].

Membrane drying is not instantaneous. The drying cell voltage takes about 15 min to go from 0.65 to 0.55 V [8], but the return to higher humidity levels is much faster with about 4 min. As with flooding, drying does not necessarily affect all cells in the stack. When drying, the cells in the stack center are the most affected because of a non-uniform and higher stack temperature in the center [9, 10]. This defect can be corrected by adjustments to the operating parameters. For example, by reducing cathodic stoichiometry to reduce water evacuation, or by increasing the current density to produce more water, or by lowering the temperature of the cell [11].

The damage described above is inevitable. They are part of the PEMFC aging. This aging can be accelerated and damage accentuated due to poor operating conditions. To optimize the life of PEMFC, these faults must be avoided. When faults appear, they must be detected and corrected before failure appear [12]. Fault tolerant control is considered a step in the implementation of recovery strategy, error management.

The main objective of this paper is to provide fuel cell model to reproduce the PEMFC behavior in event of fault. This model will allow us to apply fault tolerant control strategies in the future.

35.2 Fuel Cell System Model

Many works are proposed on the FC modeling. These models are classified in different categories: theoretical, empirical and semi-physical. Among the different existing modeling approaches, we are interested in a semi-physical model, generally used for electric vehicle applications, that provides advantages for analysis, control and predict the system behavior. It can also be easily linked to DC/DC converter models in order to develop control laws.

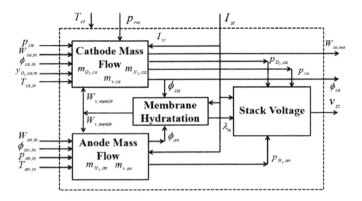

Fig. 35.1 Fuel cell stack synoptic diagram

The FC system model is based on thermodynamic and mechanical relationships. Generally, the fuel cell stack model is made up of four subsystems Fig. 35.1: stack voltage model, the anode flow model, the cathode flow model, and the membrane hydration model. In this work, only cell voltage model will be presented, in order to capture the effect of the faults on the output fuel cell voltage. While the other subsystems models can be reviewed in many references in the literature [13, 15, 16].

35.2.1 Stack Voltage Model

The stack voltage is calculated as a function of stack current, cathode pressure, reactant partial pressures, stack temperature and membrane humidity. The output voltage of a single cell can be given by [13–15]

$$v_{fc} = E - v_{act} - v_{ohm} - v_{conc} \qquad (35.1)$$

where E is the open circuit voltage, v_{act} is the activation overpotential, caused by the activation of the anode and the cathode, v_{ohm} is the ohmic overpotential resulted from the resistance of the polymer membrane to the protons transfer and the resistance of the electrode and the collector plate to the electrons transfer, v_{conc} is the concentration overpotential resulted from the drop in concentration of the reactants as they are consumed in the reaction. v_{fc} represents the voltage of a single fuel cell, the total voltage of the stack v_{st} is calculated by multiplying the v_{fc} and the total number of cells n, and the current density i is defined as cell current I_{st} per cell active area A_{fc}

$$\begin{cases} v_{st} = n \times v_{fc} \\ i = \frac{I_{st}}{A_{fc}} \end{cases} \qquad (35.2)$$

E is the function of stack temperature T_{fc}, pressure of hydrogen p_{H_2} and pressure of oxygen p_{O_2} [13, 14]

$$E = 1.229 - 0.85 \times 10^{-3}(T_{fc}-298.15)$$
$$+ 4.3085 \times 10^{-5} T_{fc} \left[\ln\left(\frac{p_{H_2}}{1.01325}\right) + \frac{1}{2} \ln\left(\frac{p_{O_2}}{1.01325}\right) \right] \quad (35.3)$$

Activation voltage loss v_{act} is calculated as follows

$$v_{act} = v_0 + v_a\left(1 - e^{-c_1 i}\right) \quad (35.4)$$

where v_0 is the voltage drop at zero current density, v_a and c_1 are constants

$$v_0 = 0.279 - 0.85 \times 10^{-3}(T_{fc}-298.15)$$
$$+ 4.3085 \times 10^{-5} T_{fc} \left[\ln\left(\frac{p_{ca} - p_{sat}}{1.01325}\right) \right.$$
$$\left. + \frac{1}{2} \ln\left(\frac{0.1173(p_{ca} - p_{sat})}{1.01325}\right) \right] \quad (35.5)$$

$$v_a = \left(-1.618 \times 10^{-5} T_{fc} + 1.618 \times 10^{-2}\right)\left(\frac{p_{O_2}}{0.1173} + p_{sat}\right)^2$$
$$+ \left(1.8 \times 10^{-4} T_{fc} - 0.166\right)\left(\frac{p_{O_2}}{0.1173} + p_{sat}\right)$$
$$+ \left(-5.8 \times 10^{-4} T_{fc} + 0.5736\right) \quad (35.6)$$

Ohmic voltage loss v_{ohm} is calculated as follows

$$v_{ohm} = i \cdot R_{ohm} \quad (35.7)$$

The ohmic resistance is a function of membrane conduction σ_m in the following form

$$R_{ohm} = \frac{t_m}{\sigma_m} \quad (35.8)$$

where t_m is the thickness of membrane, σ_m is the membrane conductivity and calculated as follows

$$\sigma_m = b_1 \exp\left(b_2\left(\frac{1}{303} - \frac{1}{T_{fc}}\right)\right) \quad (35.9)$$

in which b_2 is constant [14], b_1 is function of water content of membrane λ_m

$$b_1 = (b_{11}\lambda_m - b_{12}) \quad (35.10)$$

The empirical values of b_{11} and b_{12} for the Nafion 117 membrane are given in [14]

Concentration voltage loss v_{conc} is calculated as follows

$$v_{conc} = i\left(c_2 \frac{i}{i_{max}}\right)^{c_3} \qquad (35.11)$$

where c_2, c_3 and i_{max} are constant [14].

The calculation of parameters in this Equations requires the knowledge of cathode pressure p_{ca}, oxygen partial pressure p_{O_2}, and fuel cell temperature T_{fc}. The pressures are calculated from the cathode model. The temperature can be determined based on the stack heat transfer model. The membrane conductivity that is calculated in the membrane hydration model. Therefore, these models can be consulted in the reference [13].

The above equations represent the Stack voltage model for a healthy fuel cell. Concerning the model that includes the water management problems, the accumulation of water in the cell blocks gas access at the catalytic sites, thus reducing the active zone available [15]. A flooding is simulated by the reduction of the cell active zone with:

$$i = \frac{I_{st}}{A_{fc} * K} \qquad (35.12)$$

with $K = 0.8$ the flood coefficient.

In case of insufficient humidification, the membrane starts to dry. This is indicated by a continuous increase in cell resistance [10]. Drying is simulated by increasing the membrane resistance R_{ohm} (35.8) to 1.5 compared to the rated value.

35.3 Simulation Results and Discussion

Several operating conditions influence the level of the FC hydration, such as Relative humidity of gas, gas flow rates, fuel cell temperature, outlet gas pressures, fuel cell geometry. And the current density. Figure 35.2 illustrates the effect of different pressures and temperatures on the fuel cell voltage. The ideal voltage is defined as the maximum voltage that each cell can produce at a given temperature and pressure. In general, to ensure proper operation of the FC, the operating parameters must remain within a relatively narrow operating range.

Flooding or drying can still occur due to uneven local conditions inside the cell. In this study, the conditions of flooding and drying are generated by applying the following assumptions: Too much water causes flooding, i.e. a blockage of the porous passages which in turn reduces the rate of transport of reagents to the catalyst site.

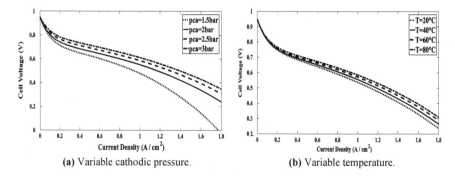

Fig. 35.2 Cell characteristics at different operating conditions

This phenomenon results in a lower active zone [15]. Too little water causes the membrane to dry, which in turn leads to an increase in resistance [10].

Figure 35.3 shows the different water content effects on PEMFC and associated degradation types. Graph (A) shows the cell polarization curve in healthy state. (B) illustrates the cell voltage of the same cell in the flooding state, this process causes a voltage drop, this voltage drop is interpreted as the result of an accumulation of liquid water in the gas diffusion layer (GDL), thus reducing the active zone available [15]. (C) displays the cell voltage in the drying condition, this process creates a net loss of water from the fuel cell. As a result, the cell membrane begins to dry over time. Finally, a voltage drop is observed over time.

During flooding or drying, fuel cell performance can be recovered by adjusting the operating parameters that affect the FC system conditions or by using a Fault-Tolerant Control (FTC) applied to the associated DC/DC converter, Fig. 35.4 shows the global diagram of this process. After evaluating the fuel cell performance and once the failure is detected, The FTC mechanism consists of reconfiguring the control signal applied to the converter in order to bring the fuel cell into optimal operation,

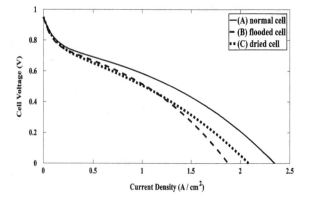

Fig. 35.3 Cell voltage according to the state of health

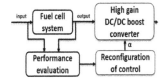

Fig. 35.4 Fault tolerant system global scheme

and adapt the output voltage to the load used. This solution makes it possible to avoid stopping the FC system.

35.4 Conclusion

Water management is one of the most complex and critical aspects of the successful operation of the PEMFC. In this paper, a model of a PEMFC to implement fault tolerant control strategies was introduced and its Simulation is implemented in the MATLAB/Simulink environment. the Simulation is done by varying the resistance and active area in the fuel cell model to obtain different cell voltage profiles, results showed the model ability to reproduce the PEMFC behavior in event of water management fault (flooding and drying). This model can therefore be used to implement fault tolerant control strategies by acting on the FC associated DC-DC power converter.

References

1. T. Ous, C. Arcoumanis, Degradation aspects of water formation and transport in proton exchange membrane fuel cell: a review. J. Power Sources **240**, 558–582 (2013)
2. M. Song, P. Pei, H. Zha, H. Xu, Water management of PEM fuel cell based on control of hydrogen pressure drop. J. Power Sources **267**, 655–663 (2014)
3. W. Schmittinger, A. Vahidi, A review of the main parameters influencing longterm performance and durability of PEM fuel cells. J. Power Sources **180**(1), 1–14 (2008)
4. S. Ge, C.Y. Wang, Liquid water formation and transport in the PEFC anode. J. Electrochem. Soc. **154**(10), B998–B1005 (2007)
5. K.J. Huang, S.J. Hwang, W.H. Lai, The influence of humidification and temperature differences between inlet gases on water transport through the membrane of a proton exchange membrane fuel cell. J. Power Sources **284**, 77–85 (2015)
6. M.A. Rubio, A. Urquia, S. Dormido, Diagnosis of performance degradation phenomena in PEM fuel cells. Int. J. Hydrogen Energy **35**(7), 2586–2590 (2010)
7. L. Boulon, K. Agbossou, D. Hissel, A. Hernandez, A. Bouscayrol, P. Sicard, M.C. Péra, Energy management of a fuel cell system: influence of the air supply control on the water issues, in *ISIE 2010*. Bari, Italy, pp. 161–166 (2010)
8. F. Barbir, H. Gorgun, X. Wang, Relationship between pressure drop and cell resistance as a diagnostic tool for PEM FC. J. Power Sources **141**(1), 96–101 (2005)
9. N. Yousfi-Steiner, P.H. Moçotéguy, D. Candusso, D. Hissel, A. Hernandez, A. Aslanides, A review on PEM voltage degradation associated with water management: impacts, influent factors and characterization. J. Power Sources **183**(1), 260–274 (2008)

10. Y. Park, J. Caton, Development of a PEM stack and performance analysis including the effects of water content in the membrane and cooling method. J. Power Sources **179**(2), 584–591 (2008)
11. X. Liu, H. Guo, C. Ma, Water flooding and two-phase flow in cathode channels of proton exchange membrane fuel cells. J. Power Sources **156**(2), 267–280 (2006)
12. E. Dijoux, N.Y. Steiner, M. Benne, M.C. Pera, B.G. Perez, A review of fault tolerant control strategies applied to PEMFCs. J. Power Sources **359**, 119–133 (2017)
13. J. Pukrushpan, A. Stefanopoulou, H. Peng, *Control of Fuel Cell Power Systems: Principles, Modeling and Analysis and Feedback Design* (Springer, New York, 2004)
14. J. Amphlett, R. Baumert, R. Mann, B. Peppley, J. Roberge, Performance modeling of the Ballard Mark IV solid polymer electrolyte fuel cell. J. Electrochem. Soci. **142**(1), 9–15 (1995)
15. A.J. del Real, A. Arce, C. Bordons, Development and experimental validation of a PEM fuel cell dynamic model. J. Power Sources **173**(1), 310–324 (2007)
16. D. Rezzak, F. Khoucha, M. Benbouzid, A. Kheloui, A. Mamoune, A DC-DC converter-based PEMFCs emulator, in *Proceedings of International Conference Power Engineering, Energy and Electrical Drives (POWERENG 2011)*, Malaga, 11–13 May 2011)

Chapter 36
Design of a Microbial Fuel Cell Used as a Biosensor of Pollution Emitted by Oxidized Organic Matter

Amina Benayyad, Mostefa Kameche, Hakima Kebaili, and Christophe Innocent

Abstract Microbial Fuel Cell (MFC) is an electrochemical device that converts polluted organic matter into electricity by using microorganisms as biocatalyst. Indeed, MFC can produce low renewable electric energy but preserve the environment. The aim of this study was to design a MFC-based biosensor using household wastes, in which anaerobes contained in the anolyte, are separated from the catholyte with a cation exchange membrane. This biosensor has been used to analyze biodegradable waste organic matter, where anaerobes act as biocatalyst for their oxidation and the transfer of electrons to the anode. Sodium acetate solution was used as substrate to obtain maximum energy. The MFC-based biosensor system was optimized by discharging it into an external circuit and phosphate solution buffer mixed with sodium chloride solution used as catholyte in the aerobic compartment. The anaerobic compartment was kept at room temperature to promote bacterial growth. It turned out that the MFC's tension increased with the concentration of sodium acetate. It varied linearly with the substrate concentration, on a semi-logarithmic scale, thus making it possible to determine the minimum and maximum concentration thresholds. In addition, the electrochemical characterization by cyclic voltammetry of the bio-anode, revealed the oxidation current due to the degradation of the organic matter, which varied linearly with the scanning speed, highlighting the adsorption phenomenon on the surface of the porous electrode carbon felt. The MFC with a renewable aerobic bacterial source could therefore be used as a biosensor for on-line detection of pollution, in this case oxidized organic matter.

Keywords Microbial fuel cell · Biosensor · Pollution

A. Benayyad (✉) · M. Kameche · H. Kebaili
Laboratoire de Physico-Chimie des Matériaux, Catalyse et Environnement, Université des Sciences et de la Technologie d'Oran Mohammed BOUDIAF, Bir El Djir, Algérie
e-mail: benayyad1amina@gmail.com

C. Innocent
Institut Européen Des Membranes, Université de Montpellier, Montpellier, France

36.1 Introduction

For several decades now, the Microbial Fuel Cells (MFCs) have been a key solution for degrading and oxidizing organic matter and thus reducing the pollution of household waste and industrial waste discharged into the environment [1]. This oxidation makes it possible both to purify the wastewater and to produce exploitable electrical energy [2]. The present work will focus on improving the biofuel to make it more cost-effective and environmentally friendly to cope with the growing industrial and household waste that keeps pace with the speed of economic development and population growth. The use of the biological fuel cell is currently used to the depollution of wastewater because the development of this device remains conditioned by the policy of the scientific community which aims to guide all research towards sustainable development to the extent not to solve a problem to create another [3]. It can be used for public lighting in isolated and offshore areas.

36.2 Materials and Methods

In the present research, we have conceived a two compartment MFC: anode compartment containing the household waste leachate (anolyte) and the cathode compartment containing a 50 mmol/l NaCl buffered solution (catholyte), separated by a Nafion 115 cation exchange membrane. Carbon Felt was used as a base material for bioanode and abiotic cathode. Leachate was prepared from liquid household waste recovered from landfill trucks. The two compartments were connected by a 10 kΩ electric resistance to deliver the electrical energy thus produced. For this purpose, a multimeter was placed in parallel to measure the voltage. The ratio of the voltage to the resistance gave the current density by considering the effective area of bioanode brought into contact with the lixiviat. The fuel (sodium acetate) was added on a regular basis (1 ml per day) at different concentrations successively and decreasingly. The biopile worked for 3 months; the generation of electric current proved the oxidation of the fuel and the transfer of electrons between the electro active biofilm and the bioanode. The anolyte was isolated anaerobically (absence of oxygen) to avoid oxidation, however. As for the catholyte, it was stirred continuously to reinforce the contact with oxygen (aerobic) and consequently reduce the protons resulting from oxidation. The reduction of the oxidant (oxygen) being as important, we used as biodegradable electrolyte a solution of NaCl (25 mmol) in 50 mmol of phosphate buffer solution. The voltage between the two electrodes has been raised regularly to reach stable values. The fuel was administered when required to maintain a stable voltage over time.

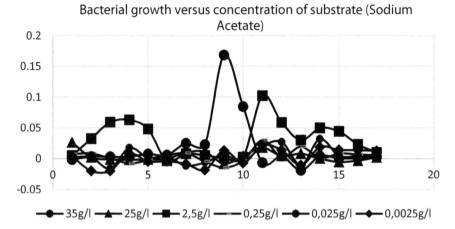

Fig. 36.1 Curves of bacterial growth in the presence of different concentrations of fuel (sodium acetate)

36.3 Results and Discussions

36.3.1 Bacterial Growth

The fuel concentration allowing optimal bacterial growth was 2.5 g/l of sodium acetate, since it made it possible to visualize a growth in two or three successive growth cycles (Fig. 36.1). It was followed by 0.025 g/l, which gave maximum growth compared to the others, after 9 h (a single peak).

36.3.2 Electrochemical Characterization

Cyclic voltammetry Cyclic Voltammetry (CV) was used to test the efficiency of the electroactive biofilm (biocatalyst) electroactivity on the bioanode as well as the oxidation of sodium acetate [4]. This technique is generally used because of its simplicity of implementation and its speed for the initial characterization of an active redox system. It provides an estimate of the redox potential of electro-active systems capable of exchanging electrons with the electrode. It also provides information on the electron transfer mechanisms between the electrode and the biofilm, as well as the stability of the analyte in the oxidation states within the chosen range of potentials. As shown in Table 36.1, this technique reveals that the electrical current delivered by the PCM decreases as the concentration increases, reaching the maximum current at 0.25 g/l (25.39 mA/cm2). Moreover, in order to demonstrate the phenomenon involved in the oxidation of the organic material by the MFC, the oxidation current as a function of the scanning speed has been represented. A straight line was obtained,

Table 36.1 Values of the oxidation peaks obtained with the fuel concentrations using the scanning speed 50 mV/s

Concentration (g/l)	Current density (mA/cm^2)
0.0025	10.81
0.025	18.17
0.25	25.39
2.5	5.99
25	4.95
35	3.40

highlighting the adsorption phenomenon, which is quite normal in our case, because of the use of Carbon Felt as an electrode (Fig. 36.2).

Electrochemical Impedance Spectroscopy (EIS) This technique has been used to explain the charge transfer between the electro-biofilm and the electrode. As shown in Fig. 36.3, the Nyquist diagrams consist of semicircles and straight lines at the high and low frequencies respectively. The semicircle and linear portions respectively represent the electron transfer and limited diffusion processes. As described by Kumar et al. [5], the inter-facial charge transfer resistance is defined from the diameter of the semicircle, representing the resistance of the electrochemical reactions on the electrode and determines the transfer kinetics of charge transfer (electrons of the redox couple at the interface of the electrode). Obviously, the best semicircle is obtained with the lowest concentration of substrate (0.0025 g/l), showing a high load transfer resistance. The low electrical conductivity is responsible for this increase in

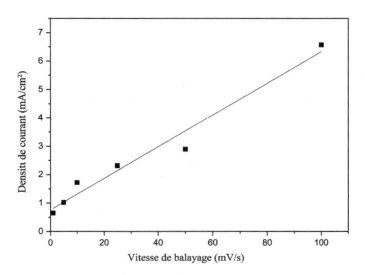

Fig. 36.2 Oxidation current versus speed (in the case of adsorption phenomenon)

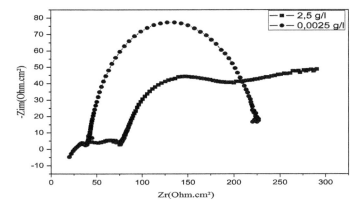

Fig. 36.3 Nyquist diagrams of the CF electrode using different fuel concentrations

electrical resistance. However, this resistance was reduced by significantly increasing the concentration of the substrate (2.5 g/l), which is 1000 times higher.

36.3.3 Power and Polarization Curves of MFC-CF

The power curves generally have the same parabolic appearance. Indeed, when the current density increases, the power density also increases and reaches a maximum and finally decreases. In our case, it reaches the maximum value at the concentration 35 g/l, followed by the other two concentrations 0.25 and 0.025 g/l. As shown in Fig. 36.4, these results corroborate those of Electrochemical Impedance Spectroscopy (EIS). The preliminary results of this electrochemical technique highlight a progressive lowering of the resistance to charge transfer (Rct) of the interface, due directly to the production of catalytic current since it involves the oxidation current.

36.4 Conclusion

As a result of this investigation, we have demonstrated the possibility of treating liquid household wastewater by using the electrochemical device of a Microbial Fuel Cell (MFC) with inexpensive materials and non-toxic substances. We followed the MFC device for more than three months and then characterized the bio-anode by chemical (bacterial growth) and electrochemical methods (cyclic voltammetry and Electrochemical Impedance Spectroscopy). The preliminary results of the SIE prove a progressive lowering of the resistance to charge transfer (Rct) of the interface, due directly to the production of catalytic current since it is it that brings the oxidation current. The bio-anode of carbon felt was the subject of all the study carried out.

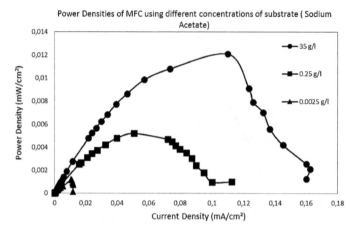

Fig. 36.4 Comparison between the power curves obtained with the different concentrations of sodium acetate

The microbial biofilm formed on the surface of the anode significantly improves the electrical conductivity of the material (carbon felt) and reduces the resistance of the electrolyte thanks to its catalytic properties that effectively oxidize organic matter.

References

1. Y. Holade, M. Oliot, A. Bergel, K. Servat, Biopiles enzymatiques et microbiennes. L'actualité chimique **400–401**, 81–82 (2015). https://www.lactualitechimique.org/Biopiles-enzymatiques-et-microbiennes
2. M. Booki, B.E. Logan, Continuous electricity generation from domestic wastewater and organic substrates in a flat plate microbial fuel cell. Environ. Sci. Technol. **38**(21), 5809–5814 (2004)
3. H. Liu, R. Ramnarayanan, B.E. Logan, Production of Electricity during Wastewater Treatment Using a Single Chamber Microbial Fuel Cell. Environ. Sci. Technol. **38**(7), 2281–2285 (2004)
4. F.S. Ketep, *Piles à combustible microbiennes pour la production d'électricité couplée au traitement des eaux de l'industrie papetière* (Université de Grenoble, Thèse de doctorat, 2012)
5. G.G. Kumar, C.J. Kirubaharan, S. Udhayakumar, C. Karthikeyan, K.S. Nahm, Conductive polymer/graphene supported platinum nanoparticles as anode catalysts for the extended power generation of microbial fuel cells. Ind. Eng. Chem. Res. **53**(43), 16883–16893 (2014)

Chapter 37
Bioelectricity Production from *Arundo Donax*-MFC and *Chlorophytum Comosum*-MFC

L. Benhabylès, **Y. M. Azri, I. Tou, and M. Sadi**

Abstract The increase of both energy demand and pollution makes essential the use of renewable energy. Green electricity production is one of numerous challenges of researches and the target is to obtain an economic clean energy with high performance and no emission of pollutants. Hydrogen fuel cell is an electrochemical process in which chemical energy is transformed into electrical energy. This reaction occurs by using hydrogen. A biological fuel cell is one of numerous types of fuel cell using living organisms to generate electricity. In a microbial fuel cell (MFC) the catalyst is the microorganisms and the most often bacteria. In plant microbial fuel cell (P-MFC), the rhizospheric microorganisms are the catalyst. Plants produce carbohydrates through photosynthesis and feed with rhizodeposits the rhizospheric microorganisms which generate electricity. This paper is a presentation of 2 P-MFC using two different plant species *Arundo donax* L. and *Chlorophytum comosum* L in the same experimental conditions. During 43 days we measured the electrical potential produced in *Arundo donax*-MFC and *Chlorophytum comosum*-MFC and we compare the 2 P-MFC system to evaluate plant species performances. Results show a different electrical potential for the two plants. The maximum tension registered for *Arundo donax* L. is +145.2 mV and for *Chlorophytum comosum* L. is + 155.3 mV.

Keywords P-MFC · *Arundo donax L* · *Chlorophytum comosum L* · Electricity · Fuel cell · Hydrogen

37.1 Introduction

Fuel cell is an old technology that produces energy by an electrochemically process. Unlike an ordinary battery that stocks energy, it produces electricity by converting fuels without interruption as long as these latter are provided [1, 2].

L. Benhabylès (✉) · Y. M. Azri · I. Tou · M. Sadi
Centre de Développement Des Energies Renouvelables, CDER, 16340 Algiers, Algeria
e-mail: lbenhabyles@gmail.com

© The Editor(s) (if applicable) and The Author(s), under exclusive license to Springer Nature Singapore Pte Ltd. 2021
A. Khellaf (ed.), *Advances in Renewable Hydrogen and Other Sustainable Energy Carriers*, Springer Proceedings in Energy,
https://doi.org/10.1007/978-981-15-6595-3_37

Type of fuels employed in a fuel cell depend on its requirements and its application [3, 4]. In a Hydrogen Fuel Cells, dihydrogene is combusted producing simultaneously electricity, water and heat. It consists of two electrodes covered by a thin layer of catalyst generally platinum, where the reaction producing electricity occurs. The electrodes are separated by an electrolyte that permit the passage of ions and blocks the electrons that pass by an external electrical circuit. Hydrogen protons migrate through the electrolyte to react with oxygen and electrons at the cathode. Even if the operating principal remains the same, different fuel cells exist. Their differences depend on influencing parameters such as: temperature, electrolyte and electrodes [3].

The use of pure hydrogen in a fuel cell produces a high output of electricity and release only water and heat, which makes a major asset in the environment protection. Otherwise, hydrogen use present disadvantages because of its unavailability in nature, high costs, high inflammability, and its difficult storage [5].

In a Microbial fuel cell (MFC), microbes use biochemicals from organics degradation to generate electricity. The MFC combined with a plant, Plant-Microbial Fuel Cell (P-MFC) is a biological fuel cell that produces bioelectricity by transforming solar energy into electric energy. Green plants produce organic matter by converting solar energy into biochemical energy through photosynthesis. Plant exudates are a part of these biochemicals mainly composed of carbohydrates. They are excreted into rhizosphere through plants roots.

Like for fuel cell, the anode is the negative electrode put at the rhizosphere under anaerobic conditions Eq. (37.37.1):

$$C_6H_{12}O_6 + 6H_2O \rightarrow 6CO_2 + 24 H^+ + 24e^- \qquad (37.1)$$

The cathode is the positive electrode put at the soil surface in contact with oxygen Eq. (37.37.2):

$$6O_2 + 24H^+ + 24e^- \rightarrow 12H_2O \qquad (37.2)$$

The catalyst, in this case, is microorganisms and are called electrochemically active microorganisms (EAM). The EAM break down organics and give electrons that are captured by the anode. The passage of electrons from the anode to the cathode generate bioelectricity.

The P-MFC presents many benefits like the low coast of the material used, continuous energy generation, creating green spaces in urban environments and no emission of greenhouse gases. Bacterial strain can also be isolated for hydrogen production under anaerobic conditions.

In this paper, we present our study in which we measured the electric potential produced in *Arundo donax*-MFC and *Chlorophytum comosum*-MFC and we compare the two P-MFC system to evaluate plant species performances.

37.2 Materials and Methods

37.2.1 Plant Material

Chlorophytum comosum L. var. Variegatum spider plants is an ornamental and indoor plant from family of asparagaceae or Liliaceae according to authors. Herbaceous, evergreen plant with tuberous roots and tuft appearance, very extensive.

Arundo donax L. Giant reed is a large perennial grass in the Mediterranean region. It is an invasive plant from family of Poaceae, it expands mainly via rhizomes.

Both species are easy to grow because of their high plasticity to environmental factors.

37.2.2 Experimental Set up

The potting of young *chlorophytum comosum* L. and *Arundo donax* L. were carried out at the end of March and the experiment lasted 43 days. In pots of 30 cm, spider plant and giant reed were placed in a loamy sandy soil without any additives, exposed to soft sunlight for 3 h a day at the morning. Light intensity varied, depending on the presence of clouds or not, from 520 to 15,000 lx and moderate temperatures ranging from 17 to 28 °C. They were watering according to their needs in order to provide them the right conditions for a good growth.

Two Graphite electrodes were placed, one at the root level "the anode" and a second one at soil surface "the cathode", for each pot.

Measurements have been done with a multimeter, every day at the morning.

37.3 Results and Discussion

Electricity production in a P-MFC is directly related to plant species because of the rhizosphezric consortium that is different from a plant species to another.

Results of measurements of *Arundo donax*-MFC and *Chlorophytum comosum*-MFC electrical potential are reported in Figs. 37.1 and 37.2.

As shown in Figs. 37.1 and 37.2, the electrical energy production in *Arundo donax*-MFC and *Chlorophytum comosum*-MFC is different.

In the present study, *Arundo donax* L. measurements show very large fluctuations and a lot of negative values. The 4th first days, the electrical tensions are negative, this is relative to the time adaptation of the plant. From the 5th to the 19th day, fluctuating values are registered with very low voltage. From 20th to 26th day, the voltage fluctuate between 0 and negative values to rise at the 27th day and fluctuate with positive values. The maximum tension registered is $+145.2$ mV at the 36 day of the experiment and it suddenly drop again to negative values during 4 days and rise

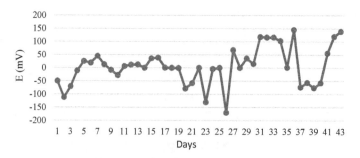

Fig. 37.1 Electrical potential of *Arundo donax* L. during 43 days of experiments

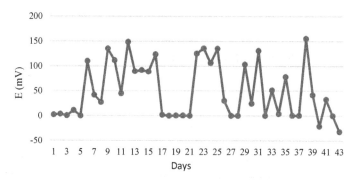

Fig. 37.2 Electrical potential of *Chlorophyum comosum* L. during 43 days of experiments

again. *Arundo donax* L. has a fast growth rate with a high biomass production [6]. It's also known for its C3 photosynthetic pathway that compete with other C3 plants and C4 grasses supposed to be more productive [7, 8]. In addition, *Arundo donax* L. forms a dense and deep root system [9]. All these characteristics might allow us to suppose that this species generate a high electrical energy. In [7], *Arundo donax* L. showed an instable electricity production but in a different experimental set up. The fluctuating tensions recorded in our work is probably due to the big size of the rhizome and the fast growth rate of the plant. It has been reported by [10], that plants react differently depending on experimental conditions.

For *Chlorophytum comosum* L. measurements show many positive values fluctuating between 0 and a maximum of +155.3 mV. During the 4th first days, the voltage was very low with positive values which corresponds to the plant adaptation time. It suddenly rise from 0 to 110 mv at the 6th day and fluctuate with positive values till the 15th day. The voltage suddenly decrease at the 16th to very low values and fluctuate positively till the 39th day and drop to negative values. *Chlorophytum comosum* L. is a C3 plant [11] with a high biomass production [12]. It forms a dense root system of fleshy roots of about 10 to 15 cm long and fine roots [13]. The best tensions recorded, during the present experimentation, are for *Chlorophytum comosum* L. while in previous work, with same conditions, tensions reached +636 mV [14].

It is already known that Photosynthesis pathway influence the bioprocess of converting solar energy into electrical energy [10]. Despite of being C3 plants, *Arundo donax* L. and *Chlorophytum comosum* L. are particular by having a high photosynthetic potential comparable to C4 species [4]. Several aspects have to be explored in order to determine the reasons of low tensions; the main ones are rhizodeposition and microorganism. The structure and activity of the rhizospheric microorganisms depend on the plant species, root morphology and rhizodeposits composition. It is also reported that structure and activity of microbiome vary through plant life cycle [15]. The challenge for an efficient P-MFC system is to produce a continuous electrical tensions, in this perspective, the rhizospheric microorganisms must be electrochemically active.

37.4 Conclusion

Both of Hydrogen fuel cell and P-MFC are technologies in performing and developing. In our work, we have presented a simple experimental device of P-MFC to show the differences with a hydrogen fuel cell. Through this experience, we have tested the production of electrical energy by *Chlorophytum comosum* L. and *Arundo donax* L. The two plant species seem to have the appropriate criteria for efficient operating of a P-MFC. However, they present different electrical potentials. The experiments are in progress investigating different parameters to perform the P-MFC system. Plant eco-physiology, the operating conditions and P-MFC system design must be coordinate to provide the optimum conditions to different plant species for producing high levels of electrical energy.

References

1. B. Cook, Introduction to fuel cell and hydrogen technology. Eng. Sci. Educ. J. **11**(6), 205–216 (2002)
2. O'Hayre, R. P., Cha, S.W., Colella, W. G, Prinz, F. B.: Fuel cell fundamentals. 3rd edn. Wiley (2016)
3. Larminie, J., Dicks, A.: Fuel cell systems explained. 2nd edn, Wiley (2003)
4. Webster, R. J., Driever, M. S., Kromdijk, J., McGrath, J., Leakey, A. D., Siebke, K., Demetriades-Shah, T., Bonnage, S., Peloe, T., Lawson, T., Long, S. P.: High C3 photosynthetic capacity and high intrinsic water use efficiency underlies the high productivity of the bioenergy grass *Arundo donax*. Sci. Rep. **6**, 20694 (2016)
5. Shah, R. K.: Introduction to fuel cells. In: Basu, S. (ed.) Recent Trends in Fuel Cell Science and Technology. Anamaya Publishers, New Delhi, India (2007)
6. Angilini, L. G., Ceccarinia L., Bonarib, E.: Biomass yield and energy balance of giant reed (*Arundo donax* L.) cropped in central Italy as related to different management practices. Europ. J. Agronomy **22**(4), 375–389 (2005)
7. M. Helder, D.P. Strik, H.V. Hamelers, A.J. Kuhn, C. Blok, C.J. Buisman, Concurrent bioelectricity and biomass production in three plant-microbial fuel cells using *Spartina anglica*, *Arundinella anomala* and *Arundo donax*. Bioresour. Technol. **101**(10), 3541–3547 (2010)

8. Y. Wang, J. Tao, J. Dai, Lead tolerance and detoxification mechanism of *Chlorophytum comosum*. Afr. J. Biotechnol. **10**(65), 14516–14521 (2011)
9. Bell, G. P.: Ecology and management of *Arundo donax*, and approaches to riparian habitat restoration in southern California. In: Plant invasions: Studies from North America and Europe. Leiden, The Netherlands Blackhuys, pp. 103–113 (1997)
10. R. Nitisoravut, R. Regmi, Plant microbial fuel cells: A promising biosystems engineering. Renew. Sustain. Energy Rev. **76**, 81–89 (2017)
11. Kerschen, E. W., Garten, C., Williams, K. A., Derby, M. M.: Evapotranspiration from spider and jade plants can improve relative humidity in an interior environment. Hort Technol. **26**(6), 803–810 (2016)
12. D. Sokic-Lazic, R.L. Arechederra, B.L. Treu, S.D. Minteer, Oxidation of biofuels: Fuel D versity and effectiveness of fuel oxidation through multiple enzyme cascades. Electroanalysis **22**(7–8), 757–764 (2010)
13. A. Braria, A. Shoaib, S.L. Harikumar, *Chlorophytum Comosum* (Thunberg) Jacques: A review. In. Res. J. Pharm **5**(7), 546–549 (2014)
14. Y.M. Azri, I. Tou, M. Sadi, L. Benhabyles, Bioelectricity generation from three ornamental plants: *Chlorophytum comosum, Chasmanthe floribunda* and *Papyrus diffusus*. Int. J. Green Energy **15**(4), 254–263 (2018)
15. L. Philippot, J.M. Raaijmakers, P. Lemanceau, W.H. Van der Putten, Going back to the roots: The microbial ecology of the rhizosphere. Nat. Rev. Microbiol. **11**(11), 789–799 (2013)

Chapter 38
Implementation of Fuel Cell and Photovoltaic Panels Based DC Micro Grid Prototype for Electric Vehicles Charging Station

N. Benyahia, S. Tamalouzt, H. Denoun, A. Badji, A. Bousbaine, R. Moualek, and N. Benamrouche

Abstract Today, electric vehicle (EV) appears as an evident solution for the future automotive market. The introduction of EV will lead to the reduction of greenhouse gas emissions and decrease the travelling cost. However, electric vehicle is truly an ecological solution only if the production of electricity necessary for its operation is produced from sustainable energy sources. In this paper, an Electric Vehicle Charging Station (EVCS) through sustainable energy sources via a DC micro-grid system has been proposed. The proposed system includes a fuel cell (FC), photovoltaic (PV) panels, storage battery and possibility of a connection to the grid. In this work a low power prototype of a micro-grid based EVCS has been first validated using a numerical simulation under Matlab/Simulink using variable irradiance and number of recharging vehicles. In the second part of this paper, an EVCS prototype has been realized in the laboratory. The tests are realized using an emulator of the PEM fuel cell with the concept of the hardware-in-the-loop (HIL). The objective of this emulation is to evaluate the performances of the whole system without the need for a real fuel cell. The whole system is implemented on the dSPACE 1103 platform and the results of the tests are discussed.

Keywords Fuel cell · Photovoltaic · Electric vehicle · Charging station

---

N. Benyahia (✉) · H. Denoun · A. Badji · R. Moualek · N. Benamrouche
LATAGE Laboratory, Mouloud Mammeri University, BP 17 RP, 15000 Tizi-Ouzou, Algeria
e-mail: benyahia.ummto@yahoo.fr

S. Tamalouzt
LT2I Laboratory, Abderrahmane Mira University, BP 17 RP, 06000 Bejaia, Algeria

A. Bousbaine
College of Engineering and Technology, University of Derby, DE22 3AW Derby, UK

© The Editor(s) (if applicable) and The Author(s), under exclusive license to Springer Nature Singapore Pte Ltd. 2021
A. Khellaf (ed.), *Advances in Renewable Hydrogen and Other Sustainable Energy Carriers*, Springer Proceedings in Energy,
https://doi.org/10.1007/978-981-15-6595-3_38

38.1 Introduction

Today, electric vehicles appear as a real solution for the future automotive sector [1]. This solution reduces greenhouse gas emissions and decreases the travel cost. However, electric vehicle is truly an ecological solution only if the production of electricity necessary for its operation is produced from sustainable energy sources. The demand of electrical energy in an EVCS is not constant and depends particularly on the number of the vehicles connected to the charging system [2]. Two solutions are proposed in the literature: EVCS's based on photovoltaic and battery storage and EVCS's that are connected to the grid. The lack of the first solution is energy efficiency, especially when the batteries are discharged. In these systems, the batteries are oversized to compensate for the photovoltaic power during low irradiance periods and to avoid the deep discharge of batteries. Therefore, the cost and the space needed by this storage system are relatively large. On the other hand, those connected to the grid, peak units must be used during periods of high demand [3] and therefore pollution can be increased.

The EVCS based Micro-Grid (MG) proposed in this paper includes a fuel cell (FC), photovoltaic (PV) panels, storage battery and possibility of an integration to the grid. The advantage of combining the FC with PV panels is that the FC has a high efficiency at full load and this efficiency is retained as the load decrease [4]. The benefit of the proposed EVCS-MG is threefold: the recharge of the EV's is generated locally using renewable energy sources, the energy efficiency is improved by adding the fuel cell and the extended parking time of the EV's offers the possibility to use the concept of vehicle to grid technology which the EV acts as a controllable thank for MG.

In this paper a low power prototype of a micro-grid based EVCS has been realized in the laboratory. The proposed MG prototype consists of fuel cell, photovoltaic panels and battery. The implementation of the MG system has been carried out using dSPACE 1103. A serial multi-cellular converter is used as a recharger converter. This structure of DC-DC converter is developed to improve some shortcomings of buck, boost and buck-boost DC-DC converters. It can also be used instead of the classical ones, allowing for an improved vehicle battery recharge performances compared to classical converters. This topology presents many advantages including: less voltage constraints on the switching components, low switching frequency, low current and voltage ripples, and the capacitor voltages balancing can be made naturally [5]. In addition, the serial multi-cellular converter offers the possibility to obtain very high energy efficiency over a wide range of power. This constitutes an important advantage for EVCS applications. The main contributions of this research work are summarized below: (1) Implementation of the proposed MG-EVCS using Matlab/Simulink SimPowerSystems Toolbox. In this implementation, the MPPT technique is used to extract the maximum power from photovoltaic panels. (2) Development of an experimental setup prototype based on MG-EVCS conversion system.

This work is organized as follows: In Sect. 38.2 and 38.3, the system's configuration and modelling of the PEM fuel cell and PV panel have been presented.

In Sect. 38.4, the simulation of the MG-EVCS conversion system has been realized using Matlab/Simulink based on SimPowerSystems Toolbox (SPS). To validate the proposed system, a real-time experimental prototype platform has been developed and the experimental results have been presented in Sect. 38.5. Finally, the conclusions and perspectives of this work have been provided in Sect. 38.6.

38.2 System Description

Figure 38.1 shows the proposed topology of the MG-EVCS system. It includes a photovoltaic array generator, a PEM fuel cell and a battery storage system. The PV panels are connected to the DC bus through a boost DC-DC converter controlled using P&O maximum point power tracking, MPPT, algorithm and the PEM fuel cell is controlled as a current source through another boost DC-DC converter. A multicellular DC-DC converter is used as vehicle battery recharge converter. This topology presents many advantages: less voltage constraints on the switching components, low switching frequency, low current and voltage ripples, and the capacitor voltages balancing can be made naturally [1, 2]. The electrical circuit of the boost DC-DC converter and serial multicellular converter used in this section have been modeled as shown in Fig. 38.1. The PV model used in this work is based on a single diode model [4].

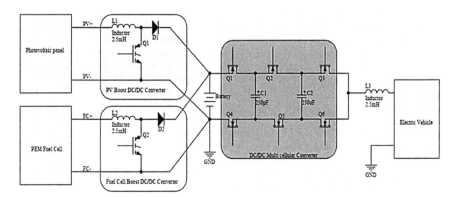

Fig. 38.1 The structure of the proposed system

38.3 System Modeling

38.3.1 PV Panel Model

In this model, the equivalent circuit of the PV cell is represented by a current source in parallel with a diode and series/parallel resistances. The PV panel model is given by Eq. (38.1):

$$I = I_{ph} - I_0[\exp((V + R_s I)/V_t \cdot a) - 1] - (V + R_s I)/R_{sh} \tag{38.1}$$

Where the V_t is the thermal voltage of the cell, I_{ph} is the photocurrent which depend on solar irradiation and is influenced by the temperature of cell, I_0 is the diode saturation current which varies with the temperature, a is the diode ideality factor. The Parameters of the PV panel used in this simulation are given in [4].

38.3.2 PEM Fuel Cell Model

PEM fuel cells are a promising technology for electric vehicle recharging station applications thanks to their higher efficiency, low emissions and direct production of electricity [6]. The output voltage of a single cell can be defined as follows [7].

$$E_{cell} = E_{Nernest} - E_{Act} - E_{Con} - E_{Ohm} \tag{38.2}$$

In Eq. (38.2), E_{Nernst} is the thermodynamic potential of the cell representing its reversible voltage, E_{Act} is the activation voltage drop, E_{Con} is the concentration voltage drop, E_{Ohm} is the Ohmic voltage drop.

38.4 Simulation Tests

The components shown in Fig. 38.1 are tested using SimPowerSystems components. Three PV panels of 240 W and PEM fuel cell of 500 W have been connected to the DC bus through two separated boosts DC-DC converters. The PV boost DC-DC converter is controlled by a classical P&O MPP technique and it is used to increase the PV output voltage from $V_{mpp} = 48.5$ VDC to 100.8–10.4 (Fig. 38.2).

The FC boost DC-DC converter is controlled using current controlled mode and it is used to increase PEMFC output $V_{fc} = 37.5$–34 VDC voltage to 100.8–110.4 VDC.

The storage system consists of 8 acid lead batteries (12 V, 7 Ah). To test the operation of the EVCS-MG, the PV panel is subjected to variable step irradiance (Fig. 38.3).

38 Implementation of Fuel Cell and Photovoltaic Panels …

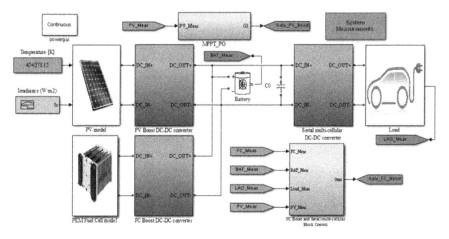

Fig. 38.2 Matlab/simulink block scheme of the proposed system

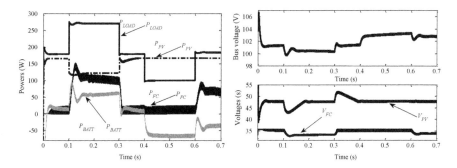

Fig. 38.3 Simulation results of the EVCS-MG

The irradiance varies according to these time settings: $t < 0.1\ s$ and $t > 0.3\ s\ E = 600\ \text{W} \bullet \text{m}^{-2}$, $0.1\ s < t < 0.3\ s\ E = 400\ \text{W} \bullet \text{m}^{-2}$. During the interval [0.1–0.3 s], the number of electric vehicle recharging is increased and the PEM fuel cell compensate for the deficiency of power from the photovoltaic panels. Under these conditions, the recharging station is fed jointly from the solar panels and the fuel cell. However, during the interval [0.4–0.6 s], the number of electric vehicle recharging is decreased, the photovoltaic power P_{PV} is sufficient, in this case, and the DC bus voltage increases slightly. It also shows that the photovoltaic voltage is fixed by the MPPT algorithm at its $V_{mpp} \simeq 48$ V under these conditions and the power delivered by the photovoltaic panel which depends on the level of the irradiation.

38.5 Experimental Validation

The experimental set up of the EVCS system is shown in Fig. 38.4. It is composed of a fuel cell emulator (1), a serial multicellular converter (2) and two boost dc-dc converters. To reproduce the behavior of the PEM fuel cell when the fuel cell current I_{fc} change, the fuel cell voltage V_{fc} must be regulated with respect to to its reference value V_{fc_ref} [8]. The reference value of PEM fuel cell current I_{fc_ref} is calculated from I_{bus} and I_{pv} and compared with the measured fuel cell current I_{fc}. The experimental set up consists of three PV panels of 240 W and a fuel cell emulator of 500 W connected to the dc bus through a boost DC-DC converters. The PV boost DC-DC converter is controlled using a classical P&O-MPPT technique and the FC boost DC-DC converter is controlled using the current control mode. The electric vehicle is replaced by a resistive load.

The power supply output current is measured and sent to the dSPACE card (1), and taken as the input of the fuel cell model. The P&O-MPPT technique is developed using the boost DC-DC converter realized in the laboratory (2). The currents are measured with LEM LA25-NP. The stabilized powers (3) are used to supply the LEM LV25-P and LEM LA25-NP sensors.

Figure 38.5 shows the evolution of the photovoltaic panel voltage, this voltage is about 52 V and it varies with the external temperature. It also shows that the photovoltaic voltage is fixed by the MPPT algorithm at its V_{mpp} value with a ripple of 10 V under these conditions. This ripple of PV voltage is due to the search for the maximum power point by the P&O-MPPT algorithm. Moreover, the drawback of this method is the oscillations of the PV output voltage around the maximum point. Figure 38.5 shows also that the DC bus voltage is not perfectly constant but depends

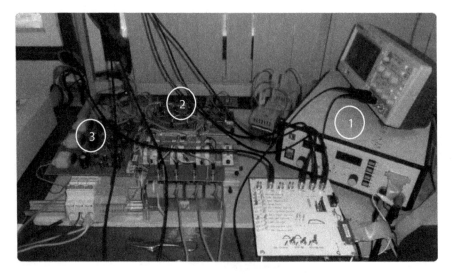

Fig. 38.4 Experimental test bench setup

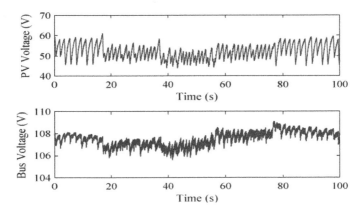

Fig. 38.5 PV and bus voltages

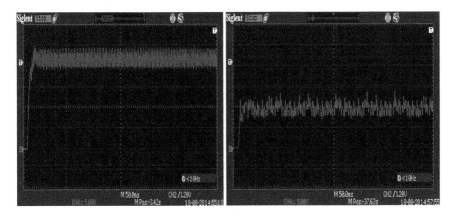

Fig. 38.6 Voltages V_{c1} and V_{c2} across capacitor in multicellular converter

on the state of the battery storage. The storage system consists of 8 lead acid batteries (12 V, 7A h) connected in series on the DC bus. In Fig. 38.6 the floating voltages V_{c1} and V_{c2} across the capacitors were shown.

38.6 Conclusion

The present paper is focused on the simulation and experimental validation of a low power prototype of a micro-grid based EVCS. The proposed MG prototype consists of fuel cell, photovoltaic panels and battery. In this system, a serial multi-cellular converter is used as a charger converter. From the simulation results, it can be seen that PV and fuel cell hybrid system present a better performance compared to the classical stand-alone PV system. The implementation of the EVCS system has been carried out using dSPACE 1103. The management of the EVCS is evaluated under sudden solar irradiation changes and load variation, the obtained simulation an experimental results indicate better performances.

In conclusion from this work, the proposed recharging electric vehicle system can bring improvements in terms of recharging power because the serial multicellular converter ensures stress free distribution in voltage on the various semiconductor components, therefore a significant recharge power.

References

1. Ashish, K. K., Sujit, R., Md. Raiu, A.: Analysis of the impact of electric vehicle charging station on power quality issues. In: 2nd International Conference on Electrical, Computer and Communication Engineering (ECCE), IEEE, Bangladesh (2019)
2. E.-A. Mehdi, C. Kent, S. Jason, Rapid-charge electric-vehicle stations. IEEE Trans. Power Delivery **3**(25), 1883–1887 (2010)
3. M. Zeinab, A. Iftekhar, H. Daryoush, V.-P. Quoc, Smart charging strategy for electric vehicle charging stations. IEEE Trans. Transp. Electrification **1**(4), 76–88 (2010)
4. Djoudi, H., Badji, A., Benyahia, N., Zaouia, M., Denoun, H., Benamrouche, N.: Modeling and power management control of the photovoltaic and fuel cell/electrolyzer system for stand-alone application. In: Proceedings of 3th International conference on Electrical Engineering, IEEE, Algeria (2015)
5. Hamida, M-L., Denoun, H., Fekkik, A., Benyahia, N., Zaouia, M., Benamrouche, N.: Cyclic reports modulation control strategy for a five cells inverter. In: Proceedings of 6th International Conference on Electrical Sciences and Technologies in Maghreb (CISTEM), IEEE, Algeria (2018)
6. N. Benyahia, H. Denoun, M. Zaouia, T. Rekioua, N. Benamrouche, Power system simulation of fuel cell and supercapacitor based electric vehicle using an interleaving technique. Int. J. Hydrogen Energy **40**(45), 15806–15814 (2015)
7. N. Benyahia, H. Denoun, A. Badji, M. Zaouia, T. Rekioua, N. Benamrouche, D. Rekioua, MPPT controller for an interleaved boost dc–dc converter used in fuel cell electric vehicles. Int. J. Hydrogen Energy **39**(27), 15196–15205 (2014)
8. N. Benyahia, T. Rekioua, N. Benamrouche, A. Bousbaine, Fuel cell emulator for supercapacitor energy storage applications. Electr. Power Components Syst. **41**(6), 569–585 (2013)

Chapter 39
Application of Hydrotalcite for the Dry Reforming Reaction of Methane and Reduction of Greenhouse Gases

Nadia Aider, Fouzia Touahra, Baya Djebarri, Ferroudja Bali, Zoulikha Abdelsadek, and Djamila Halliche

Abstract The Double layered hydroxides (LDH) called anionic clays are the most studied laminated materials. The CoAl-LDH and CoFe-LDH hydrotalcite-like compounds were successfully synthesized by co-precipitation method with $Co^{2+}/Al^{3+} = 2$ and pH = 12. After calcination under air at 800 °C. The solids were characterized by means of XRD, BET area, N_2 adsorption and desorption, TPR-H_2, chemical analysis by ICPs. Reforming of methane with carbon dioxide to synthesis gas, which is also referred to as dry reforming of methane represents an industrially relevant process that meets the criteria of green chemistry and of environmental protection: In this respect, CO_2 and CH_4, both considered as one of the main greenhouse gases responsible for out planet's global warming phenomenon, are converted to furnish a more useful mixture of gas containing H_2 and CO, syngas. After reduction, the catalysts were evaluated in the reforming of methane reaction under continuous flow with CH_4/CO_2 ratio equal to 1, at atmospheric pressure and a temperature range [500–700 °C]. The catalytic activity was tested in a fixed bed reactor.

Keywords Hydrotalcite · Cobalt · Fer

N. Aider (✉)
Département de Chimie, Faculté Des Sciences, (UMMTO), Tizi-Ouzou, Algérie
e-mail: n_aider@yahoo.fr

F. Bali · Z. Abdelsadek · D. Halliche
Laboratoire de Chimie Du Gaz Naturel, Université Des Sciences et Technologie Houari Boumediene (USTHB), Bab Ezzouar, Algérie

F. Touahra
Centre de Recherches Scientifiques (CRAPC), BP 248, 16004 Alger, Algérie

B. Djebarri
Département de Chimie, Faculté Des Sciences, Université de M'hamed Bougara, Avenu. de l'indépendance, 35000 Boumerdès, Algérie

© The Editor(s) (if applicable) and The Author(s), under exclusive license to Springer Nature Singapore Pte Ltd. 2021
A. Khellaf (ed.), *Advances in Renewable Hydrogen and Other Sustainable Energy Carriers*, Springer Proceedings in Energy,
https://doi.org/10.1007/978-981-15-6595-3_39

39.1 Introduction

Carbon deposition during the production of syngas, a mixture containing good ratios of hydrogen to carbon monoxide gas which is widely implicated in the Fischer Tropsch synthesis, still remains one of the most challenging aspects in the reaction of CO_2 reforming of methane. This reaction has been established and already widely commercialized. The target several recent studies on natural gas reforming is shifted to the catalyst development for carbon dioxide reforming also named dry reforming of methane(DRM; Eq. 39.1). So far there is no established commercially industrial process due to problem associated with this reaction comes in the form of sintering of the active phase and carbon formation [1].

This latter could be formed by direct decomposition of methane Eq. (39.2); inverse Boudouard reaction Eq. (39.3) or following the direct reduction of CO with H_2 Eq. (39.4) [2].

$$CH_4 + CO_2 \leftrightarrow 2CO + 2H_2 \quad (39.1)$$

$$CH_4 \leftrightarrow C + 2H_2 \quad (39.2)$$

$$2CO \leftrightarrow C + CO_2 \quad (39.3)$$

$$H_2 + CO \leftrightarrow C + H_2O \quad (39.4)$$

Nickel or cobalt catalysts are best suited for this reaction because they allow high activity, satisfactory stability over time and have the advantage of being inexpensive. However, the problem of catalysts based on Ni is essentially the carbon deposition originating mainly from the dissociation reaction of carbon monoxide and/or from the methane decomposition reaction [3]. Carbon formation negatively affects catalytic performance by blocking active sites [4, 5]. For this reason, the primary concern of research is to achieve efficient materials during several catalytic cycles [6, 7]. This work has notably shown that a good dispersion of the reactive species can contribute to the reduction of carbon deposition [8]. Double lamellar hydroxides (HDLs) called anionic clays, are the most studied laminate materials [9]. HDLs are quite rare in nature but relatively inexpensive and simple to synthesize in the laboratory [10, 11].

39.2 Experimental

39.2.1 Catalyst Synthesis

CoAl-LDH and CoFe-LDH solids were synthesized easily, according to the co-precipitation method described elsewhere [12]. The mixture of metals nitrate solution were added drop wise to a vigorously stirring solution of NaOH (2 M) and Na2CO3 (0.4 M) at room temperature while the pH was maintained constant at 12.

39.2.2 Material Characterization

Inductively coupled plasma emission spectrometer (ICP-ES) was employed to the determination of elemental chemical analysis of each solids, using Thermo Jarrel Ash ICAP 957. The specific surface areas (BET) were determined by nitrogen adsorption-desorption isotherms using Micromeritics Tristar 3000 at 196 °C. XRD patterns of all samples were performed with BRUKER D8 Advance diffractometer equipped with Cu-Ka radiation. Hydrogen temperature programmed reduction H2-TPR.

39.3 Results and Discussion

39.3.1 Catalyst Characterization

Hydrotalcite-type clay can be expressed by the following formula:

$$M_n^{2+}M_m^{3+}(OH)_{2(n+m)}^{m+}; A_{m/x}^{x-} \cdot yH_2O \qquad (39.5)$$

where: M2 + and M3 + are di- and trivalent cations respectively, A is the interlayer anion, x = charge of the anion, $n > m$, and y = number of water molecules of the interlayer layer [13]. The results obtained for the chemical composition and the structural formula of the samples are summarized in Table 39.1.

The results of the elemental analysis by ICP show the presence of minor anomalies is mainly associated with deficiencies in the total incorporation of the Fe cation within the brucite-like layer.

The reduction behavior of CoAlcal and CoFecal catalysts was studied by H2-TPR Fig. 39.1 depicts the temperature-programmed reduction (H2-TPR) analysis of CoAlcal-R and CoFecal-R. The H2-TPR profile of the CoAlcal-R samples shows two reduction peaks. The first peak observed at 380 °C can be assigned to the reduction of Co3O4 spinel in CoO, while the second observed at 710 °C is attributed to reduction

Table 39.1 Chemical composition and specific surface

Samples	Proposed formula	Molar ratio (M^{2+}/M^{3+})	SBET (m^2/g)		
			Precursor	Calcined	Reduced
CoAl-HDL	$[Co_{0,68}.Al_{0,32}(OH)_2]$ $[(NO_3)_{0,014}(CO_3)_{0,187}.,0,40 H_2O]$	2,19	87	116	87
CoFe-HDL	$[Co_{0,69}.Fe_{0,31}(OH)_2]$ $[(NO_3)_{0,01}(CO_3)_{0,182}.,0,85 H_2O]$	2,22	43	82	50

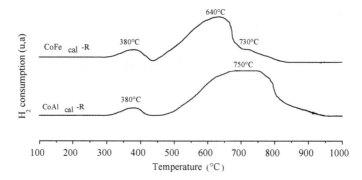

Fig. 39.1 Reduction temperature profiles of the solids $CoAl_{cal}$-R and $CoFe_{cal}$-R

of CoO species to Co0, and/or reduction of Co_2 in $CoAl_2O_4$ spinel or Co_2AlO_4, according to some studies [14].

The H2-XRD patterns of the two catalysts following H_2 reduction at 750 °C over the course of 1 h are represented in Fig. 39.2. In case of CoAlcal-R catalyst, the

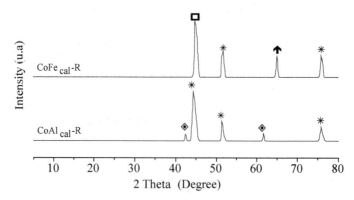

Fig. 39.2 DRX diffract grams of reduced solids at 750 °C; ◈ CoO, ✲Co^0, ↑Fe^0 and ☐ γ-CoFe

spectra showed the presence of species Co0 and CoO. The presence of Co0, Fe0 and γ-CoFe alloy were observed in case of CoFecal-R catalyst.

39.3.2 Catalytic Activity Tests

The catalysts are tested in a dry reforming reaction of CH_4, at atmospheric pressure in the range [400–700 °C]. Under the reaction mixture, the catalysts are heated from ambient to 700 °C with a speed of 5 °C/min.

The results of the catalytic activity of the CoAlcal-R and CoFecal-R catalysts, as a function of the RSM temperature are illustrated in Fig. 39.3.

In the case of two catalysts, in the temperature range [400–500 °C], the CO_2 conversion rate is generally higher than that of CH_4, while H_2/CO ratio was lower to 1 (Fig. 39.3c). This suggests the participation of the reverse water gas shift reaction (RWGS), which tends to increase CO_2 conversion and the production of CO, which becomes higher than that of H_2. At high temperatures [500–700 °C], the H_2/CO ratio is greater than the theoretical value 1 in the presence of CoAlcal-R. In addition, the CO_2 conversion is higher than the CH_4 conversion and the carbon balance obtained was lower than 100%. This suggests that the carbon deposition occurs via the inverse Boudouard reaction. At the same temperatures [500–700 °C], in the case of CoFecal-R catalyst, CO_2 conversion was always higher than that of CH_4 while the ratio H_2/CO is close to 1, indicates the occurrence of the reverse water gas shift reaction (RWGS) favored by the presence of iron [15]. The stability of our catalyst was also studied over a prolonged period.

39.4 Conclusion

CoAlcal-R and CoFecal-R catalyst were synthesized using co-precipitation method at pH = 11 followed by calcination and subsequent reduction. Various characterization techniques such as ICP, XRD, N2 adsorption and desorption, H2-TPR, were utilized to successfully identify the structure and physico-chemical properties of the solid. The catalyst was successfully applied to the synthesis of syngas from methane and carbon dioxide.

The CoAlcal-R catalyst exhibited rather high catalytic activity and stability during the reaction of CO_2 reforming of methane compared to CoFecal-R catalyst. The low reactivity of the iron-based catalysts is correlated with the re-oxidation of the active phase by the water formed via the reverse water-gas shift reaction (RWGS) favored by the presence of iron.

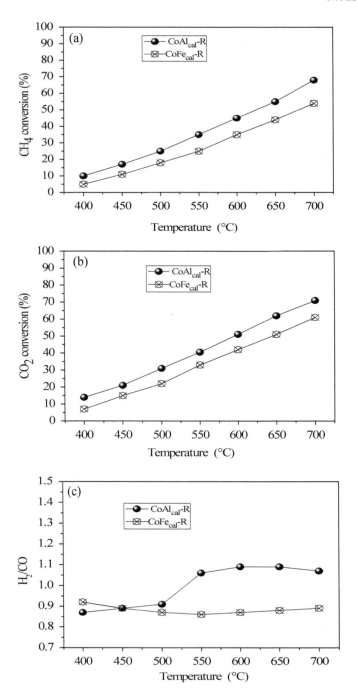

Fig. 39.3 a CH_4 conversion, b CO_2 conversion and c H_2/CO ratio obtained during DRM

References

1. Y. Cao, M. Lu, J. Fang, L. Shi, D. Zhang, Hexagonal boron nitride supported meso SiO_2-confined Ni catalysts for dry reforming of methane. Chem. Commun. **54**(53), 7549–7552 (2017)
2. M. Kogler, E.-M. Köock, B. Klöotzer, T. Schachinger, W. Wallisch, R. Henn, C.H. Huck, C. Hejny, C. Penner, High-temperature carbon deposition on oxide surfaces by CO disproportionation. J. Phys. Chem. C **3**(120), 1795–1807 (2016)
3. X. Zhao, M. Lu, H. Li, J. Fang, L. Shi, D. Zhang, In situ preparation of Ni nanoparticles in cerium–modified silica aerogels for coking- and sintering-resistant dry reforming of methane. New J. Chem. **12**(41), 4869–4878 (2017)
4. B. Djebarri, V.M. Gonzalez-Delacruz, D. Halliche, K. Bachari, A. Saadi, A. Caballero, J.P. Holgado, O. Cherifi, Promoting effect of Ce and Mg cations in Ni/Al catalysts prepared from hydrotalcites for the dry reforming of methane. React. Kinet, Mech. Catal. **111**(1), 259–275 (2014)
5. X. Zhang, N. Wang, Y. Xu, Y. Yin, S. Shang, A novel Ni–Mg–Al-LDHs/γ-Al_2O_3 catalyst prepared by in-situ synthesis method for CO_2 reforming of CH_4. Catal. Commun. **45**, 11–15 (2014)
6. D. San-José-Alonso, J. Juan-Juan, M.J. Illan-Gomez, M.C. Roman-Martinez, Ni, Co and bimetallic Ni-Co catalysts for the dry reforming of methane. Appl. Catal. A **371**, 54–59 (2009)
7. X. Zhang, C. Yang, Y. Zhang, Y. Xu, S. Shang, Y. Yin, Ni-Co catalyst derived from layered double hydroxides for dry reforming of methane. Int. J. Hydrog. Energy **46**(40), 16115–16126 (2015)
8. C. Wang, N. Sun, N. Zhao, W. Wei, Y. Zhao, Template-free preparation of bimetallic mesoporous Ni-Co-CaO-ZrO_2 catalysts and their synergetic effect in dry reforming of methane. Catal. Today **2**(281), 268–275 (2016)
9. Cavani, F., Trifiro, F., Vaccari, A.: Hydrotalcite-type anionic clays: Preparation, properties and applications. Catal. Today (11), 173–301
10. Y. Zhu, S. Zhang, B. Chen, Z. Zhang, C. Shi, Effect of Mg/Al ratio of NiMgAl mixed oxide catalyst derived from hydrotalcite for carbon dioxide reforming of methane. Catal. Today **15**(264), 163–170 (2016)
11. Y. Shiratori, M. Sakamoto, T. Uchida, H. Le, T. Quang-Tuyen, K. Sasaki, Hydrotalcite-dispersed paper-structured catalyst for the dry reforming of methane. Int. J. Hydrog. Energy **40**(34), 10807–10815 (2015)
12. F. Touahra, W. Ketir, D. Halliche, M. Trari, Photocatalytic hydrogen evolution over the heterojunction $CoAl_2O_4$/ZnO. React. Kinet., Mech. Catal. **2**(111), 805–816 (2014)
13. H. Caldararu, A. Caragheorgheopol, A. Corma, F. Rey, V. Fornes, One-electron donor sites and their strength distribution on some hydrotalcite and MgO surfaces as studied by EPR spectroscopy. J. Chem. Soc. Faraday Trans **1**(90), 213–218 (1994)
14. Rudolf, C., Dragoi, B., Ungureanu, A., Chirieac, A., Royer, S., Nastro, A., Dumitriu. E.: NiAl and CoAl materials derived from takovite-like LDHs and related structures as efficient chemoselective hydrogenation catalysts. Catal. Sci. Technol. **1**(4), 179–89 (2014)
15. O. James, S. Maity, Temperature programme reduction (TPR) studies of cobalt phases in-alumina supported cobalt catalysts. J. Pet. Technol. Altern. Fuels **7**(1), 1–12 (2016)

Chapter 40
Processing of CO$_x$ Molecules in CO$_2$/O$_2$ Gas Mixture by Dielectric Barrier Discharge: Understanding the Effect of Internal Parameters of the Discharge

L. Saidia⊙, A. Belasri⊙, and S. Baadj⊙

Abstract Environmental pollution has become a major issue due to the rapid growth of industrial and technological developments that requires a high consumption of fossil energy. A new route of treatment of pollutant molecules bases on the use of non-equilibrium thermodynamic reactive plasmas generated by electrical discharges at atmospheric pressure to neutralize or transform toxic oxides as CO$_2$ [1–6]. This type of non-equilibrium reactive plasma can be used for the decontamination of gaseous effluents and is generally generated by a pulsed discharge which constitutes a chemically very active medium of low energy consumption. Our work will be based on a zero-dimensional model, to study the reduction of CO$_x$ in the CO$_2$/O$_2$ gas mixture by dielectric barrier discharge of non-equilibrium plasma under typical operating conditions of the discharge. A model allows to calculate the temporal evolutions of chemical characteristics. The influence of certain discharge parameters such as the applied electric voltage, the gas pressure, the capacity of the dielectric, the discharge frequency and the concentration of oxygen in the gaseous mixture on the density variations of CO and CO$_2$ compared to the initial density of CO$_2$ in the gas mixture of the discharge have been analyzed.

Keywords CO$_2$/O$_2$ gas mixture · CO$_x$ · Pulsed DBD discharge

40.1 Introduction

Mitigating greenhouse gas emissions represents an important problem in today's world. While still being the main propellant for the industrial progress, burning fossil fuels gives an ever-increasing rate of CO$_x$ emission in the atmosphere. As a result, finding technological solutions for CO$_x$ reduction became a rapidly growing

---

L. Saidia (✉) · A. Belasri · S. Baadj
Laboratoire de Physique des Plasmas, des Matériaux Conducteurs et leurs Applications (LPPMCA), Département de Physique Energétique, Faculté de Physique, Université des Sciences et de la Technologie d'Oran Mohamed Boudiaf USTO-MB, BP 1505, El M'naouer, 31000 Oran, Algeria
e-mail: saaidialarbi@live.fr

© The Editor(s) (if applicable) and The Author(s), under exclusive license to Springer Nature Singapore Pte Ltd. 2021
A. Khellaf (ed.), *Advances in Renewable Hydrogen and Other Sustainable Energy Carriers*, Springer Proceedings in Energy,
https://doi.org/10.1007/978-981-15-6595-3_40

research topic in many scientific fields. Plasmas are increasingly being used for gas conversion in both research and industrial applications [7–11], such as the destruction of large hydrocarbons, volatile organic compounds (VOCs). The constantly reducing usability of fossil fuels, in combination with the need to decrease greenhouse gas emissions, has given rise to the necessity of developing sustainable energy sources through greenhouse gas as raw materials [12–17].

40.2 Description of the Discharge Model

The electrical circuit used to deposit energy in the plasma is shown in Fig. 40.1, the dielectric layers are represented by two capacitances connected in series and Cd is their equivalent capacitance.

The applied voltage through the discharge is given by the following formula:

$$Vapp(t) = Vg(t) + Vd(t) \tag{40.1}$$

where Vg(t) and Vd(t) are the plasma and the dielectric voltages, respectively.

The voltage across the dielectrics is given by the relation:

$$V_d(t) = \frac{1}{C_d} \int I(t)dt \tag{40.2}$$

where Cd is the dielectric capacitance.

The relation between the current I and voltage Vg across the gap is:

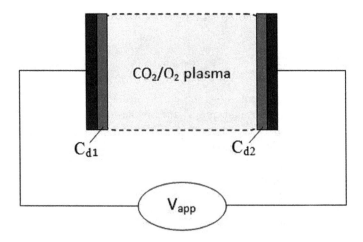

Fig. 40.1 Discharge model scheme used in the present calculations

$$I(t) = \frac{V_g(t)}{R_g(t)} \qquad (40.3)$$

The time dependence of the gap resistor Rg(t) of plasma is obtained by the following relation:

$$R_g(t) = \frac{d}{A \cdot e \cdot n_e(t) \cdot \mu_e(t)} \qquad (40.4)$$

where e, $n_e(t)$ and $\mu_e(t)$ are the electron charge, the time dependent electron density and mobility, respectively. A represents the discharge cross section in the plane of the electrodes and d is the separation between the discharge electrodes.

The system of equations describing the electrical circuit and the plasma kinetic is solved as follows: for a given time t, the plasma kinetic equations coupled with the electric circuit equations are solved with the classical GEAR method [18], between the instants t and t + dt.

In order to describe the electrical and chemical properties of the CO_2/O_2 plasma mixture, we have established a full set of processes involving twenty-one (21) species: e, O, O_2, O_3, C, CO, CO_2, C_2O, O^+, O_2^+, CO^+, CO_2^+, O^-, O_2^-, O_3^-, CO_3^-, CO_4^-, O(1D), $O_2(a)$, $O_2(b)$, CO_2^* regrouped in 113 chemical reactions. The rate coefficients of electron–molecule collisions, depending on the reduced electric field (E/N), are tabulated by solving the homogenous electron Boltzmann equation, and it can be obtained by the equation:

$$K_i = \sqrt{\frac{2e}{m}} \int_0^\infty \sigma_i \varepsilon f(\varepsilon) d\varepsilon \qquad (40.5)$$

where the parameters e, m, ε, and σ_i are the electron charge, the electron mass, electron energy and electron impact cross section of the process i, respectively, and $f(\varepsilon)$ is the electron energy distribution function (EEDF).

40.3 The Effect of Internal Parameters

In this subsection, we examine the effect of some discharge parameters, which are indicated below, on the CO concentration in the discharge and on the CO_2 conversion factor. The CO_2 conversion is defined as follows:

$$C_{CO_2}(\%) = \frac{[CO_2]_0 - [CO_2](t)}{[CO_2]_0} \times 100 \qquad (40.6)$$

where $[CO_2]_0$ is the initial concentration of CO_2 in the gas mixture (without plasma) and $[CO_2](t)$ is the concentration of CO_2 with plasma and at the instant t.

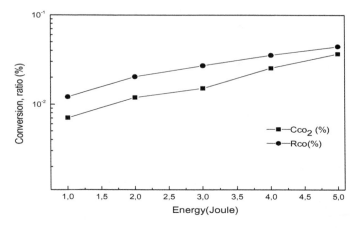

Fig. 40.2 Variations of carbon monoxide ratio and carbon dioxide conversion versus the energy deposed in the plasma

The ratio of carbon monoxide which was created during the discharge is defined by the following formulas:

$$R_{CO}(\%) = \frac{[CO](t)}{[CO_2]_0} \times 100 \qquad (40.7)$$

where $[CO](t)$ is the concentration of carbon monoxide at the instant t and N_T is the total density of the gas mixture.

40.3.1 Effect of the Applied Voltage

In Fig. 40.2, we have calculated the carbon dioxide conversion and carbon monoxide ratio under different values of applied voltage, for CO_2/O_2 (4%) gas mixture, frequency 50 kHz, gas pressure 500 Torr, dielectric capacitance 230 pF, and gas temperature 300 K. The obtained results indicate that the variations of these rates are slightly increased with the rising in the applied voltage. For voltage amplitude of 8 kV, the CO_2 conversion reaches a typical value of about 0.036%.

40.3.2 Effect of the Frequency

In order to see the influence of discharge frequency on the time evolutions of discharge behavior, we performed calculations in this subsection for the following parameters: $V_{app} = 6$ kV, $C_d = 230$ pF, gas pressure 500 Torr and gas temperature 300 K, and

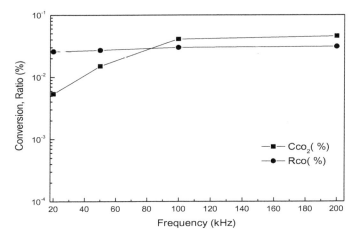

Fig. 40.3 Variations of carbon monoxide ratio and carbon dioxide conversion versus the frequency

4% of O_2 in CO_2. We have varied the frequency from 20 to 200 kHz. This range of frequency is appropriate in this study. The results reported in Fig. 40.3 represent the variation of CO_2 conversion and the CO ratio versus the discharge frequency. In addition, the ratio of CO is almost remains constant during the variation of frequency value. Moreover, the conversion of CO_2 is directly proportional to the frequency and its growth is quick at the discharge beginning (until 100 kHz). This conversion takes the greater value at highest frequency (200 kHz).

40.3.3 Effect of the Dielectric Capacitance

The calculations presented in this subsection are carried for the room temperature, frequency 50 kHz, gas pressure 500 Torr, Vapp = 4 kV, and 4% of O_2 in gas mixture.

In Fig. 40.4, we plotted the conversion of CO_2, CO ratio versus the dielectric capacitance. The evolutions of these rates are directly proportional to the increasing in the dielectric capacitance value. The CO_2 conversion takes the most value of around 0.145% at 2000 pF.

40.3.4 Effect of the Gas Pressure

The effect of the gas pressure under the following conditions: frequency 100 kHz, Vapp = 7 kV, Cd = 230 pF, gas temperature 300 K, and 4% of O_2 in CO_2, on CO_2 conversion and CO ratio is presented in Fig. 40.5. We clearly see that the conversion of CO_2 and CO ratio decrease with the increasing of the gas pressure value. The

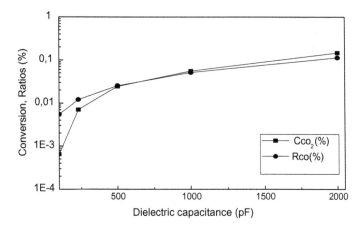

Fig. 40.4 Variations of carbon monoxide ratio and carbon dioxide conversion as a function of the dielectric capacitance

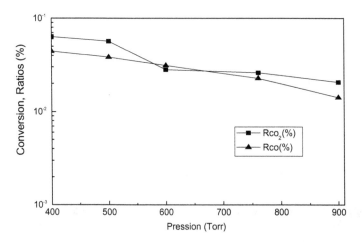

Fig. 40.5 Variations of the carbon monoxide ratio and carbon dioxide conversion as a function of the total gas pressure

most conversion of CO_2 was gained at lower pressure (400 Torr). For this pressure, the conversion of CO_2 is about 0.063%.

40.4 Conclusion

This work presents an electric and kinetic approach to study a homogeneous pulsed discharge in CO_2/O_2 gas mixture with typical operating conditions. It is based on a spatially homogeneous model (zero-dimensional model). Also, it is shown the effects of different plasma processing parameters such as applied voltage, gas pressure, dielectric capacitance, concentration of molecular oxygen in carbon dioxide, and frequency, on the discharge behavior. As well we have analyzed the effect of these parameters on the CO_2 conversion and the CO ratio. According to the calculated results and presented in this paper, on may conclude the following:

1. Higher conversion of CO_2, and CO ration can be obtained by increasing in the values of applied voltage, dielectric capacitance, and/or decreasing gas pressure.
2. CO_2 conversion is directly proportional to the rise of discharge frequency.
3. The discharge frequency indicates a weak influence on the ratio of carbon monoxide. In addition, CO_2 conversion is strongly affected by the dielectric capacitance.

References

1. L. Saidia, A. Belasri, S. Baadj, Z. Harrache, Physico-Chemical Investigation of Pulsed Discharge in CO_2/O_2 Gas Mixture. Plasma Phys. Rep. **5**(45), 501–516 (2019)
2. R. Aerts, T. Martens, A. Bogaerts, Influence of Vibrational States on CO_2 Splitting by Dielectric Barrier Discharges. Physical Chemistry C **116**(44), 23257–23273 (2012)
3. S. Ponduri, M.M. Becker, S. Welzel, M.C.M. Van de Sanden, D. Loffhagen, R. Engeln, Fluid modelling of CO_2 dissociation in a dielectric barrier discharge. Appl. Phys. **119**(9), 093301 (2016)
4. T. Kozàk, A. Bogaerts, Splitting of CO_2 by vibrational excitation in non-equilibrium plasmas: a reaction kinetics model. Plasma Sources Sci. Technol. **23**(4), 045004 (2014)
5. M.S. Moss, K. Yanallah, R.W.K. Allen, F. Pontiga, An investigation of CO_2 splitting using nanosecond pulsed corona discharge: effect of argon addition on CO_2 conversion and energy efficiency. Plasma Sources Sci. Technol. **26**(3), 035009 (2017)
6. G. Horvath, J.D. Skalny, N.J. Mason, FTIR study of decomposition of carbon dioxide in dc corona discharges. Physics. D: Appl. Physics. **41**(22), 225207 (2008)
7. H.H. Kim, Nonthermal plasma processing for air-pollution control: a historical review, current issues, and future prospects. Plasma Processes Polym. **1**(2), 91–110 (2004)
8. J. Van Durme, J. Dewulf, C. Leys, H. Van Langenhove, Combining non-thermal plasma with heterogeneous catalysis in waste gas treatment: a review. Appl. Catal. B **78**(3–4), 324–333 (2008)
9. U. Kogelschatz, B. Eliasson, W. Egli, From ozone generators to flat television screens: history and future potential of dielectric-barrier discharges. Pure Appl. Chem. **71**(10), 1819–1828 (1999)
10. U. Kogelschatz, Dielectric-barrier: discharges: their history, discharge physics, and industrial applications. Plasma Chem. Plasma Process. **23**, 1–46 (2003)
11. R. Snoeckx, A. Bogaerts, Plasma technology – a novel solution for CO_2 conversion? Chem. Soc. Rev. **46**, 5805–5863 (2017)
12. J. Han, L. Zhang, H.J. Kim, Y. Kasadani, L.Y. Li, T. Shimizu, Fast pyrolysis and combustion characteristic of three different brown coals. Fuel Process. Technol. **176**, 15–20 (2018)
13. J. Han, L. Zhang, B. Zhao, L.B. Qin, Y. Wang, F.T. Xing, The N-doped activated carbon derived from sugarcane bagasse for CO_2 adsorption. Ind. Crops Prod. **128**, 290–297 (2019)
14. B. Zhao, Y. Liu, Z. Zhu, H. Guo, X. Ma, Highly selective conversion of CO_2 into ethanol on Cu/ZnO/Al_2O_3 catalyst with the assistance of plasma. CO_2 Utilization **24**, 34–39 (2018)

15. G.R. Kale, S. Doke, A. Anjikar, Process thermoneutral point in dry autothermal reforming for CO_2 utilization. CO_2 Utilization **18**, 318–325 (2017)
16. A.A. Khan, M. Tahir, Recent advancements in engineering approach towards design of photoreactors for selective photocatalytic CO_2 reduction to renewable fuels. CO_2 Utilization **29**, 205–239 (2019)
17. J.O. Pou, C. Colominas, R. Gonzalez-Olmos, CO_2 reduction using non-thermal plasma generated with photovoltaic energy in a fluidized reactor. CO_2 Utilisation **27**, 528–535 (2018)
18. C.W. Gear, *Numerical Initial Value Problems in Ordinary Differential Equations* (Prentice-Hall, Englewood Cliffs, NJ, 1971)

Chapter 41
Performance Comparison of a Wankel SI Engine Fuelled with Gasoline and Ethanol Blended Hydrogen

Fethia Amrouche, P. A. Erickson, J. W. Park, and S. Varnhagen

Abstract The main interest of the Wankel rotary engine is its higher power density as compared to similar conventional engine. However, the unusual geometry of the Wankel engine affects negatively the engine economy and the exhaust emissions. The knowledge of the rotary Wankel engine drawbacks and fuel characteristics can help to choose which fuel can achieve the best performance with that kind of engine. The aim of this paper is to compare the experimental results of the performance of a monorotor Wankel engine fuelled with gasoline, ethanol and then after hydrogen addition to each fuel used, at lean operating condition and full load regime. Testing were carried out under constant engine speed of 3000 rpm, fixed spark timing of 15° BTDC, at lean to ultra lean and full load conditions. The test results have shown that pure ethanol burns more efficiently than gasoline. Moreover, the addition of hydrogen helps to achieve better brake mean effective pressure, thermal efficiency and reduce the fuel consumption for both fuel, however, this improvement is more significant for gasoline engine.

Keywords Lean burn · Ultra lean · Wankel rotary engine · Hydrogen enrichment · Gasoline · Ethanol

41.1 Introduction

The rotary Wankel engine is an alternative of the reciprocating engine [1]. The Wankel engine can be used in conventional vehicle as well as in Hybrid Electric Vehicles (HEV) for range extended hybrid, designed for much lighter vehicles [2]. However, the uncommon geometry of the Wankel engine affects negatively the engine economy and the exhaust emissions [3]. Indeed, the large surface to volume ratio

---

F. Amrouche (✉)
Centre de Développement des Énergies Renouvelables, CDER, 16340 Algiers, Algeria
e-mail: fethia.amrouche@gmail.com

P. A. Erickson · J. W. Park · S. Varnhagen
Mechanical and Aerospace Engineering Department, UC Davis, CA 95616 Davis, USA

of the combustion chamber slowdown the spread of the flame front [1, 4], which expands the quenching area and causes high heat transfer rates [5]. This slower the combustion rate and reduce the efficiency of the rotary Wankel engine as compared to in the reciprocating engine. The characteristics of the fuel used to run an engine, define the performance and emissions of the engine [6, 7]. Indeed, the high adiabatic flame speed, heat of vaporization and octane number help to improve the anti-knock characteristics, which allows a larger compression ratio to be used. These characteristics also helps to enhance the thermal efficiency and potentially increase the power output. Hydroxyl group of alcohol fuels benefits a complete combustion that enhance the engine efficiency and reduce the CO and HC emissions [8]. Using ethanol instead of gasoline in a Wankel rotary engine could be a way to overcome the Wankel engine drawbacks, which are reduced thermal efficiency and high levels of hydrocarbons emissions [9].

In addition, lean burn is a strategy commonly used to improve the engine economy and reduce exhausted pollution of the reciprocating engine [10]. Hydrogen's combustion proprieties are helpful to improve the engine's lean burn ability. Consequently, the use of the lean burn strategy through hydrogen enrichment could potentially helpful to improve and even extend the engine lean burn capability [6, 7].

In the literature, many researches have been conducted on the engines operating with hydrogen-blended fuels at lean and ultra-lean burn conditions [1, 6, 7]. Amrouche et al. [3, 9] have already investigated the effect of hydrogen enrichment on performance, combustion characteristics and emissions of gasoline and ethanol Wankel engine respectively, at lean and ultra-lean burn and full load. However, a small number of papers have compared some of these fuels together within the same Wankel engine. And none of them compared gasoline and ethanol enriched hydrogen fulled Wankel engine. Therefore, the aim of this paper is to compare the experimental results of the performance and emissions of a monorotor Wankel engine at lean to ultra lean burn and full load regime, while fuelled separately with gasoline and ethanol, and then after hydrogen addition to each fuel used.

41.2 Experimental Procedure

The engine used in this test bench was manufactured by Outboard Marine Corporation, USA. It is a 0.530L single rotor, air cooled Wankel engine, using a single spark plug. The spark timings, injection timings and durations of gasoline, ethanol and hydrogen were controlled via a hybrid electronic control unit (HECU) developed in the laboratory. Various concentrations of hydrogen blends 0–18% for ethanol and 0–10% for gasoline by energy were analyzed. To achieve real time control over the engine air/fuel mixture preparation as well as hydrogen addition, two fuel injection systems were added, one for gasoline/ethanol and the other for pure hydrogen. A calibrated Micromotion CMF010M Coriolis flow meter with an RFT9739 transmitter with an accuracy of 0.10% of rate for flow rates of 0–23 g/s was used to meter the gasoline/ethanol fuel flow. The hydrogen used in this study was Bottled industrial

Table 41.1 Tests operating conditions

Parameters	Value
Ignition timing (degree crank angle BTDC)	15°
Engine speed (rpm)	3000
Throttle position	Wide open
Hydrogen energy fraction	Gasoline: 0%, %2, 4%, 5%, 7%, 10% Ethanol: 0%, 3%, 5%, 6%, 8%, 9%, 12%, 15%, 18%
Equivalence ratio	Gasoline: 0.77 Ethanol: 0.47

hydrogen (99.95% purity) that is regulated and metered via a calibrated Aalborg differential pressure mass flow controller, model GFC 47, with an accuracy of ±3% (0–20% full scale), ±1.5% (20–100% full scale). The air mass flow was metered via a designed orifice plat.

The energy fraction of hydrogen in the total intake gas is defined as energy fraction, calculated as follow [3]:

$$\%H_2 = \left[\frac{(\dot{m}_{H_2} \times LHV_{H_2})}{(\dot{m}_f \times LHV_f) + (\dot{m}_{H_2} \times LHV_{H_2})} \right] \times 100 \quad (41.1)$$

where $\dot{m}_{H_2}$, $\dot{m}_f$ are respectively the mass flow rate of Hydrogen and Fuel (gasoline/ethanol) (g/s). LHV_{H_2}, LHV_f are the lower heating value respectively of Hydrogen and fuel (gasoline/ethanol) (MJ/Kg), such as $LHV_{H_2} = 120.1$, $LHV_g = 43.5$; $LHV_E = 26.9$.

A Telma CC100 eddy current dynamometer was coupled to the engine to control and measure the engine speed and torque output.

The operating condition used are listed in Table 41.1.

41.3 Results and Discussion

The brake mean effective pressure is one of the important aspects that determine the performance of an engine. Indeed, this parameter reflected the power output Fig. 41.1 shows the effect of fuel used and hydrogen enrichment on engine brake mean effective pressure "BMEP". The BMEP are higher for ethanol than for gasoline. However, with hydrogen enrichment, the increased BMEP is more evident for gasoline than for ethanol (see Fig. 41.2).

The increase of the BMEP for ethanol is due to ethanol's heat of evaporation that is higher compared to that of gasoline, thus providing air-fuel charge cooling and increasing the density of the charge. Therefore, a higher power output is obtained.

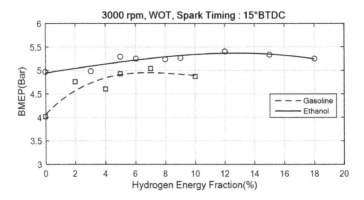

Fig. 41.1 BMEP of gasoline and ethanol enriched hydrogen

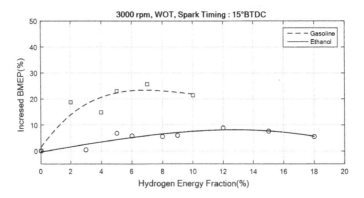

Fig. 41.2 Increased BMEP for gasoline and ethanol enriched hydrogen

This can be verified through the volumetric efficiency graph. Indeed, according to Fig. 41.3, that presents the volumetric efficiency of Gasoline and Ethanol blended hydrogen, Ethanol shows higher volumetric efficiency compared to that of gasoline.

Also according to Fig. 41.3, the addition of hydrogen decrease slightly the volumetric efficiency for both fuels. This is because hydrogen displaces part of the intake air. Thus, reduce the present air in the combustion chamber. However, at these operating condition, this does not affects the BMEP, this is because the engine run at lean condition. Lean mixture ignite hardly and the addition of hydrogen help to improve the ignition and burning rate of the lean mixture.

Engine economy is reflected through thermal efficiency. Figure 41.4 shows the changes of the increased brake thermal efficiency for gasoline, ethanol and with hydrogen enrichment. It can be observed that the increase of brake thermal efficiency is more significant with hydrogen addition to gasoline than to ethanol. However, it is important to indicated that the thermal efficiency of pure ethanol is higher by more than 26% than that of pure gasoline. This can be explained by better combustion

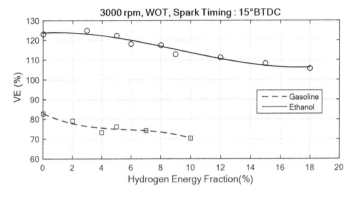

Fig. 41.3 Volumetric efficiency for gasoline and ethanol enriched hydrogen

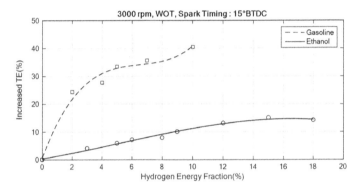

Fig. 41.4 Increased thermal efficiency for gasoline and ethanol enriched hydrogen

efficiency of oxygenated fuels and the improvement of the heat transfer within the Wankel engine elongated chamber.

Figure 41.5 shows the variations of the BSFC for ethanol, gasoline and with hydrogen enrichment. As shown in this figure, the BSFC increased for pure ethanol, and it decrease with hydrogen addition for both fuel. However, the specific fuel consumption of ethanol was highest compared to that of gasoline for all hydrogen enrichment. Moreover, hydrogen has greater effect on the reduction of gasoline consumption than for ethanol. This is described with heating value, and stoichiometric air-fuel ratio that is the smallest for ethanol, which means that for specific air-fuel equivalence ratio, more fuel is needed. Quetz de Almeida et al. [11] have observed that the addition of small amount of hydrogen gives no significant change on fuel consumption for operation at a specific mixture strength from stoichiometric to lean conditions.

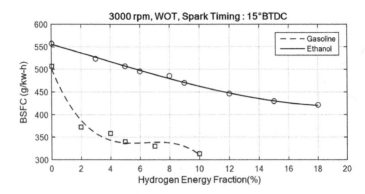

Fig. 41.5 BSFC for gasoline and ethanol enriched hydrogen

41.4 Conclusion

The purpose of the present paper is to demonstrate the influence of the fuel used (ethanol, gasoline) blended hydrogen on Wankel spark-ignition engine performance. The experimental data collected through this work, have shown that the use of pure ethanol can improve the engine performance compared to gasoline. However, ethanol is a high fuel consumer. The addition of hydrogen at lean condition improve the combustion characteristics, and thus the BMEP and the engine economy that is reflected over the thermal efficiency and BSFC. However, through this work, it can be noticed that hydrogen affect more significantly gasoline as fuel than ethanol.

Acknowledgements The authors thank Moller International for their donation of the Wankel research engines. The University of California, Davis, Green Transportation Laboratory and the Energy Research Laboratory and all their associated members made this work possible.

References

1. F. Amrouche, P.A. Erickson, S. Varnhagen, J.W. Park, An experimental analysis of hydrogen enrichment on combustion characteristics of a gasoline Wankel engine at full load and lean burn regime. Int. J. Hydrogen Energy **43**(41), 19231–19242 (2018)
2. H.T. Arat, Alternative fuelled hybrid electric vehicle (AF-HEV) with hydrogen enriched internal combustion engine. Int. J. Hydrogen Energy **44**(34), 19005–19016 (2019)
3. F. Amrouche, P. Erickson, J. Park, S. Varnhagen, An experimental investigation of hydrogen-enriched gasoline in a Wankel rotary engine. Int. J. Hydrogen Energy **39**(16), 8525–8534 (2014)
4. E. Chouinard, F.J. Hamady, H.J. Schock, Airflow visualization and LDA measurements in a motored rotary engine assembly: part 2. SAE paper 900031 (1990)
5. T. Su, C. Ji, S. Wang, L. Shi, J. Yang, X. Cong, Effect of spark timing on performance of a hydrogen-gasoline rotary engine. Energy Convers. Manage. **148**, 120–127 (2017)

6. F. Amrouche, P.A. Erickson, J.W. Park, An experimental evaluation of ultra-lean burn capability of a hydrogen enriched ethanol-fuelled Wankel engine at full load condition. Int. J. Hydrogen Energy **41**(42), 19231–19242 (2016)
7. F. Amrouche, P.A. Erickson, J.W. Park, S. Varnhagen, Extending the lean operation limit of a gasoline Wankel rotary engine using hydrogen enrichment. Int. J. Hydrogen Energy **41**(32), 14261–14271 (2016)
8. B. Zhang, C. Ji, S. Wang, Performance of a hydrogen-enriched ethanol engine at unthrottled and lean conditions. Energy Convers. Manage. **114**, 68–74 (2016)
9. F. Amrouche, P.A. Erickson, S. Varnhagen, J.W. Park, An experimental study of a hydrogen-enriched ethanol fueled Wankel rotary engine at ultra-lean and full load conditions. Energy Convers. Manage. **123**, 174–184 (2016)
10. D. Dunn-Rankin, *Lean Combustion: Technology and Control, 1st edn*. Academic Press (2007)
11. L. Quetz de Almeida, L.C. Monteiro Sales, J.R. Sodré, Fuel consumption and emissions from a vehicle operating with ethanol, gasoline and hydrogen produced on-board. Int. J. Hydrogen Energy **40**(21), 6988–6994 (2015)

Chapter 42
Parametric Study of SO_3 Conversion to SO_2 in Tubular Reactor for Hydrogen Production *via* Sulfuric Cycle

F. Lassouane

Abstract Hybrid sulfur (HyS) cycle or Westinghouse cycle is among the main candidates for full-scale solar hydrogen production. This cycle is one of the most attractive and simplest thermochemical processes, because it comprises only two global reaction steps and has only fluid reactants. The decomposition of sulfuric acid is a vital step in the sulfuric hybrid cycle process for the production of hydrogen from water. For this, the temperature needed for SO_3 to be converted into SO_2 and O_2 is very high. It requires operating temperatures above 800 °C. Thus, SO_2 compound resulting from the decomposition reactor is the key compound used to feed the electrolyser where it is oxidized to regenerate H_2SO_4 and produce H_2. This work focused on the sulfuric acid decomposition step in suggested tubular plug flow reactor with possible use of solar energy. A parametric study on the effects of temperature and pressure on the thermodynamics conversion of SO_3 to SO_2 was investigated. The theoretical sulfur trioxide conversion calculation was applied where a plug flow reactor model was assumed.

Keywords Hydrogen · Hybrid sulfuric cycle · Plug flow reactor · Solar energy

42.1 Introduction

Hydrogen production process by splitting water cycles is the viable alternative for substitution of fossil fuels in large-scale. This process is efficient and cost-effective, non-pollutant and safe.

The sulfuric cycles is well-known processes proposed in the first half of the twentieth century such as the hybrid sulfur cycle (Westinghouse cycle), the sulfur-iodine cycle and the sulfur bromine hybrid cycle [1]. In fact, these cycles have in common the decomposition of sulfuric acid session, which is the crucial step in the process. The hybrid sulfuric cycle investigated in this study requires a high

F. Lassouane (✉)
Centre de Développement des Energies Renouvelables CDER, BP. 62, Route de l'Observatoire, 16340 Alger, Algeria
e-mail: flassouane@gmail.com

temperature heat source to decompose the sulfuric acid in the sulfur dioxide SO_2. The latter compound will supply the electrolyser unit to produce hydrogen [2].

The use of solar energy is one of the most promising energy technologies that can considerably contribute to a sustainable energy supply in the future. Several research groups have focused on reactor concepts for sulfuric acid decomposition by solar route such as the General Atomics groups, Sandia National Laboratories in USA and JRC-Ispra in Italy [3, 4]. Likewise, the German Aerospace center (DLR) developed a solar receiver-reacor within the European project HYTHEC (HYdrogen THErmochemical Cycles) [5].

In this paper, the hybrid sulfuric thermochemical cycle concept are described by focusing on the decomposition of sulfur acid with the solar energy in the tubular reactor. A study of thermal decomposition of sulfuric acid, the influence of temperature and pressure on the SO_3 conversion to SO_2 was carried out.

42.2 Description of the Hybrid Sulfuric Cycle Acid Decomposition Section

The decomposition of sulfuric acid is the most endothermic reaction that takes place at high temperatures (800–900 °C). Indeed, the decomposition reaction occurs in two steps: the non-catalytic thermal decomposition of sulfuric acid to sulfur trioxide and water vapor and Thermo-catalytic reduction of sulfur trioxide to sulfur dioxide and oxygen. The first reaction step is fast and takes place spontaneously, while the reduction of SO_3 in SO_2 is very low even at high temperatures. In order to increase the reaction rate, the presence of catalyst is needed [4].

Banerjee et al. [6] describe the decomposition of sulfuric acid as the "most endothermic step" of researched sulfur-based cycles for hydrogen production, which is also observed in the standard heats of the reactions.

Kondamudi et al. [7] states that two reactions can take place simultaneously, wherein the sulfuric acid is first vaporized and decomposed into sulfur trioxide and water, followed by a second stage where the sulfur trioxide is decomposed into sulfur dioxide and oxygen.

42.2.1 SO_3 Decomposition in Tubular Reactor

The decomposition of pure sulfuric acid take place in two simultaneous steps:

$$H_2SO_4(l) \rightarrow SO_3(g) + H_2O(g) \tag{42.1}$$

$$SO_3(g) \rightarrow SO_2(g) + 1/2\, O_2(g) \tag{42.2}$$

After H_2SO_4 is vaporized and superheated, the decomposition of gaseous sulfuric acid into sulfur dioxide (first reaction) is spontaneously occurs at the temperature of about 400 °C (673 K). In the second reaction, the sulfur trioxide is converted on sulfur dioxide and oxygen at high temperature (>800 °C, 1073 K) in the presence of catalyst [4].

42.2.2 Reactor Designs Used for the SO_3 Decomposition

Several reactor designs have been developed and used for sulfur trioxide decomposition. Thomey et al. [4] developed a reactor for decomposition of sulfuric acid by concentrated solar radiation, tested in the solar furnace of DLR in Cologne using ferric oxide (Fe_2O_3), and mixed oxides ($CuFe_2O_4$) catalysts. Sander et al. [8] have been used a laboratory scale fixed bed reactor for the decomposition of sulfur trioxide with a supported platinum and palladium-based catalyst. Solar reactor system has been investigated for SO_3 conversion using ferric oxide (Fe_2O_3) on an Al_2O_3 support within a temperature range of 1050–1200 K [9].

Choi et al. [10] have been modelled the conventional shell and tube reactors operated with helium. Another reactor design has also been developed to decompose the SO_3 which include the bayonet type decomposer that were successfully tested [11].

42.3 Tubular Reactor for Thermodynamic Study of SO_3 Decomposition

Solar thermochemical processes are thermodynamically favorable ways for producing solar hydrogen because of the good potential for converting the sustainable solar energy into chemical energy efficiently.

The conversion of SO_3–SO_2 (X) can be obtained in proposed plug flow tubular reactor fed (Fig. 42.1) by concentrated solar radiation.

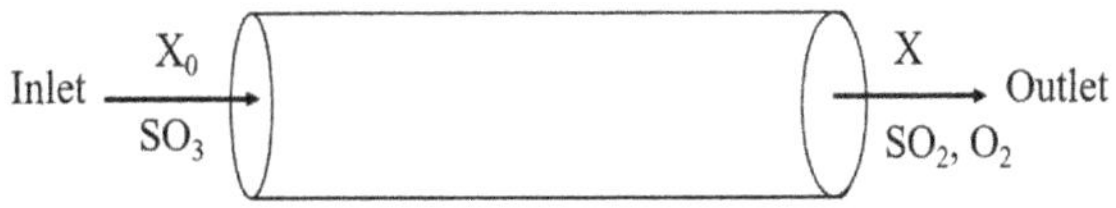

Fig. 42.1 Schematic of plug flow tubular reactor

42.3.1 Standard Gibbs Energy ΔG_r^0 of the Reaction

The relation of the standard Gibbs energy (J.mol.$^{-1}$.K^{-1}) for the reaction is given by:

$$\Delta G_r^\circ = \Delta H_r^\circ - T \Delta S_r^\circ \tag{42.3}$$

To evaluate the enthalpy of the reaction, first the change in the enthalpy, entropy of the individual species entering in the reaction are estimated using the following equations:

$$\Delta H_r^\circ = \Delta H_{298}^\circ + \int_{298}^{T} \Delta C_p^\circ dT \tag{42.4}$$

$$\Delta S_r^\circ = \Delta S_{298}^\circ + \int_{298}^{T} \frac{\Delta C_p^\circ}{T} dT \tag{42.5}$$

The thermodynamics data for each compound (SO_3, SO_2 and O_2) are given in the literature [12]. Cp° is the specific heat at constant pressure of each species. The reference temperature is taken as the room temperature (298 K).

42.3.2 Equilibrium Constant K Calculation

The equilibrium constant is calculated from the standard Gibbs energy method:

$$-\ln K = \frac{\Delta G_T^\circ}{RT} \tag{42.6}$$

K is the equilibrium constant and R is the gas constant (8.314 J.K^{-1}.mol^{-1}).

The parameters are calculated in the temperature range of 300 and 1200 K with the different at Gibbs energies for each compound involved in the composition reaction.

42.3.3 Estimation of Equilibrium Conversion (X)

The relationship between the equilibrium constant K and the composition of SO_3, SO_2 and O_2 is given by the following equation [13]:

$$\left(\frac{P}{P^\circ}\right)^\gamma K = \prod (Y_i \varphi_i)^{\gamma_i} \tag{42.7}$$

The molar fractions for each component (Y_{SO_3}, Y_{SO_2}, Y_{O_2}) were determined with the following equations:

$$Y_{SO_3} = \frac{1-X}{1+0.5X} \quad (42.8)$$

$$Y_{SO_2} = \frac{X}{1+0.5X} \quad (42.9)$$

$$Y_{O_2} = \frac{0.5X}{1+0.5X} \quad (42.10)$$

P is the operating pressure (bar), $P°$ is the pressure in standard state taken for one bar and Yi is the molar fraction (%) of each chemical.

φi, γ, γ_i are the fugacity coefficient for each compound (SO_3, SO_2 and O_2), global stoichiometric Coefficient and stoichiometric Coefficient for each element, respectively.

Equation (42.7) relates K to fugacities of the reacting species, as they exist in the real equilibrium mixture. These fugacities reflect the nonidealities of the equilibrium mixture and are functions of temperature, pressure, and composition. They have been calculated from the data given in the literature [13].

The standard pressure was set at 1 bar while the operating pressure was set at values of 1, 3, 6, 9, 12 and 20 bar.

42.4 Results

The molar fractions of SO_3, SO_2 and O_2 at 1200 K were presented in Fig. 42.2. In this temperature of the SO_3 decomposition, the highest molar fraction of SO_2 was obtained at 1 bar.

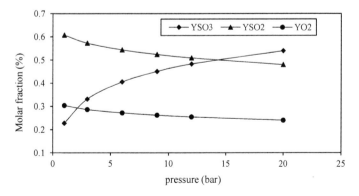

Fig. 42.2 Molar fraction variation of SO_3 to SO_2 and O_2 at different pressures

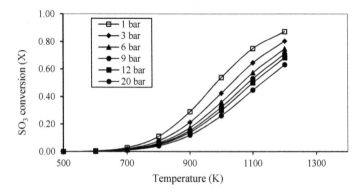

Fig. 42.3 Estimation of equilibrium conversion of SO_3 to SO_2 as function of temperature at different pressure

The equilibrium constant values for the temperature interval between 300 and 1200 K at different operating pressure are presented in Fig. 42.3. The plot clearly shows the significant influence of pressure on the equilibrium conversion. Indeed, the lowest operating pressure increase the equilibrium conversions of SO_3 to SO_2.

The study of Coetzee. [14] was compared with conversions obtained at the same conditions in order to validate the calculations that were done. It can be seen that the results were almost identical.

The pressure of 1 bar allows for conversion of 87.17% at temperature of 1200 K.

42.5 Conclusion

We have presented the thermodynamics of sulfuric trioxide SO_3 decomposition reaction and given a parametric study showing the effects of temperature, pressure on the SO_3 conversion. Thermodynamic results show that the conversions on SO_2 decrease with increasing pressure.

Thermodynamic data of SO_3 reduction to SO_2 in the calculated temperature range of 300–1200 °C shows that the reaction is favored by increasing the temperature.

References

1. L.M. Gandia, G. Arzamedi, P.M. Dieguez, *Renewable hydrogen technologies: Production, purification, storage, applications and safety* (Elsevier, USA, 2013)
2. Banerjee, A. M., Shirole, A. R., Pai, M. R., Tripathi, A. K., Bharadwaj, S. R., Das, D., Sinha, P. K.: Catalytic activities of Fe_2O_3 and chromium doped Fe_2O_3 for sulfuric acid decomposition reaction in an integrated boiler, preheater, and catalytic decomposer. Appl. Catal, B-ENVIRON **127**, 36–46 (2012)

3. General Atomics. Decomposition of sulfuric acid using solar thermal energy, GA-A17573. General Atomics. (1985)
4. D. Thomey, L. de Oliveira, J.P. Sack, M. Roeb, C. Sattler, Development and test of a solar reactor for decomposition of sulphuric acid in thermochemical hydrogen production. Int. J. Hydrog. Energ. **37**(21), 16615–16622 (2012)
5. A. Le Duigou, J.M. Borgard, B. Larousse, D. Doizi, R. Allen, B.C. Ewan et al., HYTHEC: An EC funded search for a long-term massive hydrogen production route using solar and nuclear technologies. Int. J. Hydrog. Energ. **32**(10), 1516–1529 (2007)
6. A.M. Banerjee, M.R. Pai, S.S. Meena, A.K. Tripathi, S.R. Bharadwaj, Catalytic activities of cobalt, nickel and copper ferrospinels for sulfuric acid decomposition: The high temperature step in the sulphur based thermochemical water splitting cycles. Int. J. Hydrog. Energ. **36**(8), 1–13 (2011)
7. K. Kondamudi, S. Upadhyayula, Kinetic studies of sulfuric acid decomposition over $Al-Fe_2O_3$ catalyst in the sulfur-iodine cycle for hydrogen production. Int. J. Hydrog. Energ. **37**(4), 3586–3594 (2012)
8. Stander, B. F., Everson, R. C., Neomagus, H. W. J. P., Van der Merwe, A. F., Tietz, M. R.: Sulphur trioxide decomposition with supported platinum/palladium on rutile catalyst: 2. Performance of a laboratory fixed bed reactor. Int. J. Hydrog. Energ. **4**(6), 2493–2499 (2015)
9. Brutti, S., De Maria, G., Cerri, G., Giovannelli, A., Brunetti, B., Cafarelli, P., Semprin, E., Barbarossa, V., Ceroli. A.: Decomposition of H_2SO_4 by direct solar radiation. Ind. Eng. Chem. Res **46**, 6393–6400 (2007)
10. J.H. Choi, N. Tak, Y.J. Shin, C.S. Kim, K.Y. Lee, Numerical analysis of a SO_3 packed column decomposition with alloy RA330 structural material for nuclear hydrogen production using the sulfur iodine process. Nucl. Eng. Technol. **41**(10), 1275–1284 (2009)
11. V. Nagarajan, V. Ponyavin, Y. Chen, M.E. Vernon, P. Pickard, A.E. Hechanova, CFD modelling and experimental validation of sulfur trioxide decomposition in bayonet type heat exchanger and chemical decomposer with porous media zone and different packed bed designs. Int. J. Hydrog. Energ. **33**(6), 6445–6455 (2009)
12. Kotz, J. C., Treichel, P. M., Townsend, J. R.: Chemistry and chemical reactivity, 8th edn, Brooks/Cole, Cengage learning, Appendix L, A-31. USA (2012)
13. J.M. Smith, H.C. VanNess, M.M. Abbott, *Introduction to chemical engineering thermodynamics*, 6th edn. (Mc Graw Hill, New York, 2001)
14. Coetzee, MD.: The chemical reactor for the decomposition of sulphuric acid for the hybrid sulphur process. Ph.D. Dissertation, North West University (2008)

Chapter 43
Preparation and Physical/Electrochemical Characterization of the Hetero-System 10% NiO/γ–Al$_2$O$_3$

I. Sebai, R. Bagtache, A. Boulahaouache, N. Salhi, and Mohamed Trari

Abstract 10% NiO/γ–Al$_2$O$_3$ was synthesized by wet impregnating γ–Al$_2$O$_3$ with Ni(NO$_3$)$_2$, 6H$_2$O solution. After evaporation, the sample was calcined in air at 450 °C. The insulator γ–Al$_2$O$_3$ was used as support to offer a large distribution of NiO, leading to a higher specific surface area, an enhanced photosensibility. The sensitizer NiO was characterized by physical and photoelectrochemical techniques. The XRD pattern exhibits the peaks of both γ–Al$_2$O$_3$ and NiO. The particle size of NiO (18 nm) was calculated from the full width at half maximum. The optical gap was found at 1.51 eV and the transition is directly permitted. The capacitance measurement indicates n-type semiconductor and the potential of conduction band (-0.35 V$_{SCE}$), is more cathodic than the H$_2$O/H$_2$ level (~ -0.3 V$_{SCE}$) at pH $\sim$13. Such condition generates a direct water reduction under visible light illumination. The hydrogen yield of 10% NiO/γ–Al$_2$O$_3$ was compared with those produced with NiO alone.

Keywords NiO · γ–Al$_2$O$_3$ · Photo-electrochemical · Wet impregnation · Hydrogen · Visible light

43.1 Introduction

The demand of hydrogen as clean fuel increased over the recent years. Currently, it is mainly obtained from polluting non-renewable process [1] and the catalytic steam reforming of hydrocarbon takes the most important industrial route of H$_2$ production [2, 3]. The conversion of solar energy into chemical energy in the form of hydrogen in presence of semiconductor has become an attractive alternative [4,

---

I. Sebai (✉) · N. Salhi
Faculty of Chemistry, USTHB, LCGN, BP 32, 16111 Algiers, Algeria
e-mail: ibtissamsebai.ch@gmail.com

R. Bagtache · M. Trari
Faculty of Chemistry, USTHB, LSVER, BP 32, 16111 Algiers, Algeria

A. Boulahaouache · N. Salhi
Faculty of Sciences, University of BLIDA1, LCPMM, BP 270 Soumaa Road, Blida, Algeria

© The Editor(s) (if applicable) and The Author(s), under exclusive license to Springer Nature Singapore Pte Ltd. 2021
A. Khellaf (ed.), *Advances in Renewable Hydrogen and Other Sustainable Energy Carriers*, Springer Proceedings in Energy,
https://doi.org/10.1007/978-981-15-6595-3_43

5]. Therefore, a great attention has been focused on the photoelectrochemical water splitting as resource of hydrogen. Nowadays, this renewable and sustainable procedure is considered as a cleanest strategy which reduces considerably the emission of greenhouse gases [6].

Many semiconductors were synthesized, characterized by photo-electrochemistry and successfully tested to hydrogen production [7, 8]. Among these materials, oxides have received a particular attention due to their low cost, easy synthesis and high photo-catalytic performance [5]. NiO is found to be an effective semiconductor owing to its chemical stability over a wide pH range (3–14). It can work as electrons pump and the position of its conduction band, made up of cationic orbital Ni^{2+} 3d, is more cathodic than of the H_2O/H_2 level with a high reducing ability, yielding spontaneous hydrogen formation. Also, NiO has an optical gap (E_g) near to the ideal value for photo-electrochemical applications [9, 10].

The photocatalyst-support interface interaction is beneficial for the charge separation and transfer and offer large metal distribution in surface provides more active sites [9]. It is important that the photocatalyst does not diffuse deeply in the support matrix, that will weaken the visible light to excite the photocatalyst. Also, previous works suggested that the semiconductor loading on an insulator or a wide band gap semiconductor, can possibly prevent the loss by charge recombination [11, 12].

In this respect, 10% $NiO/\gamma-Al_2O_3$ was prepared by impregnation method and evaluated for the H_2 formation under visible light irradiation. NiO works as sensitizer, it was characterized by physical and photo-electrochemistry methods. The $\gamma-Al_2O_3$ was used as support to improve remarkably the photoactivity of hydrogen production of NiO, by offering a large distribution of NiO, thus inducing a higher specific surface area. In addition, using 10% of NiO instead of pure NiO is economically extremely favorable.

43.2 Experimental

The hetero-system was prepared by wet impregnation of $Ni(NO_3)_2$, $6H_2O$ (1.7 M, Merck) on $\gamma-Al_2O_3$ (Aldrich) in proportion 10%–90% respectively. The solutions were stirred during 1 h and then evaporated at 80 °C. The solids were dried at 100 °C (12 h), calcined at 450 °C (4 h) with a heating rate of 5 °C min^{-1}.

The phases were identified by X-ray diffraction (XRD) equipped with CuK_α anticathode ($\lambda = 0.15418$ nm) over the 2θ range (15–80°). The morphology was investigated by the scanning electron microscopy (JEOL JSM6360-LV). N_2 adsorption–desorption isotherms were used to determine the BET surface areas and pore size distribution (Micromeritics type apparatus at 77 K).

The forbidden band (E_g) of NiO was measured with a Jasco 650 UV-Vis spectrophotometer in presence of $BaSO_4$ as blank.

The electrical contact on the pellet, sintered at 500 °C, was made with silver cement. The photoelectrochemical properties were studied in a standard cell containing Pt auxiliary electrode and a saturated calomel electrode (SCE). The electrode potential was piloted by a computer aided Versa STAT 3 potentiostat while the capacitance was measured at 10 kHz.

The photocatalytic procedure was performed with 250 mL of solution of Na_2SO_4 (10^{-3} M) in the photocatalytic reactor connected to a thermostatic bath (50 °C) in order to maintain constant temperature during irradiation. The dispersion of 100 mg of catalyst (NiO and 10% NiO/γ–Al_2O_3) was maintained by continuous magnetic stirring (100 rpm) with a magnetic stirrer, whilst three tungsten lamps provided the intense light source (2.07×10^{19} photons/s).

43.3 Results and Discussion

The XRD peaks of NiO and γ–Al_2O_3 are indexed according to the standard JCPDS 47-1049 and 29-0063. 10% NiO/γ–Al_2O_3 indicating the highly crystallized NiO phase, and clearly shows mixed phases, Fig. 43.1.

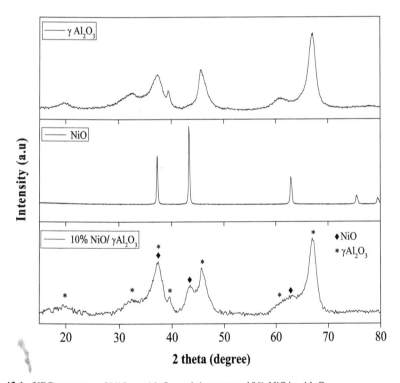

Fig. 43.1 XRD patterns of NiO, γ–Al_2O_3 and the system 10% NiO/γ–Al_2O_3

Fig. 43.2 SEM micrographs of 10% NiO/γ–Al$_2$O$_3$ and γ–Al$_2$O$_3$

The SEM images of γ–Al$_2$O$_3$ (Fig. 43.2) displays relatively a uniform shape while 10% NiO/γ–Al$_2$O$_3$ shows rather agglomeration of particles with small amount of NiO.

The specific surface area, the pore diameter and the volume of pores of the synthesized system 10% NiO/γ–Al$_2$O$_3$ are respectively 83.53 m^2/g, 59 Å and 0.147 cm^3/g.

The forbidden band (E$_g$) of the sensitizer NiO is of high importance in photocatalytic applications and the diffuse reflectance spectrum is used for the determination of the nature of the transition and its value through the Pankov formula:

$$(\alpha h\nu)^{2/k} = Cst \times (h\nu - E_g) \qquad (43.1)$$

where Cst is the proportionality constant, α the optical absorption coefficient (cm^{-1}) and hν (eV) the photon energy; k = 2 or 0.5 respectively for direct or indirect transitions.

The extrapolation of the linear part (αhν)^2 to the energy axis (hν= 0) gives a direct gap of 1.51 eV (Fig. 43.3a), due to the electronic transition: t$_{2g}$ → e$_g$, due to the crystal field splitting of Ni^{2+} in six-fold coordination octahedrally. Because of the gray color of NiO, the gap E$_g$ can be taken as the minimal energy needed to excite electrons from the valence band (VB) to the conduction band (CB). The small mobility of NiO (μ$_e$= 0.8 × 10^{-7} cm^2 V^{-1} s^{-1}), calculated from the relation (σ= e N$_D$ μ$_e$), is due to the obstruction of O^{2-} ions to the electrons jump between mixed valences Ni$^{2+/+}$ in the compact structure, in conformity with a conduction mechanism by low polarons with phonon assisted conductivity, the electron density (N$_D$) was deduced by photo-electrochemistry (see below). The conductivity determined on sintered pellet obeys to an exponential law, activation energy (E$_a$) of 0.12 eV was deduced from the slope dσ/dT^{-1}, so one can conclude that the donors are no longer ionized at room temperature.

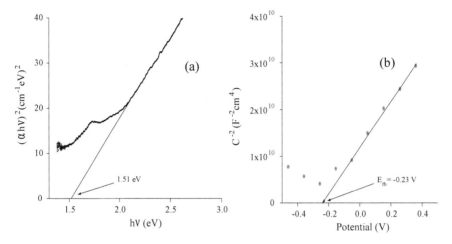

Fig. 43.3 Direct optical transition of NiO (**a**) and the Mott-Schottky plot of NiO in Na_2SO_4 (0.1 M) (**b**)

Because of the interest the hetero-system 10% $NiO/\gamma-Al_2O_3$ in photocatalysis, photo-electrochemistry analyses were undertaken. The latter underlines the role of the crystal structure which imposes not only the position of the electronic bands (VB and CB) but also the width of the gap through the covalence degree of the chemical bond.

NiO exhibits a good stability in alkaline solution; the cyclic voltammetry plotted at pH ~13. The flat band potential (E_{fb}) is determined from the intercept of the straight line to the infinite capacitance ($C^{-2} = 0$) according to the relation:

$$C^{-2} = (2/\varepsilon\varepsilon_o N_A)(E-E_{fb}) \qquad (43.2)$$

where ε the permittivity of NiO (~20) and ε_o the permittivity of vacuum. The positive slope characterizes n type conduction (Fig. 43.3b) and the free electrons (majority carriers) arise from oxygen vacancies whose associated levels are positioned at ~0.12 eV below the conduction band ($NiO \rightarrow NiO_1 -_\delta + 2\delta e_{CB}^- + 0.5\delta O_2$), the reaction is written according to the Kröger-Vink notation ($O_o \leftrightarrow 0.5\ O_2 + V_{\ddot{o}} + 2\ e^-$). The straight-line C^{-2} versus E is characteristic of a classical semiconductor. The intersection of the line to infinite capacity ($C^{-2}= 0$) and the slope give respectively a potential E_{fb} of -0.23 V and a density N_D of 7.06×10^{19} cm^{-3}. The potential E_{fb} measured at different pHs does not change significantly, and this indicates that the both the bands VB and CB derive from the same cationic orbital; the energy of CB ($E_{CB}= -0.35$ V) is given by:

$$E_{CB} = 4.75 + e\ E_{fb} - E_a \qquad (43.3)$$

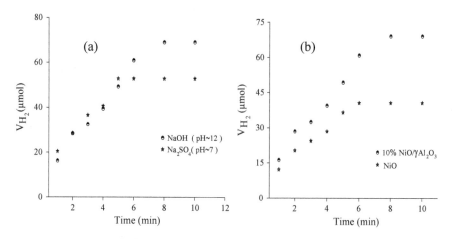

Fig. 43.4 Volumes of evolved H_2 versus illumination time on 10% $NiO/\gamma-Al_2O_3$ at different pHs (**a**) and comparison between the hetero-system 10% $NiO/\gamma-Al_2O_3$ and NiO (**b**)

The bands of NiO are pH insensitive while the level H_2O/H_2 changes by -0.6 V pH^{-1}. The best photoactivity was obtained at pH ~12. This property was judiciously exploited to have an optimal band bending and an enhancement of 25% was obtained at pH ~12. In the aim to improve the photoactivity of NiO through spreading it onto $\gamma-Al_2O_3$ with large specific surface area (111 m^2/g) to extend its active surface area and simplify the accessibility of light absorption. 10% $NiO/\gamma-Al_2O_3$ was prepared and tested in the same conditions as those of NiO. The photoactivity on the hetero-system increases by 42% compared to NiO alone (Fig. 43.4).

43.4 Conclusion

10% $NiO/\gamma-Al_2O_3$ was prepared by wet impregnation. NiO and 10% $NiO/\gamma-Al_2O_3$ were characterized by physico-chemical and photoelectrochemical techniques. The X-ray diffraction showed mixed phase ($\gamma-Al_2O_3/NiO$) and the average particle size of NiO is 18 nm. The diffuse reflectance shows a direct band gap appropriately matched to the solar spectrum. The capacitance measurement indicates n-type behaviour where the conduction band, is more negative than the level of the couple H_2O/H_2 at pH ~13, thus providing a better hydrogen evolution under visible light illumination. The photoactivity of 10% $NiO/\gamma-Al_2O_3$ was more effective with 42% than that obtained with NiO alone, since $\gamma-Al_2O_3$ offer a large distribution of NiO with an increased active surface area.

References

1. F. Valle, J.L.G. Fierro, R.M. Navarro, M.C. Sa, Hydrogen production from renewable sources : Biomass and photocatalytic opportunities, pp. 35–54 (2009)
2. D. Li, Y. Nakagawa, K. Tomishige, Methane reforming to synthesis gas over Ni catalysts modified with noble metals. Appl. Catal. A **408**, 1–24 (2011)
3. N. Salhi, A. Boulahouache, C. Petit, A. Kiennemann, C. Rabia, Steam reforming of methane to syngas over $NiAl_2O_4$ spinel catalysts. Int. J. Hydrog. Energy **36**, 11433–11439 (2011)
4. A. Belhadi, S. Boumaza, M. Trari, Photoassisted hydrogen production under visible light over NiO/ZnO hetero-system. Appl. Energy **88**, 4490–4495 (2011)
5. X. Chen, S. Shen, L. Guo, S.S. Mao, Semiconductor-based photocatalytic hydrogen generation. Chem. Rev. **110**, 6503–6570 (2010)
6. H. Ahmad, S.K. Kamarudin, L.J. Minggu, M. Kassim, Hydrogen from photo-catalytic water splitting process: A review. Renew. Sustain. Energy Rev. **43**, 599–610 (2015)
7. Y. Wang, Q. Wang, X. Zhan, F. Wang, M. Safdar, J. He, Visible light driven type II heterostructures and their enhanced photocatalysis properties: A review. Nanoscale **5**, 8326–8339 (2013)
8. X. Li, J. Yu, J. Low, Y. Fang, J. Xiao, X. Chen, Engineering heterogeneous semiconductors for solar water splitting. J. Mater. Chem. **A6**, 2485–2534 (2015)
9. J. Zou, C. Liu, Y. Zhang, Control of the metal-support interface of NiO-loaded photocatalysts via cold plasma treatment, pp. 2334–2339 (2006)
10. S. Boumaza, A. Belhadi, M. Doulache, M. Trari, Photocatalytic hydrogen production over NiO modified silica under visible light irradiation. Int. J. Hydrogen Energy **37**, 4908–4914 (2012)
11. M.K. Arora, N. Sahu, S.N. Upadhyay, A.S.K. Sinha, Alumina-supported cadmium sulfide photocatalysts for hydrogen production from water: role of dissolved ammonia in the impregnating solution. Ind. Eng. Chem. Res. **38**, 4694–4699 (1999)
12. N. Sahu, M.K. Arora, S.N. Upadhyay, A.S.K. Sinha, Phase transformation and activity of cadmium sulfide photocatalysts for hydrogen production from water: Role of adsorbed ammonia on cadmium sulfate precursor. Ind. Eng. Chem. Res. **37**, 4682–4688 (1998)

Chapter 44
Synthesis and Characterization of the Double Perovskite La$_2$NiO$_4$-Application for Hydrogen Production

S. Boumaza, R. Brahimi, L. Boudjellal, Akila Belhadi, and Mohamed Trari

Abstract Ternary oxides are practically employed in most technological fields and continue to attract much interest because of their low cost and simple preparation. In this work, we have synthesized La$_2$NiO$_4$ by sol-gel method and characterized by X-ray diffraction, BET surface area, SEM analysis, laser granulometry and electrical conductivity. The XRD pattern shows that the pure La$_2$NiO$_4$ is obtained beyond 900 °C. The BET measurements give relatively a small surface area (10 m^2 g^{-1}). The elemental chemical analysis by laser granulometry, confirmed the agglomerated nature of the synthesized powder observed by SEM analysis and attributed to the fine powder obtained by sol-gel method. The diffuse reflectance gives an optical gap of 1.3 eV, in agreement with the dark color. The transport properties show a semiconductor behavior with *p*-type conductivity and activation energy of 44 meV. The results of the absorption analysis show that La$_2$NiO$_4$ exhibits an excellent chemical stability in the pH range (6–14). The capacitance measurement (C^{-2}-E) in basic electrolyte reveals a linear behavior from which a flat band potential of 0.1 V$_{SCE}$ is obtained. La$_2$NiO$_4$ is tested successfully as photocatalyst for the hydrogen production upon visible light and the best performance is obtained in alkaline solution (NaOH 0.1 M, Na$_2$S$_2$O$_3$ 10^{-3} M) with an average rate of 23.6 µmol mn^{-1} (g catalyst)$^{-1}$. The system shows a tendency toward saturation whose deceleration is the result of the competitive reduction of the end product S$_4$O$_6^{2-}$.

Keywords K$_2$NiF$_4$-type oxide · La$_2$NiO$_4$ · Hydrogen

S. Boumaza (✉) · L. Boudjellal · A. Belhadi
Laboratory of Chemistry of Natural Gas, Faculty of Chemistry (USTHB), BP 32, 16111 Algiers, Algeria
e-mail: boumazasouhila@gmail.com

R. Brahimi · M. Trari
Laboratory of Storage and Valorization of Renewable Energies, Faculty of Chemistry (USTHB), BP 32, 16111 Algiers, Algeria

© The Editor(s) (if applicable) and The Author(s), under exclusive license to Springer Nature Singapore Pte Ltd. 2020
A. Khellaf (ed.), *Advances in Renewable Hydrogen and Other Sustainable Energy Carriers*, Springer Proceedings in Energy,
https://doi.org/10.1007/978-981-15-6595-3_44

44.1 Introduction

Oxides with perovskite structure have attracted considerable interest due to their importance technological properties and applications [1, 2]. The general formula for a stoichiometric perovskite oxide can be written AMO_3. A related structure to that of perovskite is the K_2NiF_4 structure in which perovskite-type layers are separated by rock salt layers [3]. A wide range of materials with this structure-type have been studied including La_2CuO_4 [4, 5]. The general formula of these compounds can be written A_2MO_4.

Several techniques were used for the preparation of perovskite oxides with nanoparticles powder structure like hydrothermal route, solid-state reactions, co-precipitation, sol-gel, Pechini method, gas phase preparations and colloidal chemistry route [6–8]. The sol-gel process has proved to be efficient to prepare ultra-fine particles. In this work, we are synthesized La_2NiO_4 by the citrate sol-gel method and characterized by different techniques analyses; the obtained powder is tested successfully for the hydrogen production under visible light.

44.2 Experimental

La_2NiO_4 was prepared by sol gel method. The precursors $La(NO_3)_3,6H_2O$ (Merck 99%) and $Ni(NO_3)_2,6H_2O$ (Biochem 98%) were separately dissolved in distilled water and mixed; then a solution of citric acid was added drop wise and the solution was evaporated under vacuum in a rotavapor. The gel was denitrified and the obtained powder was grounded and heated at 950 °C (3 °C min^{-1}). The crystalline phase was confirmed by X-ray diffraction (XRD) using XPERT-PRO diffractometer with $Cu_{K\alpha}$ radiation ($\lambda = 0.15406$ nm).

The specific surface area was determined by BET method on ASAP2010 micromeritics apparatus using N_2 gas as adsorbent at 77 K. The diffuse reflectance data were collected with a UV-VIS spectrophotometer (Shimadzu 1800) equipped with integration sphere. The morphology was analyzed by scanning electron microscopy (SEM) using JSM Jeol 6360L microscope. The size of the particles were dispersed in water, exposed to ultrasound and characterized by laser granulometry (Malvern Mastersizer 2000/3000).

The Mott Schottky characteristic is plotted in alkaline solution (NaOH, 0.1 M) in a standard electrochemical cell containing the working electrode, Pt emergency electrode and a saturated calomel electrode (SCE). The electrode potential was monitored by a potentiostat Voltalab PGZ301.

The photoactivity was measured through the volume of evolved hydrogen. The experiments were realized in a closed Pyrex reactor at 50 °C. The powder (200 mg) was maintained in suspension by magnetic agitation (210 rpm) in 200 mL of the prepared solution ($Na_2S_2O_3$, 10^{-3} M) and deoxygenated by N_2 bubbling for 30 min. The light source consisted of tungsten lamps (3 × 200 W Sonelec, Algeria) with a total flux of 29 mW cm^{-2}. Hydrogen formation was confirmed by gas chromatography (Agilent Technologies 7890A GC System).

44.3 Results and Discussion

The XRD profile of La$_2$NiO$_4$ (Fig. 44.1) shows a single phase crystallizing in K$_2$NiF$_4$ type structure. All peaks are indexed in an orthorhombic symmetry (space group: Fmmm) (JCPDS N°89-3460) with lattice constants: $a = 5.4462$ Å; $b = 5.4828$ Å; $c = 12.6566$ Å.

The active surface area (10 m^2 g^{-1}) was determined experimentally from the BET measurement and is in good agreement with that cited in the literature [9].

The SEM micrographs of La$_2$NiO$_4$ (Fig. 44.2) show a porous homogeneous structure where the grains appear spherical and agglomerated.

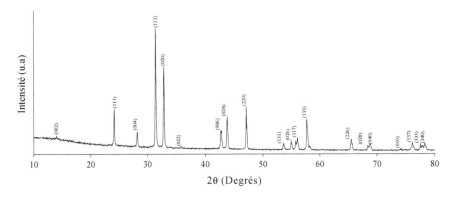

Fig. 44.1 X-ray diffraction pattern of La$_2$NiO$_4$ calcined at 950 °C

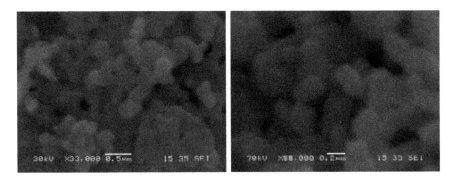

Fig. 44.2 SEM micrographs of La$_2$NiO$_4$

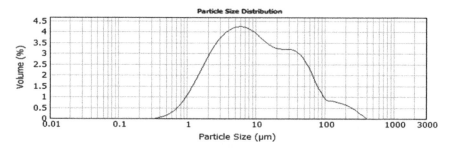

Fig. 44.3 Laser granulometry of La_2NiO_4 powder

La_2NiO_4 is formed by dispersed grains with small sizes and the laser granulometry of the powder confirms these observations with a large size distribution of grains (0.2–400 μm). Some grains sizes exceed 100 μm, while others do not reach 10 μm (Fig. 44.3).

The relationship between the absorption coefficient (α) and the energy of the incident photons ($h\nu$) is expressed by the relation:

$$(\alpha h\nu)^n = A(h\nu - E_g) \tag{44.1}$$

A is a constant and n = 2 and ½ for direct and indirect transition respectively. Figure 44.4 represents the $(\alpha h\nu)^2$ as a function of $h\nu$. The intercept of the linear plot $(\alpha h\nu)^2$ with the $h\nu$-axis gives a direct transition at 1.31 eV; this value is close to those reported in literature [10].

Figure 44.5 shows the thermal dependence of the electrical conductivity (σ) of La_2NiO_4 recorded in the range (300–450 K), the conductivity increases with increasing temperature indicating a semiconducting behavior. An activation energy (E_a) of 44 meV is determined from the slope $d(\log\sigma)/d(T^{-1})$.

The flat band potential E_{fb} was determined from reciprocal capacitance (C^{-2}) as a function of the potential (E) in alkaline solution (pH ~ 13) using the relation:

$$C^{-2} = \pm(2/e\varepsilon\varepsilon_o N_A)\{E - E_{fb}\} \tag{44.2}$$

where ε_o is the permittivity of vacuum (8.85 × 10^{-12} F m^{-1}) and ε (~10^5) is the permittivity of La_2NiO_4 [11]. The potential E_{fb} (= 0.10 V) and the holes concentration ($N_A = 10^{19}$ cm^{-3}) are deduced by extrapolation of the linear part and the slope respectively (Fig. 44.6); the negative slope confirms the p-type conduction.

The H_2 volume on La_2NiO_4 was studied at two pHs in aqueous solution containing $S_2O_3^{2-}$ as hole scavenger (Fig. 44.7).

The best performance was obtained in basic medium and this can be explained as follows: as pH increases from 7 to 13, the H_2O/H_2 level which varies by − 0.6 V/pH approaches the conduction band and reaches the optimal band bending, which enhances the electron transfer, thus, producing an increase in the volume of

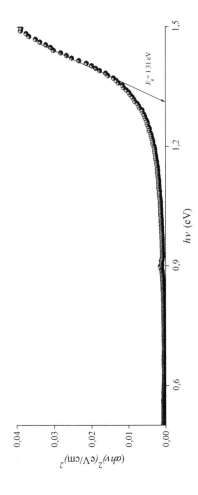

Fig. 44.4 Direct optical transition of La_2NiO_4

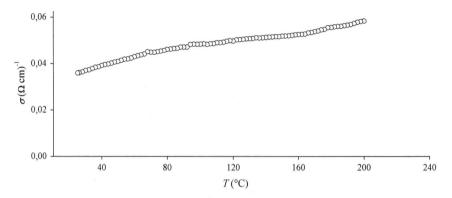

Fig. 44.5 Electrical conductivity (σ) of La_2NiO_4 as function of temperature

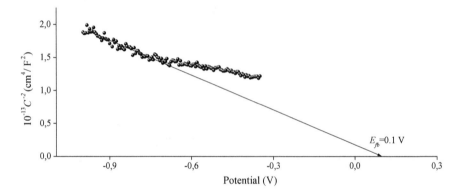

Fig. 44.6 The Mott-Schottky plot of La_2NiO_4/NaOH (0.1 M)/Pt plotted in N_2 bubbled solution at 10 kHz

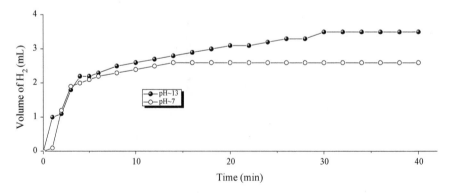

Fig. 44.7 Volume of evolved H_2 as a function of illumination time of La_2NiO_4, $S_2O_3^{2-}$ (10^{-3} M)

evolved hydrogen. The process of the hydrogen photo-production is given below:

$$\textbf{Anodic pole:} \quad La_2NiO_4 + h\upsilon \rightarrow e^-_{CB} + h^+_{VB} \qquad (44.3)$$

$$2S_2O_3^{2-} + 2h^+ \rightarrow S_4O_6^{2-} \qquad (44.4)$$

$$\textbf{Cathodic pole:} \quad 2H_2O + 2e^- \rightarrow H_2 + 2OH^- \qquad (44.5)$$

44.4 Conclusion

The double perovskite La_2NiO_4 was synthesized by citrate sol-gel route; The X-ray diffraction shows that the phase is formed at 900 °C. The optical transitions (1.31 eV) make La_2NiO_4 attractive for the solar energy conversion. In addition, it is photo-electrochemically active and has environmental friendly characteristic. The capacitance measurement showed p type behaviour. La_2NiO_4 is stabilized by holes capture via the reducing specie $S_2O_3^{2-}$ and the best photoactivity (23.6 µmol mn^{-1} (g catalyst)$^{-1}$) was obtained in the alkaline medium using $S_2O_3^{2-}$ as scavenger holes, is in good agreement with that obtained with the homologous La_2CuO_4 [4].

References

1. Y. Mao, H. Zhou, S. Wong, Synthesis, properties and applications of perovskite-phase metal oxide nanostructures. Material Matters **5**(2), 50 (2010)
2. L. Boudjellal, A. Belhadi, R. Brahimi, S. Boumaza, M. Trari, Physical and photoelectrochemical properties of the ilmenite NiTiO$_3$ prepared by wet chemical method and its application for O$_2$ evolution under visible light. Mater. Sci. Semicond. Process. **75**, 247–252 (2018)
3. I.B. Sharma, D. Singh, Solid state chemistry of Ruddlesden-Popper type complex oxides Bulltin. Mater. Sci. **21**(5), 363–374 (1998)
4. H. Lahmar, M. Trari, Photocatalytic generation of hydrogen under visible light on La$_2$ CuO$_4$. Bull. Mater. Sci. **38**(4), 1043–1048 (2015)
5. A. Tsukada, T. Greibe, M. Naito, Phase control of La 2 CuO 4 in thin film synthesis. Phy. Rev. B. **66**(18), 184515 (2002)
6. A. Ecija, K. Vidal, A. Larrañaga, L. Ortega-San-Martín, M.I. Arriortua, Synthetic methods for perovskite materials—structure and morphology, in *Advances in Crystallization Processes*, ed. by Dr. Y. Mastai (InTech, 2012). ISBN: 978-953-51-0581-7
7. L. Pan, G. Zhu, Perovskite materials: synthesis, characterisation, properties, and applications. Janeza Trdine. ISBN: 978-953-51-2245-6 (2016), (Chapter 4)
8. P. Kanhere, Z. Chen, A review on visible light active perovskite-based photocatalysts. Molecules **19**, 19995–20022 (2014)
9. L. Bouyssiéres, R. Schifferli, L. Urbina, P. Araya, J.M. Palacios, Study of perovskites obtained by the sol-gel method. J. Chil. Chem. Soc. **50**(1), 407–412 (2005)

10. C.A. Silva, J.B. Silva, M.C. Silva-Santana, P. Barrozo, N.O. Moreno, Influence of the Zn dopent in structural and electrical properties of the $La_2Ni_{1-x}Zn_xO_4$. Adv. Mater. Res. **975**, 75–85 (2014)
11. X.Q. Liu, C.L. Song, Y.J. Wu, X.M. Chen, Giant dielectric constant in $Nd_2NiO_{4+\delta}$ ceramics obtained by spark plasma sintering. Ceram. Int. **37**(7), 2423–2427 (2011)

Chapter 45
Optimal Design and Comparison Between Renewable Energy System, with Battery Storage and Hydrogen Storage: Case of Djelfa, Algeria

Ilhem Nadia Rabehi

Abstract Algeria's energy mix is almost exclusively based on fossil fuels (Meriem in Renewable Energy in Algeria Reality and Perspective, pp. 1–19, 2018) [1], especially natural gas. However, the country has enormous renewable energy potential, mainly solar, which the government is trying to harness by launching an ambitious renewable energy and energy efficiency programs (Ministry of Energy and Mining of Algeria in Renewable Energy and Energy Efficiency Program, 2011) [2]. Despite being a hydrocarbon-rich nation, Algeria is making efforts to harness its renewable energy potential. The renewable energies could represent an economic solution for the case of isolated sites, but their intermittency needs a storage system, that could be either by the use of batteries or hydrogen technologies. However, these two storage systems still face challenges, especially economic ones. This study deals with an economic study of several configurations of renewable energy systems, it aims to compare between the conventional storage systems and the new technologies of the hydrogen. In this study, HOMER will be used to simulate three configurations for a school on the high land region of Algeria named Had-Saharry. Many configurations will be simulated using HOMER in order to have an over view about the techno-economic feasibility and the use of hydrogen for the storage. The system has been designed according to the school's load profile. Then compare between the costs of the systems and their performance on the Algerian high lands weather conditions. As result the systems with batteries proved to be less expensive than the hydrogen storage, as well as, the hybrid system (PV, WECS) proved to be cost effective.

Keywords Renewable · Hybrid · Storage · Batteries · Hydrogen · HOMER

I. N. Rabehi (✉)
Pan African University Institute of Water and Energy Sciences, Tlemcen, Algeria
e-mail: ilhemnadia95@gmail.com

45.1 Introduction

Algeria is the biggest African country and among the biggest exporter of gas and it possesses the world's fifth largest natural gas reserves. Despite of being a hydrocarbon reach country Algeria has an huge renewable energy (RE) potential mainly solar and wind. In order to diversify the sources of energy used and reduce the carbon emission, Algeria has started in 2011 an ambitious renewable energy program that aims to generate 22,000 Mw from RE s by 2030 [2].

Renewable energies represents a green alternatives and cost effective system for remote and isolated areas [3], but it still face the intermittency issue that pose the storage problems. It exist to ways to store energy, either by the use of the batteries, which are not environmentally friendly, and the hydrogen storage system, which still a new technology.

45.2 Materials and Methods

HOMER (Hybrid Optimization Model for Electric Renewable) software was used in this study to model and optimize the systems because it is a techno-economic tool. HOMER software is developed by NREL (National Renewable Energy Laboratory), it determines optimal size of its components through carrying out the techno-economic analysis. It simulations possible combination of components entered and ranks the systems according to user-specified criteria such as cost of energy (COE). The software requires six types of input data for simulation and optimization including meteorological data, load profile, equipment characteristics, and search space, economic and technical data [4]. In this study, we consider three proposed configurations:

The first one is a photovoltaic (PV) system with a battery storage then with hydrogen storage system Fig. 45.1 represents the systems.

The second configuration represents a WECS (Wind Electricity Conversion System) with a batteries then hydrogen storage system Fig. 45.2 represents the systems.

The third configuration represents a hybrid (PV, WECS) system with batteries then with a hydrogen storage system, Fig. 45.3 represents the systems.

The components used on configurations are:

- **Module**: the PV modules used on this system are a polycrystalline panels with a maximum of 275 W and an efficiency of 17%.
- **Wind turbines**: a wind turbine from AWS HC 3.3 kW and a rated power of 3.3 kW, 4.65 rotor diameter and 12 m hub height.
- **Battery**: Battery bank stores the electrical energy produced by the PV, and makes the energy available at night or on dark days (days of autonomy or no-sun-days). The batteries used on this system are BAE SECURA SOLAR 9 PVV (2 V, 2.92 kWh).

45 Optimal Design and Comparison Between Renewable … 349

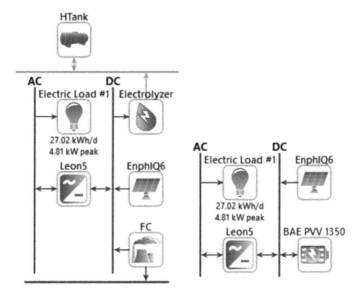

Fig. 45.1 Configuration 01 (PV system)

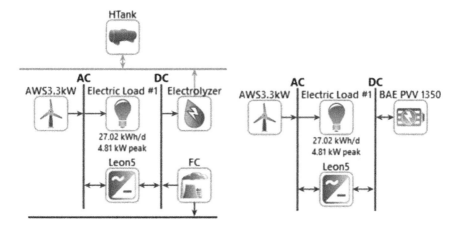

Fig. 45.2 Configuration 02 (WACS system)

- **Converter**: The inverter model used in this project is Leonics S-219Cp 5 kW it was chosen based on the power unit (5 kW).
- **Fuel cell and electrolyzer**: the choice of these components were generic, connected to the DC bus with an search space 1, 3, 5, 10, 20 kW.
- **Tank**: The tank chosen search spaces were 1, 3, 5, 10, 20 kW.

The overall summary of economic parameters of the different components according to [5–7] are presented in Table 45.1.

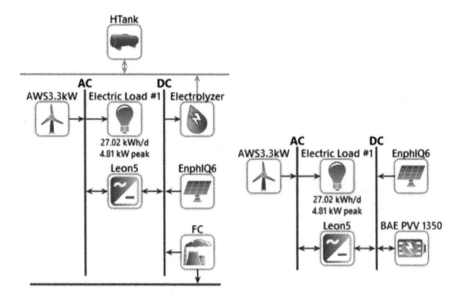

Fig. 45.3 Configuration 03 (hybrid system)

Table 45.1 Investment, replacement and O&M costs by components

	Investment ($/kw)	Replacement ($/kW)	O&M ($/kW)
PV	2500	2000	25
WECS	1000	800	100
FC	3000	2500	0.15/op.hr
Tank	1500	1200	30
Electrolyser	2000	1600	20

45.2.1 Solar Energy Potential

The geographical coordinates of the data collection site were 35°21.3' N latitude 3°21.8' E longitude and 1140 m altitude above mean sea level. From these geographic data HOMER, generate automatically the solar radiation and the wind speed on the location

The monthly average of solar radiation and clearness index of the province for 22 years obtained through HOMER. The solar radiation data for the selected remote area was estimated to range between 2.17 and 7.42 kW h/m2/day with an average annual solar radiation estimated to 4.76 kW h/m2/day. Notice that more solar irradiance can be expected from the month of May to August while less solar irradiance is to be expected from November to January.

The wind speed is around 4.6 m/s; it is almost constant during the year. The highest value is 5 m/s in the period of December–April.

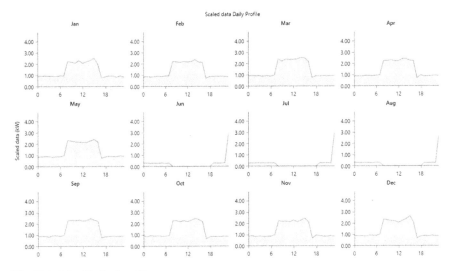

Fig. 45.4 Monthly load profile

45.2.2 Electrical Load

An important step in the design of the system is the determination of electricity load. Figure 45.4 Shows the monthly profile for the assumed electric load. The load has an average value of 27.02 kWh/day and a peak of 4.81 kW.

45.3 Results

The PV system with the batteries gives a total electricity production of 35,804 kwh/year, while the PV system with hydrogen storage gives 53.503 kWh/yr 92.4% from PV and 7.61% from FC Table 45.2 represents the COE, investment cost O&M/year and the NPV. Figure 45.5 represents the share costs of components in each system

The WACS system with the batteries gives a the total electricity production of 31,745 kwh/yr, while the WACS system with hydrogen storage gives 29,214 kWh/yr a 88.9% from WACS and 11.1% from FC Table 45.3 represents the COE, investment

Table 45.2 Costs of the configuration 01

Conf	COE ($/kWh)	Investment ($)	O&M ($/yr)	NPV ($)
(PV, batteries)	0.477	62,597	2034	110,391
(PV, FC)	1.26	113,112	7567	290,916

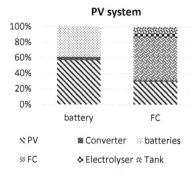

Fig. 45.5 The cost of each component on the two systems

Table 45.3 Costs of the configuration 02

Conf	COE ($/kWh)	Investment ($)	O&M ($/yr)	NPV ($)
(WECS, batteries)	0.602	50,596	3783	139,489
(WECS, FC)	1.04	57,333	7815	240,964

cost O&M/year and the NPV. Figure 45.6 represents the costs share of components in each system

The hybrid system (PV, WECS) with the batteries gives a the total electricity production of 21,336 kwh/year 40.6% from wind and 59.4% from PV, while the hybrid system with hydrogen storage gives 24,278 kWh/yr. A 30.6% from PV, 10% from FC and 95.4% from WACS. Table 45.4 represents the COE, investment cost O&M/year and the NPV. Figure 45.7 represents the share costs of components in each system.

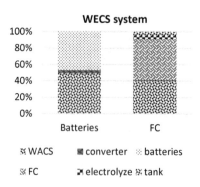

Fig. 45.6 The cost of each component on the two systems

Table 45.4 Costs of the configuration 03

Conf	COE ($/kWh)	Investment ($)	O&M ($/yr)	NPV($)
(PV, WECS, batteries)	0.397	36,295	2374	92,066
(PV, WECS, FC)	0.994	52,863	7556	230,389

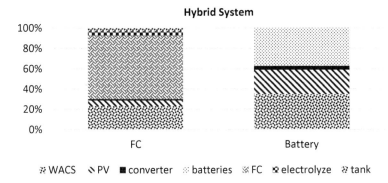

Fig. 45.7 The cost of each component on the two system

45.4 Conclusions

From the economical analysis, it is easy to notice that, the hybrid renewable system with a battery storage is the most economic way to generate electricity among the other systems. The use of hydrogen is expensive especially with the PV system even though it is more environmentally friendly than the batteries, which represent a big challenge in the recycling and waste management.

References

1. B. Meriem, Renewable Energy in Algeria Reality and Perspective, pp. 1–19 (2018)
2. D. Djalel, S. Youcef, B. Chellali, Renewable Energy and Energy Efficiency Program (2017)
3. A.B. Stambouli, Z. Khiat, S. Flazi, Y. Kitamura, A review on the renewable energy development in Algeria: current perspective, energy scenario and sustainability issues. Renew. Sustain. Energy **16**, 4445–4460 (2012)
4. S. Bahramara, M.P. Moghaddam, M.R. Haghifam, Optimal planning of hybrid renewable energy systems using HOMER : A review **62**, 609–620 (2016)
5. A. Gabour, A. Metatla, Optimal Design and Comparison Between Renewable Energy, Hybrid Energy and Non-renewable Energy Systems , pp. 677–682 (2016)
6. M.J. Khan, M.T. Iqbal, Pre-feasibility study of stand-alone hybrid energy systems for applications in Newfoundland **30**, 835–854 (2005)
7. K.M. Kabir et al., Design & simulation of hydrogen based hybrid green power system using sea water for Cox' s Bazar design and simulation of hydrogen based hybrid. Cogent **2**, 1–14 (2017)

Chapter 46
New Neural Network Single Sensor Variable Step Size MPPT for PEM Fuel Cell Power System

Abdelghani Harrag

Abstract This paper deals with the development of a single sensor neural network controller used to track the maximum power of proton exchange membrane fuel cell power system. The proposed single sensor neural network controller has been developed and trained using single sensor maximum power point tracking data obtained previously. The developed maximum power point tracking controller has been used to track the output power of the fuel cell power system composed of 7 kW proton exchange membrane fuel cell powering a resistive load via a DC-DC boost converter controlled using the proposed controller. Simulation results obtained using the developed MATLAB/Simulink model show that the proposed single sensor neural network maximum power point tracking controller can track effectively the maximum power using only one sensor compared to the classical power point tracking controllers using two sensors reducing by the way the cost and the complexity of the fuel cell maximum power tracking controller.

Keywords PEM fuel cell · MPPT · Single sensor · Neural network · NN

46.1 Introduction

In the twenty-first century, the demand for clean and sustainable energy sources has become a strong driving force in continuing economic development and hence in the improvement of human living conditions. In that respect, fuel cells have been recognized to form the cornerstone of clean energy technologies due to their high efficiency, high energy density, and low or zero emissions. Recently, fuel cells have seen explosive growth and application in various energy sectors including transportation, stationary and portable power, and micro-power. The rapid advances in fuel cell system development and deployment require basic knowledge of science and technology as well as advanced techniques on fuel cell design and analysis [1].

A. Harrag (✉)
Mechatronics Laboratory, Optics and Precision Mechanics Institute, Ferhat Abbas University, Setif 1, Cite Maabouda (ex. Travaux), 19000 Setif, Algeria
e-mail: a.harrag@univ-setif.dz

© The Editor(s) (if applicable) and The Author(s), under exclusive license to Springer Nature Singapore Pte Ltd. 2021
A. Khellaf (ed.), *Advances in Renewable Hydrogen and Other Sustainable Energy Carriers*, Springer Proceedings in Energy,
https://doi.org/10.1007/978-981-15-6595-3_46

A fuel cell is an electrochemical device that uses reverse electrochemical reactions and continuously converts the chemical energy content of the fuel into electrical energy, water, and some heat as long as fuel and oxidant are supplied. Among various types, the high temperature solid oxide fuel cell (SOFC) and the low temperature proton exchange membrane fuel cell (PEMFC) have been identified as the likely fuel cell technologies that will capture the market in the future. The proton exchange membrane fuel cell (PEMFC) uses a solid membrane that transports protons. It can operate from about 0–80 °C with the output power ranging from a few watts to several hundred kilowatts [2].

The output power of a fuel cell is not regulated, and its stability is a relevant issue. The small voltage of each individual cell is heavily influenced by changes in electric current, partial gas pressures, reactants humidity level, gas speed, stoichiometry, temperature and membrane water content. As a consequence, the extraction of maximum power from a fuel cell power source is essential for its optimum and economical utilization. However, the maximum extractable power varies dynamically during the fuel cell operation for varying load current requirements as well as for varying operating conditions which makes the maximum power extraction a challenging task [3].

The last decade has seen a huge development of the maximum power point tracking (MPPT) controllers for fuel cell power systems [4, 5], among them: perturb and observation [6], incremental conductance [7], sliding mode controller [8], fractional order filter controller [9], hysteresis controller [10], extremum seeking control [11], fuzzy logic controller [12, 13], particle swarm optimization controller [14], water cycle algorithm controller [15], unified tracker algorithm controller [16], eagle strategy controller [17], neural network controller [18, 19], etc.

In this paper, we present a new single sensor neural network controller used to track the maximum power of proton exchange membrane fuel cell power system. The proposed single sensor neural network controller has been developed and trained using single sensor maximum power point tracking data obtained previously. The developed maximum power point tracking controller has been used to track the output power of the fuel cell power system composed of 7 kW proton exchange membrane fuel cell powering a resistive load via a DC-DC boost converter controlled using the proposed controller. Simulation results show that the proposed single sensor neural network maximum power point tracking controller can track effectively the maximum power using only one sensor compared to the classical power point tracking controllers using two sensors reducing by the way the cost and the complexity of the fuel cell maximum power tracking controller.

The remainder of this paper is organized as follows. In Sect. 46.2, the PEM fuel cell modelling is described. Section 46.3 presents the proposed neural network single sensor MPPT controller. The simulation results and discussions are presented in Sect. 46.4. While Sect. 46.5 stated the main conclusions of this study.

46.2 PEM Fuel Cell Modeling

A fuel cell is a simple electrochemical device that uses the chemical energy present in hydrogen and oxygen to produce electricity in the form of direct current along with water and heat as the byproducts. More specifically, hydrogen is fed into the anode, where it is dissociated into protons and electrons with the help of a catalyst.

The electrons provide the electrical current as they pass through the external circuit and reach the cathode. The protons pass through the proton-conducting membrane and crossover into the cathode to recombine with the electrons as well as the oxygen (which is fed into the cathode) to generate water (Fig. 46.1) [20].

Oxidation of hydrogen reaction at anode:

$$H_2 \rightarrow 2H^+ + 2e^- \tag{46.1}$$

Reduction of oxygen reaction at a cathode:

$$O_2 + 4e^- \rightarrow 2O^{-2} \tag{46.2}$$

The overall hydrogen reaction is:

$$H_2 + \frac{1}{2}O_2 \rightarrow H_2O + electrical energy + heat \tag{46.3}$$

Each cell voltage can be defined by the well known expression [21]:

$$V_{FC} = E_{nernst} - V_{act} - V_{ohmic} - V_{conc} \tag{46.4}$$

E_{nernst} is the reversible open circuit voltage [22]:

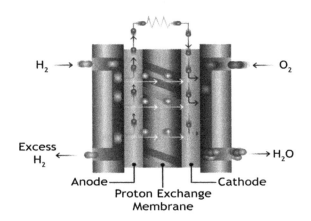

Fig. 46.1 PEM fuel cell

$$E_{nernst} = 1.229 - (8.5 \times 10^{-4})(T - 298.15)$$
$$+ (4.385 \times 10^{-5}T[\ln(P_{H2}) + 0.5\ln(P_{O2})] \quad (46.5)$$

where T is the temperature; P_{O_2} the oxygen pressure and P_{H_2} the hydrogen pressure.

V_{act} is the activation voltage drop reflecting the fact that the cell requires a certain amount of energy to start electron circulation and create/break chemical bonding [23]:

$$V_{act} = \xi_1 + \xi_2 \cdot T + \xi_3 \cdot T \cdot \ln(C_{O_2}) + \xi_4 \cdot T \cdot \ln(i_{FC}) \quad (46.6)$$

where $\xi_{i(i=1\ to\ 4)}$ are parametric coefficients for each cell; i_{FC} is the cell current and C_{O_2} is the oxygen's concentration.

V_{ohmic} is the ohmic linear voltage drop proportional to electric current representing the resistance of the polymeric membrane to proton circulation as well as to the electrical resistance of electrodes and current collectors [24]:

$$V_{ohmic} = R_{ohmic} \cdot i_{FC} \quad (46.7)$$

where i_{FC} is the cell current; R_{ohmic} is the sum of the contact resistance R_c and the membrane resistance R_m.

V_{conc} concentration voltage drop due to the changes in the concentration of reactants as they are consumed by the electrochemical reaction [25]:

$$V_{conc} = -b \cdot \ln\left(1 - \frac{i_{FC}/A}{I_{max}}\right) \quad (46.8)$$

where b is the concentration loss constant; i_{FC} is the cell current; A is the is cell active area; I_{max} is the maximum current density.

46.3 Proposed Single Sensor Neural Network MPPT Algorithm

The proposed single sensor neural network variable step size MPPT controller is developed in two steps: (1) firstly in offline mode required to test several set of neural network parameters to find the optimal architecture and (2) secondly, the optimal found parameters are used in online mode to track the maximum output power of the PEM fuel cell power system. The developed controller uses the PEM fuel cell current and old PWM duty cycle as inputs to compute the new PWM ratio, used as output.

The neural network controller has been trained using the mean squared errors (MSE) for minimizing the overall error measure between the neural networks output and the data generated previously using the conventional single sensor MPPT [26].

Once trained, the optimized neural network controller is used in the online mode to track the maximum available power under different testing scenario considering changing atmospheric and operating conditions.

46.4 Results and Discussion

The Matlab/Simulink model of the PEM fuel cell power system including the 7 kW PEM fuel cell powering a resistive load *via* a DC-DC boost converter controlled using the developed neural network single sensor MPPT controller has been implemented (Table 46.1).

To evaluate the efficiency and the performance of the proposed neural network single sensor MPPT, the implemented controller using Matlab/Simulink environment is validated considering two test scenarios: (1) temperature (T) variation; and (2) hydrogen pressure (P_{H_2}) variation. In the two considered experiments, we use a fast stepped pattern considered as strained case to validate the efficiency and the capability tracking of the proposed MPPT controller. Figure 46.2a, b shows the output power corresponding to the two considered scenarios.

From Fig. 46.2a, b, we can see that the proposed neural network single sensor MPPT track effectively the maximum available output power regarding temperature or pressure changes.

Table 46.1 PEM fuel cell parameters

Parameter	Value
Maximum power at MPP P_{MPP} (W)	7000
Cell open circuit voltage V_{OC} (V)	1.29
Number of cells N	50
Cell active surface (A/cm^2)	200
Hydrogen partial pressure P_{H_2} (bar)	2.6
Oxygen partial pressure P_{O_2} (bar)	0.3

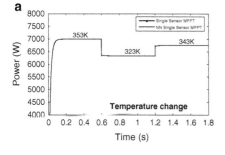
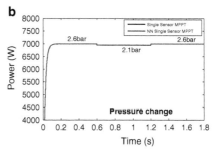

Fig. 46.2 Output power: **a** temperature (T) variation, **b** hydrogen pressure variation

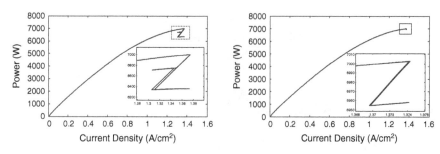

Fig. 46.3 P-I characteristics: **a** temperature variation, **b** pressure variation

Figure 46.3a, b shows the corresponding P-I characteristics according to the considered temperature and pressure variation, respectively.

46.5 Conclusion

This paper addresses the development of a neural network single sensor MPPT controller used to track the maximum PEM fuel cell power system composed of the 7kW PEM fuel cell powering a resistive load *via* a DC-DC boost converter controlled using the proposed controller. Simulation results otained using Matlab/Simulink environment show that the proposed neural network single sensor MPPT controller can effectively track the maximum power using only one sensor for the PEM fuel cell current which reduces the cost as well as the complexity of the PEM fuel cell power system. As future work, we work currently on the experimental validation of the developed single sensor neural network MPPT in the hardware in the loop mode using the STM32F4 board.

Acknowledgements The Algerian Ministry of Higher Education and Scientific Research via the DGRSDT supported this research (PRFU Project code: A01L07UN190120180005).

References

1. S. Revankar, P. Majumdar, *Fuel cells: Principles, Design and Analysis* (CRC Press, Taylor & Francis, Boca Raton, 2014)
2. Z. Qi, *Proton Exchange Membrane Fuel Cells* (CRC Press, Taylor & Francis, Boca Raton, 2014)
3. C.A. Ramos-Paja, G. Spagnuolo, G. Petrone, R. Giral, A. Romero, Fuel cell MPPT for fuel consumption optimization, in *Proceedings IEEE International Symposium on Circuits and Systems (ISCAS)*, Paris, France (2010), pp. 2199–2202
4. M. Becherif, D. Hissel, MPPT of a PEMFC based on air supply control of the motocompressor group. Int. J. Hydrogen Energy **35**(22), 12521–12530 (2010)

5. N. Karami, R. Outbib, N. Moubayed, Fuel flow control of a PEM fuel cell with MPPT, in *Proceedings IEEE International Symposium on Intelligent Control (ISIC)*, Dubrovnic; Crotia (2012), pp. 289–294
6. M. Sarvi, M.M. Barati, Voltage and current based MPPT of fuel under variable temperature conditions, in *Proceedings Universities Power Engineering Conference (UPEC)*, Cardiff, Wales (2010), pp. 1–4
7. M.Z. Romdlony, B.R. Trilaksono, R. Ortega, Experimental study of extremum seeking control for maximum power point tracking of PEM fuel cell, in *International Conference on System Engineering and Technology (ICSET)*, Bandung, Indonesia (2012), pp. 1–6
8. S.H. Abdi, K. Afshar, N. Bigdeli, S. Ahmadi, A novel approach for robust maximum power point tracking of PEM fuel cell generator using sliding mode control approach. Int. J. Electrochem. Sci. **7**, 4192–4209 (2012)
9. J. Liu, T. Zhao, Y. Chen, Maximum power point tracking of proton exchange membrane fuel cell with fractional order filter and extremum seeking control, in *Proceedings ASME/IEEE International Conference on Mechatronic and Embedded Systems and Applications (ICMESA)*, Boston, Massachusetts, USA (2015), pp. 1–6
10. J.D. Park, Z. Ren, Hysteresis controller based maximum power point tracking energy harvesting system for microbial fuel cells. J. Power Sources **205**, 151–156 (2012)
11. K. Ettihir, L. Boulon, K. Agbossou, S. Kelouwani, MPPT control strategy on PEM fuel cell low speed vehicle. in *Proceedings IEEE Vehicle Power and Propulsion Conference (VPPC)*, Seoul, South Korea (2012), pp. 926–931
12. J. Jiao, X.A. Cui, Real-time tracking control of fuel cell power systems for maximum power point. J. Comput. Inf. Syst. **9**(5), 1933–1941 (2013)
13. A. Harrag, S. Messalti, How fuzzy logic can improve PEM fuel cell performance? Int. J. Hydrogen Energy **43**(1), 537–550 (2018)
14. I. Soltani, M. Sarvi, H. Marefatjou, An intelligent, fast and robust maximum power point tracking for proton exchange membrane fuel cell. World Appl. Program. **3**(7), 264–281 (2013)
15. I.N. Avanaki, M. Sarvi, A new maximum power point tracking method for PEM fuel cells based on water cycle algorithm. J. Renew. Energy Environ. **3**(1), 35–42 (2016)
16. H. Fathabadi, Novel highly accurate universal maximum power point tracker for maximum power extraction from hybrid fuel cell/PV/wind power generation systems. Energy **116**, 402–416 (2016)
17. M. Sarvi, M. Parpaei, I. Soltani, M.A. Taghikhani, Eagle strategy based maximum power point tracker for fuel cell system. IJE Trans. A Basics **28**(4), 529–536 (2015)
18. A. Harrag, H. Bahri, Novel neural network IC-based variable step size fuel cell MPPT controller. Int. J. Hydrogen Energy **42**(5), 3549–3563 (2017)
19. M. Hatti, M. Tioursi, W. et Nouibat, Neural network approach for semi-empirical modeling of PEM fuel-cell, in *IEEE International Symposium on Industrial Electronics* (2006), pp. 1858–1863
20. C. Spiegel, *PEM Modeling and Simulation Using Matlab* (Academic Press, Burlington, MA, 2008)
21. R. O'Hayre, Fuel Cell Fundamentals (Wiley, 2009)
22. R.F. Mann, J.C. Amphlett, M.A. Hooper, H.M. Jensen, B.A. Peppley, P.R. Roberge, Development and Application of a generalized steady-state electrochemical model for a PEM fuel cell. J. Power Sources **86**(1–2), 173–180 (2000)
23. D. Yu, and S. Yuvarajan, A novel circuit model for PEM fuel cells, in *IEEE 19th Applied Power Electronics Conference and Exposition*, Anaheim, CA, USA (2004), pp. 362–366
24. J.H. Lee, T.R. Lalk, A.J. Appleby, Modelling fuel cell stack system. J. Power Sources **73**, 229–241 (1998)
25. H.J. Avelar, E.A.A. Coelho, J.R. Camacho, J.B. Vieira, L.C. Freitas, M. Wu, PEM fuel cell dynamic model for electronic circuit simulator, in *IEEE Electrical Power & Energy Conference*, Montreal, QC, Canada (2009), pp. 1–6
26. A. Harrag, H. Bahri, A novel single sensor variable step size maximum power point tracking for proton exchange membrane fuel cell power system. **19**(2), 177–189 (2019).

Chapter 47
Modified P&O-Fuzzy Type-2 Variable Step Size MPPT for PEM Fuel Cell Power System

Abdelghani Harrag

Abstract This paper proposes a modified perturb and observe based fuzzy type-2 maximum power point controller. In this study the fuzzy type-2 controller has been used to drive the variable step size of classical perturb and observe algorithm in order to track the maximum power point of the proton exchange membrane fuel cell power system. The proposed controller has been validated using the Matlab/Simulink environment where the whole fuel cell power system composed of 7 kW proton exchange membrane fuel cell powering a resistive load via a DC-DC boost converter controlled using the proposed controller have been implemented. The developed model has been investigated in case of temperature variation. Comparative simulation results obtained using the classical perturb and observe algorithm, the fuzzy type-1 algorithm and the proposed fuzzy type-2 algorithm prove the superiority of the proposed controller to track effectively the maximum power regarding used dynamic and static performance metrics.

Keywords PEM fuel cell · MPPT · Perturb and observe · P&O · Fuzzy logic · FL · Type-2

47.1 Introduction

Green energy technologies including energy storage and conversion will play the mainly role in overcoming global pollution and fossil fuel exhaustion for the sustainable development of human society. Among green energy technologies, conversion and electrochemical energy storage are considered the most feasible, sustainable and environmentally friendly. Electrochemical energy technologies such as batteries, fuel cells, supercapacitors, hydrogen generation and storage, as well as solar energy conversion have been or will be used in important application areas including transportation and stationary and portable/micro power. To respond to the increasing

---

A. Harrag (✉)
Mechatronics Laboratory, Optics and Precision Mechanics Institute, Ferhat Abbas University, Setif 1, Cite Maabouda (ex. Travaux), 19000 Setif, Algeria
e-mail: a.harrag@univ-setif.dz

© The Editor(s) (if applicable) and The Author(s), under exclusive license to Springer Nature Singapore Pte Ltd. 2021
A. Khellaf (ed.), *Advances in Renewable Hydrogen and Other Sustainable Energy Carriers*, Springer Proceedings in Energy,
https://doi.org/10.1007/978-981-15-6595-3_47

demand in both the energy and power densities of these electrochemical energy devices in various new application areas, further research and development are essential to overcome major obstacles, such as cost and durability, which are considered to be hindering their application and commercialization [1].

In fixed operating conditions, the fuel cell characteristics presents only one unique point on power-current curve representing the maximum power point (MPP) at which the fuel cell produces its maximum power. However, this curve is heavily influenced by the operating parameters: cell temperature, anode and cathode pressures, relative humidity, stoichiometry and anode and/or cathode gas mole fraction. Therefore, to supply a load with a constant voltage, a maximum power point controller is necessary to adjust the point on the P/I curve corresponding to the actual power demand [2, 3].

In this spirit, a huge development of the maximum power point tracking controllers for fuel cell power systems have emerged that can be divided in two categories: (1) the first one based on controlling the gas flow and/or the pressure [4–8]; and (2) the second one, inspired from PV control strategies, based on the use of a power converter [9–14].

This paper proposes a modified perturb and observe based fuzzy type-2 maximum power point controller. In this study the fuzzy type-2 controller has been used to drive the variable step size of classical perturb and observe algorithm in order to track the maximum power point of the proton exchange membrane fuel cell power system. The proposed controller has been validated using the Matlab/Simulink environment where the whole fuel cell power system composed of 7 kW proton exchange membrane fuel cell powering a resistive load via a DC-DC boost converter controlled using the proposed controller have been implemented. Comparative simulation results between the proposed controller and the classical perturb and observe one prove the superiority of the proposed controller to track effectively the maximum power regarding used dynamic and static performance metrics. The remainder of this paper is organized as follows. In Sect. 47.2, the PEM fuel cell modelling is presented. Section 47.3 gives the implementation methodology of the proposed P&O-based fuzzy type-1 MPPT controller. The simulation results and discussions are presented in Sect. 47.4. Finally, Sect. 47.5 draws the main conclusions of this work.

47.2 PEM Fuel Cell Modeling

A fuel cell is a simple electrochemical device that uses the chemical energy present in hydrogen and oxygen to produce electricity in the form of direct current along with water and heat as the byproducts based on the following equations [15]:

Oxidation of hydrogen reaction at anode:

$$H_2 \rightarrow 2H^+ + 2e^- \qquad (47.1)$$

Reduction of oxygen reaction at a cathode:

$$O_2 + 4e^- \rightarrow 2O^{-2} \qquad (47.2)$$

The overall hydrogen reaction is:

$$H_2 + \frac{1}{2}O_2 \rightarrow H_2O + electrical\,energy + heat \qquad (47.3)$$

Each cell voltage can be defined by the well known expression given by [16]:

$$V_{FC} = E_{nernst} - V_{act} - V_{ohmic} - V_{conc} \qquad (47.4)$$

E_{nernst} is the reversible open circuit voltage approximated by [17]:

$$E_{nernst} = 1.229 - (8.5 \times 10^{-4})(T - 298.15) \\ + (4.385 \times 10^{-5}T[\ln(P_{H_2}) + 0.5\ln(P_{O_2})] \qquad (47.5)$$

where T is the temperature; P_{O_2} the oxygen pressure and P_{H_2} the hydrogen pressure. V_{act} is the activation voltage drop approximated by [18]:

$$V_{act} = \xi_1 + \xi_2 \cdot T + \xi_3 \cdot T \cdot \ln(C_{O_2}) + \xi_4 \cdot T \cdot \ln(i_{FC}) \qquad (47.6)$$

where ξ_i (i = 1 to 4) are parametric coefficients for each cell; i_{FC} is the cell current and C_{O_2} is the oxygen's concentration.

V_{ohmic} is the ohmic linear voltage drop proportional to electric current approximated by [19]:

$$V_{ohmic} = R_{ohmic} \cdot i_{FC} \qquad (47.7)$$

where i_{FC} is the cell current; R_{ohmic} is the sum of the contact resistance R_c and the membrane resistance R_m.

V_{conc} is the concentration voltage drop approximated by [20]:

$$V_{conc} = -b \cdot \ln\left(1 - \frac{I_{FC}/A}{I_{max}}\right) \qquad (47.8)$$

where b is the concentration loss constant; i_{FC} is the cell current; A is the is cell active area; I_{max} is the maximum current density.

47.3 Proposed P&O-Based Fuzzy Logic Type-2 MPPT Algorithm

The success of type-1 fuzzy logic systems naturally led to the development of fuzzy logic systems based on type-2 fuzzy logic. The structure of a type-2 fuzzy logic systems shares the same core components of its type-1 counterpart, namely: a fuzzifier, a rule-base, an inference engine and an output processor. Compared to type-1 fuzzy system structure, in the type-2 case the output processor embraces an additional stage where the type-2 fuzzy system is firstly converted into an equivalent type-1 fuzzy system. This procedure is implemented by a type-reduction (TR) algorithm [21].

In this study, the proposed P&O-based fuzzy type-2 MPPT requires as inputs the error E and the change in error ΔE defined by:

$$E_k = \left| \frac{P_k - P_{k-1}}{I_k - I_{k-1}} \right| \quad (47.9)$$

$$\Delta E_k = E_k - E_{k-1} \quad (47.10)$$

The output is the step size change used to drive the P&O classical MPPT.

Fuzzification The universe of discourse for input variables (E_k and ΔE_k) as well as for the output variable (Δd) is divided into five fuzzy sets: mf1 (NB), mf2 (NS), mf3 (ZE), mf4 (PS) and mf5 (PB).

Inference Method In this study, we use the Sugeno inference system.

Rule Base The Fuzzy algorithm tracks the MPP point based on the rule-base consisting of 25 rules as shown in Table 47.1.

Type-Reduction In this study, we use the Begian-Melek-Mendel Method for the type reduction.

Defuzzification After applying the type reduction method, the obtained interval fuzzy set still has to be converted into a crisp number so it becomes suited to the most part of the fuzzy logic system application scenarios. The defuzzified value is obtained by simply computing the average of the interval's left and right endpoints.

Table 47.1 Fuzzy rule base

$E_k/\Delta E_k$	NB	NS	ZE	PS	PB
NB	ZE	ZE	NB	NB	NB
NS	ZE	ZE	NS	NS	NS
ZE	ZE	NP	NS	NS	PS
PS	PS	PS	PS	ZE	ZE
PB	PB	PB	PB	ZE	ZE

47.4 Results and Discussion

The developed Matlab/Simulink model of the PEM fuel cell power system including the 7 kW PEM fuel cell powering a resistive load *via* a DC-DC boost converter controlled using the developed P&O based fuzzy type-2 variable step size MPPT controller has been implemented (Table 47.2).

Figure 47.1 shows the output power corresponding to temperature variation as described below 353 K for the interval 0–0.6 s, 323 K for the interval 0.6–1.2 s and 343 K for the interval 1.2–1.8 s. The legends PO-FS, PO-FZT1 and PO-FZT2 correspond to fixed step P&O, P&O-based fuzzy type-1 and P&O-based fuzzy type-2 MPPT algorithms.

Table 47.2 PEM fuel cell parameters

Parameter	Value
Maximum Power at MPP P_{MPP} (W)	7000
Cell open circuit voltage V_{OC} (V)	1.29
Number of cells N	50
Cell active surface (A/cm^2)	200
Hydrogen partial pressure P_{H_2} (bar)	2.6
Oxygen partial pressure P_{O_2} (bar)	0.3

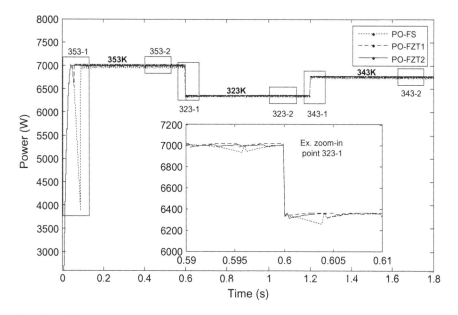

Fig. 47.1 PEM fuel cell output power

Table 47.3 Response time in case of temperature variation (in ms)

Temperature variation	PO-FS	PO-FZT1	PO-FZT2
0–353 K at 0 s	90.7	59.5	50.8
353–323 K at 0.6 s	7.5	3.5	3.5
323–343 K at 1.2 s	1.1	7.2	9.2

Table 47.4 Ripple around the maximum point in case of temperature variation (in W)

Temperature	PO-FS	PO-FZT1	PO-FZT2
353 K from 0–0.6 s	66	33	10
323 K from 0.6–1.2 s	33	31	9
343 K from 1.2–1.8 s	55	33	11

The developed PEM fuel cell power system Matlab/Simulink model has been used to evaluate and compare the dynamic and static performances of the proposed P&O-based fuzzy type-2 MPPT controller to the two others the classical P&O fixed step size and the P&O-based fuzzy type-1 MPPT controllers using response time as dynamic performance metric and the ripple as static performance metric. Tables 47.3 and 47.4 gives the response time and the ripple for the three considered MPPTs.

From Tables 47.3 and 47.4, we can confirm that the proposed P&O-based fuzzy type-2 MPPT controller outperforms both P&O fixed step size as well the P&O-based fuzzy type-1 MPPT controllers regarding dynamic or static performances. The proposed MPPT controller reduces significantly the response time and the ripple around the maximum power point leading by the way to an overall reduction of the energy losses.

47.5 Conclusion

The maximum power of fuel cell depends on operating conditions. Therefore, to supply a load with a constant voltage, a maximum power point controller is needed. This paper proposes a modified perturb and observe based fuzzy type-2 maximum power point controller. In this study the fuzzy type-2 controller has been used to drive the variable step size of classical perturb and observe algorithm in order to track the maximum power point of the proton exchange membrane fuel cell power system. The proposed controller has been validated using the Matlab/Simulink environment where the whole fuel cell power system composed of 7 kW proton exchange membrane fuel cell powering a resistive load via a DC-DC boost converter controlled using the proposed controller have been implemented. Comparative simulation results between the proposed controller and the classical perturb and observe one prove the superiority of the proposed controller to track effectively the maximum power regarding used dynamic and static performance metrics.

Acknowledgements The Algerian Ministry of Higher Education and Scientific Research via the DGRSDT supported this research (PRFU Project code: A01L07UN190120180005).

References

1. Z. Qi, *Proton Exchange Membrane Fuel Cells* (CRC Press, Taylor & Francis, Boca Raton, 2014)
2. C.A. Ramos-Paja, G. Spagnuolo, G. Petrone, R, Giral, A. Romero, Fuel cell MPPT for fuel consumption optimization, in *Proceedings IEEE International Symposium on Circuits and Systems (ISCAS)*, Paris, France (2010) pp. 2199–2202
3. M. Sarvi, M.M. Barati, Voltage and current based MPPT of fuel under variable temperature conditions, in *Proceedings Universities Power Engineering Conference (UPEC)*, Cardiff, Wales (2010), pp. 1–4
4. M. Becherif, D. Hissel, MPPT of a PEMFC based on air supply control of the motocompressor group. Int. J. Hydrogen Energy **35**(22), 12521–12530 (2010)
5. N. Karami, R. Outbib, N. Moubayed, Fuel flow control of a PEM fuel cell with MPPT, in *Proceedings IEEE International Symposium on Intelligent Control (ISIC)*, Dubrovnic, Crotia (2012), pp. 289–294
6. D. Li, Y. Yadi, J. Qibing, G. Zhiqiang, Maximum power efficiency operation and generalized predictive control of PEM fuel cell. Energy **68**, 210–217 (2014)
7. C.A. Ramos-Paja, J.J. Espinosa, A control system for reducing the hydrogen consumption of PEM fuel cells under parametric uncertainties. Tecno Lógicas **19**(37), 45–59 (2016)
8. F.X. Chen, Y. Yu, J.X. Chen, Control system design of power tracking for PEM fuel cell automotive application. Fuel Cells **17**(5), 671–681 (2017)
9. A. Harrag, H. Bahri, Novel neural network IC-based variable step size fuel cell MPPT controller. Int. J. Hydrogen Energy **42**(5), 3549–3563 (2017)
10. K.J. Reddy, N. Sudhakar, A new RBFN based MPPT controller for grid-connected PEMFC system with high step-up three-phase IBC. Int. J. Hydrogen Energy **43**(37), 17835–17848 (2018)
11. A. Harrag, S. Messalti, How fuzzy logic can improve PEM fuel cell performance? Int. J. Hydrogen Energy **43**(1), 537–550 (2018)
12. M. Derbeli, O. Barambones, J.A. Ramos-Hernanz, L. Sbita, Real-time implementation of a super twisting algorithm for PEM fuel cell power system. Energies **12**(1594), 1–20 (2019)
13. A. Harrag, H. Bahri, A novel single sensor variable step size maximum power point tracking for proton exchange membrane fuel cell power system. Fuel Cells **19**(2), 177–189 (2019)
14. M. Hatti, Neural network controller for PEM fuel cells, in *IEEE International Symposium on Industrial Electronics*, pp. 341–346 (2007)
15. C. Spiegel, *PEM Modeling and Simulation Using Matlab* (Academic Press, Burlington, MA, 2008)
16. R. O'Hayre, *Fuel Cell Fundamentals*. Wiley (2009)
17. R.F. Mann, J.C. Amphlett, M.A. Hooper, H.M. Jensen, B.A. Peppley, P.R. Roberge, Development and Application of a generalized steady-state electrochemical model for a PEM fuel cell. J. Power Sources **86**(1–2), 173–180 (2000)
18. D. Yu, S. Yuvarajan, A novel circuit model for PEM fuel cells, in *IEEE 19th Applied Power Electronics Conference and Exposition*, Anaheim, CA, USA (2004), pp. 362-366
19. J.H. Lee, T.R. Lalk, A.J. Appleby, Modelling fuel cell stack system. J. Power Sources **73**, 229–241 (1998)

20. H.J. Avelar, E.A.A. Coelho, J.R. Camacho, J.B. Vieira, L.C. Freitas, M. Wu, PEM fuel cell dynamic model for electronic circuit simulator, in *IEEE Electrical Power & Energy Conference*, Montreal, QC, Canada (2009), pp. 1–6
21. R. Antão, A. Mota, R.E. Martins, *Type-2 Fuzzy Logic Uncertain Systems' Modeling and Control* (Springer 2017)

Chapter 48
A GIS-MOPSO Integrated Method for Optimal Design of Grid-Connected HRES for Educational Buildings

Charafeddine Mokhtara, Belkhir Negrou, Noureddine Settou, Abdessalem Bouferrouk, Yufeng Yao, and Djilali Messaoudi

Abstract In this paper, an optimal design of a grid-connected PV-battery-hydrogen hybrid renewable energy system (HRES) at a University campus in Ouargla, Algeria is carried out. To achieve this goal, geographical information system (GIS), CAD software and multi-objective particle swarm optimization (MOPSO) are used. First, the rooftop's solar energy potential, optimal zones to install PV panels and selection of the PV system's best installation are determined, considering many design criteria. Thus, based on these outcomes, optimal sizing of the proposed hybrid system is then performed using MATLAB. Cost of energy, loss of power supply probability, and renewable usage are the objectives to be optimized. Here, an energy management strategy is adopted to select the most adequate storage option at each simulation time step. In this study, selling of the excess hydrogen gas has suggested instead of selling electricity to the grid. Results show that standard multi crystalline PV panels with an inclination angle of 17° is the best installation. In addition, the obtained optimal HRES, which includes PV/battery/hydrogen has a renewable usage of 90%, and cost of energy of only 0.22 $/kWh with high reliability.

Keywords Solar PV · Hybrid energy storage · Multi objective particle swarm optimization · GIS · Hydrogen

---

C. Mokhtara (✉) · B. Negrou · N. Settou · D. Messaoudi
Laboratory Promotion et Valorisation des Ressources Sahariennes (VPRS), University of Kasdi Merbah Ouargla, BP 511, 30000 Ouargla, Algeria
e-mail: mokhtara.chocho@gmail.com

C. Mokhtara · D. Messaoudi
Department of Mechanical Engineering, University of Kasdi Merbah Ouargla, Ouargla, Algeria

A. Bouferrouk · Y. Yao
Department of Engineering Design and Mathematics, University of the West of England, Bristol BS16 1QY, UK

© The Editor(s) (if applicable) and The Author(s), under exclusive license to Springer Nature Singapore Pte Ltd. 2021
A. Khellaf (ed.), *Advances in Renewable Hydrogen and Other Sustainable Energy Carriers*, Springer Proceedings in Energy,
https://doi.org/10.1007/978-981-15-6595-3_48

48.1 Introduction

In Algeria, 97% of electricity is generated from naturel gas [1]. Buildings, which are 99% grid-connected, consume a large amount (43%) of this generated electricity. Therefore, the integration of renewable sources is mandatory to reduce CO_2 emissions by reducing the use of fossil fuels. In this context, the Algerian government aims to increase renewable energy generation share to 27% of the total power generation in the country [2]. However, due to uncertainty of renewable sources like solar energy, a hybrid energy system is one of the most promising solutions at present. For Algeria where solar energy is abundant [3] especially in the southern Sahara region, a rooftop, grid-connected solar PV hybrid energy system is an attractive solution to supply energy to buildings in urban areas. However, optimal design and size optimization of such systems is a great challenge. Many research works have been conducted in this area. The majority of these efforts have focused on either evaluating energy potential of rooftop solar PV systems in urban areas using GIS-based methods [4] and CAD software, or on optimal sizing and energy management of grid-connected PV based hybrid systems. The optimal sizing problem is solved as a single-objective or multi-objective function using either classical or meta-heuristic algorithms. Meta-heuristic algorithms such as the particle swarm optimization (PSO) have been widely used by researchers. In addition, cost of energy (COE), renewable usage, CO_2 emission, and reliability indices are often the objective functions. From literature, the integration of spatial analysis with size optimization of a HRES has not been investigated extensively. In addition, a solar system with hybrid hydrogen-battery storage system has not been much discussed. In this work, an integrated approach is developed for the optimal design of a rooftop grid-connected solar PV with hybrid hydrogen-battery storage system. The main contributions of this work are: (1) to identify feasible zones and installations for rooftop PV system, and (2) to find optimal sizing of the suggested HRES.

48.2 Methodology

A GIS-Multi Objective PSO (GIS-MOPSO) method is developed for optimal design of rooftop solar PV with hybrid hydrogen-battery storage system to electrify a grid-connected University building in Ouargla, Algeria. In order to achieve the aims of the study, the following methodologies are followed. First, a map of the building (from Google-Earth) is exported, separately, to ArcGIS to perform spatial analysis, and to sketch up software to create a 3D model of the building. Based on the developed 3D model, Ecotect software is used to evaluate shading effects and sun light hours at each rooftop's zone for a one-year simulation. Hence, identifying the best zones for installing PV panels, selection of the best PV system's installation based on technical and spatial criteria, and finding the optimal sizing of the proposed HRES are carried out next.

Fig. 48.1 Map of the campus building

48.2.1 Building Description and Climatic Data

The studied educational building, located in a hot dry climate, has a total roof area of 18,209 m^2, and has an energy demand of 1485 MWh/year. Figure 48.1 shows a map of the building. In addition, the monthly load of the building (as supplied by energy provider Sonelgas) is presented in Fig. 48.2. Meteonorm 7 software was used to collect the required climatic data for the building.

48.2.2 Assessment of Rooftop Solar Energy Potential

Solar potential represents the theoretical maximum amount of PV that can be deployed on the rooftop of buildings, which depends on different factors [5]. Partial shading due to inter rows distance or neighbouring obstacles [6], wind speed, PV panel technology and panel inclination, and ambient temperature are generally reported to be the most influential factors for PV panels productivity and their set up. The assessment of rooftop solar energy potential consists of two major steps:

(1) Identifying optimal rooftop zones ArcGIS software (V. 10.2) was used to select the best rooftop zones to install PV panels considering the weights of five evaluation criteria, which are sunlight hours (30%), exposer ratio (30%), shape factor (20%),

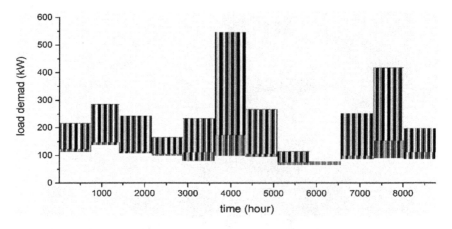

Fig. 48.2 Building monthly load demand

available zone area (15%) and height of each zone (5%). After creating the raster maps of the five criteria, a multi criteria decision-making (MCDM) calculation is performed.

(2) Selecting the best PV system installation To select the best PV system installation, many factors are considered such as PV panels' technology (including thin film of First-Solar (FS) and crystalline modules of Trina-Solar (TS)), inclination angle, and available area for the obtained optimal zones. Three inclination angles are considered: 17°, 47°, and 32°; these represent the optimum tilt angles at the studied location for summer period, winter period, and yearly average, respectively. The maximum capacity of the best rooftop PV installation is used as a constrained in the size optimization.

48.2.3 HRES Components

Solar Photovoltaic (PV) The electric output power of the PV module is evaluated using Eq. 48.1 [7].

$$P_{pv} = P_{Npv} \times \frac{G}{G_{ref}} \times \left[1 + K_t \times \left(\left[T_{amb} + \frac{NOCT - 20}{800}\right] \times G - T_{ref}\right)\right] \quad (48.1)$$

Battery Storage (BS) Batteries are used to store excess electricity from PV panels.

Hydrogen Storage Includes Electrolyzer (EL), Storage tank (ST), and Fuel cell (FC).

Grid When the PV system and storage devices are not sufficient to supply the load, the grid is used to supply the deficit power. In Algeria, the purchase price of electricity without subsidies is 0.1 $/kWh. The rest of stored hydrogen at the end of the academic year will be sold. The sell rate of hydrogen gas is set at 3.3 $/kg (0.08 $/kWh) [8], similar to the sell rate of hydrogen gas produced by conventional procedures.

Converter It is a device that converts electricity from DC to AC and vice versa.

48.2.4 Energy Management Strategy

The energy management (EM) strategy applied in this work has two modes:

Mode 1 The surplus electricity from PV panels is used to charge batteries or produce hydrogen gas. The selected storage option depends on the time of operation. In the summer vacation (from 10 July to 10 September), surplus electricity is converted to hydrogen. Out of this period, surplus electricity is used to charge batteries.

Mode 2 When the PV panels are unable to meet the required load demand, then batteries and/or fuel cells are used to supply the shortage. If there is further shortage, the grid will provide the rest of the demand. Loss of power supply is evaluated if the complete hybrid system fails to meet the load.

48.2.5 Multi Objective Optimization

In this work, cost of energy, loss of power supply probability (LPSP) and renewable usage (RU) are the objectives to be optimized. The decision variables include size of PV panels, battery bank, electrolyser, storage tank and fuel cell. The size of PV system depends on available area, the obtained optimal zones, and the selected PV system installation (results of Sect. 48.2.2). The size optimization problem is solved by multi objective PSO. Simulations are carried out in MATLAB. The simulation time step is 1 h, climatic data (temperature and solar radiation) and building energy load are defined for one entire year (8760 h). Within the project lifetime of 20 years, the optimization must proceed until the maximal iteration value is reached.

Fig. 48.3 Area suitability for installing PV panels on three buildings of the campus (from ArcGIS)

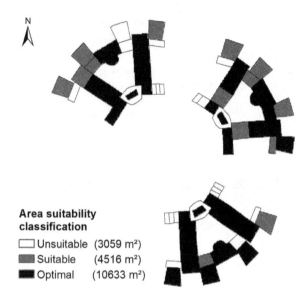

Area suitability classification
☐ Unsuitable (3059 m²)
▩ Suitable (4516 m²)
■ Optimal (10633 m²)

48.3 Results and Discussion

48.3.1 Area Suitability Classification

Results from ArcGIS of MCDM analysis for area classification within the building are presented in Fig. 48.3. Results clearly show three categories of areas, with the optimal zones (of area 10,633 m²) representing more than half of the total area of the building. Only this zone area is selected for installing PV panels. Therefore, the maximum capacity of rooftop PV depends on the available area from this optimal zone.

48.3.2 Selecting the Best PV System Installation

Based on the results of area suitability, only the optimal zone (zone 1), as seen in blue colour in Fig. 48.3, is considered for installing PV panels. Different configurations are compared to select the best one based on their yearly energy production and the allowable installed capacity. Table 48.1 provides the results of ranking, where only the first option is shown and it is the one that will be selected for the simulation of optimal sizing of the proposed HRES.

From Table 48.1, multi crystalline PV panels at 17° inclination represent the optimal installation for this case study. This is because it provides the largest yearly energy production by exploiting the entire area of the optimal zone.

Table 48.1 Ranking of PV system installations

Configuration	Required area per kW	Maximum allowable PV capacity (optimal zone) (kW)	PV output per kW (kWh/year)	PV total power (optimal zone) (MWh/year)	Rank
TS_17°	8.12	1310	1781	2333.110	1
...	...	...	...	...	...
FS_47°	1167	912	1759	1,604,208	6

Table 48.2 Result of optimal sizing of the HRES

Component	PV (kW)	BS (kWh)	EL (kW)	ST (kWh)	FC (kW)	LPSP (%)	COE ($/kWh)	NRU (%)
Size	1310	866	300	2601	80	4.5	0.225	0.099

48.3.3 Optimal Sizing of the HRES

Multi crystalline PV panels at inclination angle of 17° are selected for further optimization. Results of optimal sizing of the studied HRES are presented in Table 48.2.

The obtained HRES configuration from optimal sizing has a renewable usage of 99%, and COE of 0.22 $/kWh. Despite the COE being higher than the current price of purchase of electricity, the optimised HRES is more environmentally attractive. In addition, this final HRES configuration produces hydrogen gas, which is considered the sustainable fuel of the future, and therefore, more jobs could be created. Furthermore, with the continuous decrease of the price of PV panels and hydrogen storage components, the proposed HRES will be more cost effective than using conventional sources.

48.4 Conclusions

In this work, a rooftop PV/battery/hydrogen grid-connected HRES is optimally designed using a combined GIS and MOPSO method. Results show that the use of standard multi crystalline PV panels at an inclination angle of 17° is the most suitable configuration for the studied building. Moreover, PV/battery/Electrolyzer/storage tank/fuel cell is found to be the optimal configuration, with 90% of renewable energy usage, and COE of 0.22 $/kWh. The results have emphasized the efficiency of the proposed method in solving a complex energy problem, which relates to HRES.

References

1. C. Mokhtara, B. Negrou, N. Settou, A. Gouareh, B. Settou, Pathways to plus-energy buildings in Algeria : design optimization method based on GIS and multi-criteria decision-making. Energy Procedia **162**, 171–80 (2019) (Elsevier B.V)
2. A. Government, Renewable energy and energy efficiency development plan 2015–2030. https://www.iea.org/policiesandmeasures/renewableenergy/?country=ALGERIA (2015)
3. Z. Abada, M. Bouharkat, Study of management strategy of energy resources in Algeria. Energy Rep. **2018**(4), 1–6 (2018). (Elsevier Ltd.)
4. L. Romero, E. Duminil, J. Sánchez, U. Eicker, Assessment of the photovoltaic potential at urban level based on 3D city models : a case study and new methodological approach. Sol Energy **146**, 264–75 (2017). (Elsevier Ltd.)
5. L. Kurdgelashvili, J. Li, C.H. Shih, B. Attia, Estimating technical potential for rooftop photovoltaics in California, Arizona and New Jersey. Renew Energy **95**, 286–302 (2016). (Elsevier Ltd.)
6. A.K. Shukla, K. Sudhakar, P. Baredar, Design, simulation and economic analysis of standalone roof top solar PV system in India. Sol Energy **136**, 437–449 (2016). (Elsevier Ltd.)
7. A. Kumar, A.R. Singh, Y. Deng, X. He, P. Kumar, R.C. Bansal, Integrated assessment of a sustainable microgrid for a remote village in hilly region. Energy Convers Manag. **180**, 442–72 2019 (Elsevier; October 2018)
8. B. Negrou, N. Settou, N. Chennouf, B. Dokkar, Valuation and development of the solar hydrogen production. Int J Hydrogen Energy **6**(36), 4110–4116 (2010). (Elsevier Ltd.)

Chapter 49
A Comparison Between Two Hydrogen Injection Modes in a Metal Hydride Reactor

Bachir Dadda, Allal Babbou, Rida Zarrit, Youcef Bouhadda, and Saïd Abboudi

Abstract The optimization of hydrogen storage within metal hydride reactors is one of the main issues in the recent works. The aim of this paper is to compare between two hydrogen injection modes within a hydrogen tank in terms of heat transport. The charging process in the cylindrical tank is releasing heat because it undergoes an exothermic reaction. In order to guarantee a maximum absorption of hydrogen from the alloy ($LaNi_5$), a heat exchanger along the cylindrical walls is considered. Two hydrogen injection modes have been considered in this study: (1) injection from the top, (2) injection from the axis of the cylinder. The governing equations based on mass, heat and momentum balances are transient and two-dimensional. The results show that axial injection is more advantageous than the top one, since it ensures a better heat transfer within the hydride bed and therefore helps to absorb the maximum amount of hydrogen in a shorter time.

Keywords Metal hydride · $LaNi_5$ · Heat and mass transfer · Finite volume method

Nomenclature

T	Gas and solid temperature (K)
P	Pressure (Pa)
λ	Effective thermal conductivity (W/m K)
m	Hydrogen mass absorbed (kg/m^3 s)

B. Dadda (✉) · A. Babbou · R. Zarrit · Y. Bouhadda
Unité de Recherche Appliquée en Energies Renouvelables, URAER, Centre de Développement des Energies Renouvelables, CDER, 47133 Ghardaïa, Algeria
e-mail: dadbac@gmail.com

Y. Bouhadda
e-mail: bouhadda@yahoo.com

S. Abboudi
Département COMM, Laboratoire ICB, Univ. Bourgogne Franche-Comté, UTBM, Site de Sévenans, Belfort Cedex 90010, France

© The Editor(s) (if applicable) and The Author(s), under exclusive license to Springer Nature Singapore Pte Ltd. 2021
A. Khellaf (ed.), *Advances in Renewable Hydrogen and Other Sustainable Energy Carriers*, Springer Proceedings in Energy,
https://doi.org/10.1007/978-981-15-6595-3_49

C_a Constant
E_a Activation energy (J/mole)
P_{eq} Equilibrium pressure
ρ_s Hydride density (kg/m)
ρ_{ss} Hydride density at the saturation state
ΔH^0 Reaction heat of formation (J/kg)
ε Hydride porosity
C_p Specific heat (J/kg K)
k   Permeability (m^2)

49.1 Introduction

One of the major issues in hydrogen storage systems is the heat transfer limit in metal hydride reactors (MHRs). Heat transfer rate is the control variable that determines the rate at which hydrogen gas can be extracted from the hydride tank. Muthukumar [1] analyzed the effect of different working conditions within a tank made of metal hydrides using AB_5. He found that, fluids at low temperatures have significant effects on the storage capacity of hydrogen at low pressure, and give better absorption and desorption rates.

The optimization of hydrogen storage within MHRs has been investigated by Kikkinides et al. [2]. In their paper, authors have considered different cooling system designs by inserting heat exchangers in the tank. Bhouri et al. [3] used the numerical tool to evaluate the influence of the fin thickness and the number of heat exchanger tubes on the loading and discharging processes. Their results show that the number of heat exchanger tubes and the bed thickness are very important for the optimization of the hydrogen storage application design. Several investigations have been made to increase the heat transfer rate, by modifying the intern properties of the tank [4–7]. That is generally realized by increasing the effective thermal conductivity of the hydride bed. Brendan and Andrew [8] studied the effect of extern fins on the hydrogen release rate of a MHR. They used a one-dimensional analysis with a two-dimensional transient model.

An experimental work focused on the influence of different tank configurations has been presented by Kaplan [9]. They concluded that, hydrogen storage in MHRs is principally heat transfer management. Elhamshri and Kayfeci [10] presented a performance analysis. Their work shows the impact of using heat pipes and fins for enhancing heat transfer in MHRs at varying hydrogen supply pressures (2–15 bar).

This work aims to compare between two injection modes of hydrogen in a MHR.

49.2 Mathematical Model

The geometry of the studied reactor is cylindrical. A heat exchanger is considered in the walls in order to ensure a good reaction progress. The obtained equations are based on the mass, heat and momentum balance. We assume that:—Since the pressure and temperature are not very high, hydrogen is considered as an ideal gas. The work of compression and the viscous dissipation are neglected, because of the gaseous nature of hydrogen. Finally, the gas and solid temperatures are considered equal (local thermal equilibrium), according to Jemni and Ben Nasrallah [11] findings.

49.2.1 Energy Equation

$$(\bar{\rho}C_p)_e \frac{\partial T}{\partial t} = \frac{1}{r}\frac{\partial}{\partial r}\left(r\lambda_e \frac{\partial T}{\partial r}\right) + \frac{\partial}{\partial z}\left(\lambda_e \frac{\partial T}{\partial z}\right) - (\bar{\rho}C_p)_g V_{gr} \frac{\partial T}{\partial r}$$
$$- (\bar{\rho}C_p)_g V_{gz} \frac{\partial T}{\partial z} - m(\Delta H^0) + T\left(C_{p_g} - C_{p_s}\right) \qquad (49.1)$$

49.2.2 Momentum Equation

$$\varepsilon \frac{\partial \bar{\rho}_g^g}{\partial t} + \text{div}\left(\bar{\rho}_g^g V_g\right) = -m \qquad (49.2)$$

49.2.3 Mass Conservation Equation

For the gas phase: $\varepsilon \dfrac{M_g}{RT}\dfrac{\partial P_g}{\partial t} + \varepsilon \dfrac{M_g P_g}{R}\dfrac{\partial (1/T)}{\partial t} = \dfrac{k}{\vartheta_g}\dfrac{1}{r}\dfrac{\partial}{\partial r}\left(r\dfrac{\partial P_g}{\partial r}\right) + \dfrac{k}{\vartheta_g}\dfrac{\partial^2 P_g}{\partial z^2} - m$

$$(49.3)$$

For the solid phase: $(1-\varepsilon)\dfrac{\partial \bar{\rho}_s}{\partial t} = -m^2 \qquad (49.4)$

With $m = C_a e^{-E_a/RT} \ln\left(\dfrac{P}{P_{eq}}\right)(\rho_{ss} - \rho_s) \qquad (49.5)$

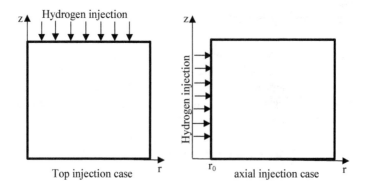

Fig. 49.1 Study area sketch for both injection cases

49.2.4 Initial and Boundary Conditions

Figure 49.1 shows the considered domain for the two cases of hydrogen injection:

At t = 0
Initially, pressure, temperature and hydride density are supposed to be constant:

$$T = T_0, P = P_0, \rho_s = \rho_0 \tag{49.6}$$

At t > 0

- for the top injection case:

$$\frac{\partial P}{\partial r} = 0, \frac{\partial T}{\partial r} = 0, \quad \text{for } r = 0, 0 \leq z \leq H \tag{49.7}$$

- for axial injection case:

$$P = P_0, T = T_0, \quad \text{for } r = r_0, 0 \leq z \leq H \tag{49.8}$$

- The reactor walls are considered impermeable:
  - for the both cases:

$$\frac{\partial P}{\partial r} = 0 \quad \text{for } r = R \text{ and } 0 \leq z \leq H \tag{49.9}$$

  - for the top injection case:

$$\frac{\partial P}{\partial z} = 0 \quad \text{for } z = H \text{ and } 0 \leq r \leq R \tag{49.10}$$

- for the axial injection case:

$$\frac{\partial P}{\partial z} = 0 \quad \text{for } z = H \text{ and } r_0 \leq r \leq R \tag{49.11}$$

- At the top of the reactor,
  - for the top injection case:

$$P = P_0, T = T_0, \quad \text{for } z = 0, 0 \leq r \leq R \tag{49.12}$$

- for the axial injection case:

$$\frac{\partial T}{\partial z} = h(T - T_f) \quad \text{for } z = 0 \text{ and } r_0 \leq r \leq R \tag{49.13}$$

- On the lateral walls and the base of the reactor, the heat fluxes are given by:
  - for the two cases:

$$\frac{\partial T}{\partial r} = h(T - T_f) \quad \text{for } r = R \text{ and } 0 \leq z \leq H \tag{49.14}$$

- for the top injection case:

$$\frac{\partial T}{\partial z} = h(T - T_f) \quad \text{for } z = H \text{ and } 0 \leq r \leq R \tag{49.15}$$

- for the axial injection case:

$$\frac{\partial T}{\partial z} = h(T - T_f) \quad \text{for } z = H \text{ and } r_0 \leq r \leq R \tag{49.16}$$

h the heat transfer coefficient between the hydride bed and the cooling fluid.
T_f the cooling temperature

To solve the mathematical model, the finite volume method has been used. The resulting nonlinear tridiagonal matrices have been solved directly with Thomas algorithm using the line-by-line technique as detailed in our previous work [12]. A CFD code has been worked out in FORTRAN language by an i5-2310, 2.90 GHz processor with 1.9 GB of RAM.

49.3 Results and Discussion

In Fig. 49.2, the contours of the isotherms within the reservoir are plotted for the two different injection modes at two different instants (injection from the top and axial injection at t = 100 s and t = 2500 s). For the second mode of injection, the cooling is applied on the top wall as well. Therefor, the boundary condition (49.16) is considered at that region. From Fig. 49.2, it can be noticed that at the beginning of the absorption process, the major part of the reactor is heated, since the reaction inside is exothermic. After a while, the temperatures start decreasing, because of the cooling fluid circulating around the tank walls. A clear difference between the isotherm shapes corresponding to the two modes of injection is observed. This difference is due to the fact that the injection of hydrogen from the axis of the cylinder induces a cooling region at that side. On the other hand, at advanced stages, it can be noticed that the bed temperature decreases much more in the case with axial injection than in the other case. Therefor, the axial injection is more efficient and favorable for the heat transfer in the hydride bed.

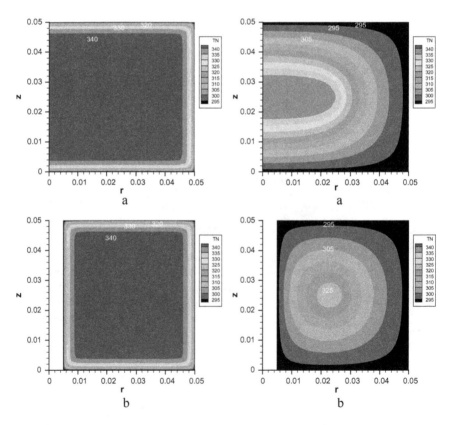

Fig. 49.2 Isotherms for two different injection modes: **a** top injection, **b** axial injection

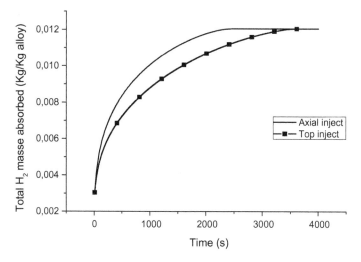

Fig. 49.3 Total absorbed hydrogen mass for the two injection modes

In order to study the influence of the hydrogen injection mode on the total absorbed hydrogen mass, the time evolution curves of this quantity are plotted in Fig. 49.3 for the two cases of injection.

The hydrogen mass absorbed by the alloy increases over the time, until it reaches the maximum value that corresponds to the saturation of the hydride bed. This value is around 12 g of H_2/(Kg of alloy). It is clear from this figure that, in the axial injection case, the tank tends to absorb the maximum amount of hydrogen in a very short time compared to the case with top injection. For the axial injection case, this maximum amount is reached at t = 2300 s. Whereas, it is reached at t = 3200 s for the second case, almost 20 min of gap, which is relatively significant.

49.4 Conclusion

This work aims to study the heat transfer enhancement within a metal hydride reactor. The cylindrical considered reactor is equipped with a cooling system in order to ensure a maximum hydrogen absorption rate. Two different modes of hydrogen injection are treated. The mathematical two-dimensional governing equations are solved by the finite volume method. The obtained numerical results show that, axial injection is more efficient in terms of heat transfer and hydrogen mass absorption.

References

1. P. Muthukumar, Experiment on a metal hydride-based hydrogen storage device. Int. J. Hydrogen Energy **30**(15), 1569–81 (2005)
2. E.S. Kikkinides, M.C. Georgiadis, A.K. Stubos, On the optimization of hydrogen storage in metal hydride beds. Int. J. Hydrogen Energy **31**(6), 737–51 (2006)
3. M. Bhouri, J. Goyette, B.J. Hardy, D.L. Anton, Sensitivity study of Alanate hydride storage system. Int. J. Hydrogen Energy **36**(1), 621–33 (2011)
4. S. Lévesque, M. Ciureanu, R. Roberge, T. Motyka, Hydrogen storage for fuel cell systems with stationary applications—I. Transient measurement technique for packed bed evaluation. Int. J. Hydrogen Energy **25**(11), 1095–1105 (2000)
5. M. Nagel, Y. Komazaki, S. Suda, Dynamic behavior of paired metal hydride systems—I. J Less Common Met. **120**(1), 45–53 (1986)
6. K.J. Kim, G. Lloyd, J.A. Razani, K.T. Feldman, Development of $LaNi_5$/Cu/Sn metal hydride powder composit. Powder Technol. **99**(1), 40–45 (1998)
7. K.J. Kim, B. Montoya, A. Razani, K. Lee, Metal hydride compacts of improved thermal conductivity. Int. J. Hydrogen Energy **26**(6), 609–613 (2001)
8. M.D. Brendan, M.R. Andrew, Impacts of external heat transfer enhancements on metal hydride storage tanks. Int. J. Hydrogen Energy **31**(12), 1721–1731 (2006)
9. Y. Kaplan, Effect of design parameters on enhancement of hydrogen charging in metal hydride reactors. Int. J. Hydrogen Energy **34**(5), 2288–94 (2009)
10. F.A.M. Elhamshri, M. Kayfeci, Enhancement of hydrogen charging in metal hydride-based storage systems using heat pipe. Int. J. Hydrogen Energy **44**(34), 18927–18938 (2018)
11. A. Jemni, S. Ben Nasrallah, Study of two dimensional heat and mass transfer during absorption in a metal-hydrogen reactor. Int. J. Hydrogen Energy **20**, 43–52 (1995)
12. B. Dadda, A. Babbou, R. Zarrit, Y. Bouhadda, S. Abboudi, Effect of boundary conditions on the hydrogen absorption in a metal hydride reactor (April 15, 2019), in *2nd National Conference on Computational Fluid Dynamics & Technology 2018 (CFD & Tech 2018)*

Chapter 50
Effect of the Complexing Agent in the Pechini Method on the Structural and Electrical Properties of an Ionic Conductor of Formula $La_{1-x}Sr_xAlO_{3-\delta}$ (x = 0, 0.05, 0.1, 0.15)

F. Hadji, F. Bouremmad, S. Shawuti, and M. A. Gulgun

Abstract The Ion conductors are used as electrolytes in high temperature Solid Oxide Fuel Cells SOFCs. The preparation route has an important role on their structural and electrical properties. In this study, we used a modified Pechini method to prepare an ionic conductor based on lanthanum aluminate doped with strontium $La_{1-x}Sr_xAlO_{3-\delta}$ (x = 0.0.05, 0.1, 0.15). The effect of two complexing agents on structural and electrical properties was studied, we used Ethylene Diamine Tetra Acetic EDTA, and tartaric acid TA as complexing agents. The perovskite phases were obtained at 900 °C and characterized by different techniques; SEM images show that grain size is in the nanometer range, XRD analysis shows that the compounds prepared by use of the two complexing agents crystallize in a perovskite structure with an orthorhombic system and an R3m space group, the doped phases prepared by EDTA have a secondary phase $LaSrAl_3O_7$ which is absent in the compounds prepared by tartaric acid. The determination of the ionic conductivity by electrochemical impedance spectroscopy shows clearly the effect of the complexing agent. Indeed we have found that the value of the ionic conductivity is higher for the phases produced by the Pichini method in the presence of tartaric acid as complexing agent.

Keywords Ionic conductivity · Impedances · Solid electrolyte

50.1 Introduction

Fuel cells convert chemical energy into electrical energy. Solid oxide fuel cells (SOFCs) are a very important family in the development of energy; they have attracted

---

F. Hadji · F. Bouremmad (✉)
Laboratoire de l'Interaction des Matériaux Environnement (LIME), Université Mohammed Seddik Ben Yahia, Jijel, Algeria
e-mail: bouremmad_farida@yahoo.com

S. Shawuti · M. A. Gulgun
Sabanci University, FENS, 34956 Orhanli Tuzla, Istanbul, Turkey

© The Editor(s) (if applicable) and The Author(s), under exclusive license to Springer Nature Singapore Pte Ltd. 2021
A. Khellaf (ed.), *Advances in Renewable Hydrogen and Other Sustainable Energy Carriers*, Springer Proceedings in Energy,
https://doi.org/10.1007/978-981-15-6595-3_50

attention due to their high-energy efficiency thanks to the flexibility in the choice of gas. However, the operating temperature range of SOFC is very high, of the order of 800 to 1000 °C. This is a condition for the solid electrolyte to provide sufficient ionic conductivity, which is a disadvantage for the use of these fuel cells [1–3].

Today, the research is focused on the study of electrolyte that could help reduce the SOFC operating temperature and thus ensure cheaper electricity generation [4].

Among the ionic conductive materials, which have attracted particular attention, those of the perovskite type; of chemical formula ABO_3, they have a degree of freedom in the creation of the ionic conductivity by making substitutions on A or B sites [5–7].

To prepare these materials, several methods are used, each having benefits and disadvantages, the method of Pechini classified as a soft chemistry method is based on two types of reactions: first a complexation and then a polymerization, basically the complexing agent used is citric acid [8, 9].

The purpose of this work is to prepare an ionic conductor $La_{1-x}Sr_xAlO_3$, substituting lanthanum by strontium for the purpose of increasing ionic conductivity, the Pechini method is used. We will also study the effect of the complexing agent on the structural and electrical properties of the samples; two complexing agents are thus used: tartaric acid and EDTA. The characterization of the $La_{1-x}Sr_xAlO_3$ phases is carried out by XRD and SEM. The electrical conductivity at high temperature is determined by the electrical impedances method.

50.2 Experimental

50.2.1 Preparation

All samples are prepared in the same way and the same method used in [9] with replacement of citric acid by tartaric acid or EDTA.

50.2.2 Characterization

Several techniques were employed to investigate the phases formed at different stages of processing. The X-ray diffraction (XRD) experiments were performed on a D8 Advance Bruker AXS diffractometer with CuKα radiation equipped with a curved graphite monochromator. The data were collected in the 2θ range of 10–80° with a step size of 0.03° and a count time of 2 s per step. Scanning Electron Microscopy (SEM) images were recorded at room temperature on a LEO Supra 35VP FEG SEM (by ZEISS, Germany) apparatus. For the complex plane impedance, measurements were performed via a Solartron 1260 impedance analyzer with dielectric interface. The measurements were done in the temperature range between 400–700 °C and the frequency range from 5 Hz to 20 MHz with a frequency step of 5 points per decade and an oscillating voltage of 100 mV. For high temperature measurements, a ProboStat setup was uzed.

50.3 Results

50.3.1 XRD

The powders prepared by the Pechini method using tartaric acid as a complexing agent are essentially a pure phase after calcination at 900 °C, as shown on (Fig. 50.1), no secondary phase is noted and all peaks are indexed in the perovskite phase.

The powders prepared with EDTA consist essentially of the perovskite phase, however only the undoped phase is pure, the others have a secondary phase at 2θ equal to 30.7°; this shows that the complexation step was critical and tartaric acid is better complexing agent than EDTA. The found perovskite phase crystallizes in the rhombohedral system with a space group R3m.

After the indexing step, the cell volume variation of the perovskite as a function of strontium rate x is represented in (Fig. 50.2). We note that the cell volume increases with increasing x values, this implies that Sr^{2+} having an ionic radius of 1.44 Å effectively replaced the La^{3+} with ionic radius of 1.36 Å.

On the other hand, it is noted that the cell volume of the perovskite phase varies according to the type of the complexing agent. The presence of secondary phase in the case of EDTA resulting from a bad complexation reaction a priori distorts the cell by decreasing its volume.

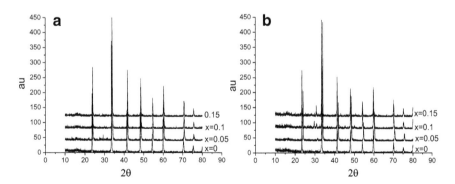

Fig. 50.1 XRD spectra of $La_{1-x}Sr_xAlO_{3-\delta}$ (x = 0; 0.05; 0.1; 0.15): **a** prepared with TA, **b** prepared with EDTA

Fig. 50.2 Effect of strontium rate on cell volume

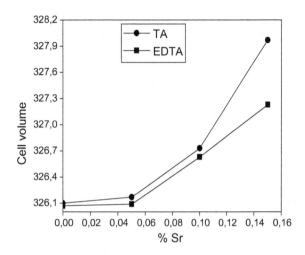

50.3.2 SEM

The SEM observations of $La_{1-x}Sr_xAlO_{3-\delta}$ structures, x takes the values: 0 and 0.1 as examples, prepared by the two complexing agents are shown in (Fig. 50.3). The SEM images reveal agglomerated grains with spherical morphology of nanometric size ranging in 100 nm to 200 nm, similar results are found by S. D. Neelapala et al., who prepared Sr-doped $LaAlO_3$ by reverse strike co-precipitation Method [10].

50.3.3 Electrical Conductivity

The electrical conductivity was studied as a function of temperature in the range of 400–700 °C, by using the electrochemical impedance method. The results are presented by the complex plane diagram (Nyquist representation associated with an RC model-circuit) which connects the imaginary part—Z'' to the real part Z'. As an indication, we present in (Fig. 50.4), the results concerning the samples ($La_{0.95}Sr_{0.05}AlO_{3-\delta}$) at 700 °C.

Figure 50.5 shows the results for electrical conductivity obtained for temperatures between 400 and 700 °C. The conductivity σ, was calculated from the relationship, σ = L/(R.S) (where L is the thickness and S is the contact area of the electrode).

For x = 0, which corresponds to the undoped $LaAlO_3$ phase, it is clear that for both complexing agents, the temperature has practically no effect on the electrical conductivity; in this case, phases are insulating. For the other doped phases, we note that the conductivity increases with temperature, its values are greater in the case of the phases prepared in the presence of tartaric acid. For all temperature values and for both agents we can see that the conductivity increases as x goes from 0 to 0.05, then it decreases for x = 0.1, finally it increases for x = 0.15, this behavior is probably

50 Effect of the Complexing Agent in the Pechini Method … 391

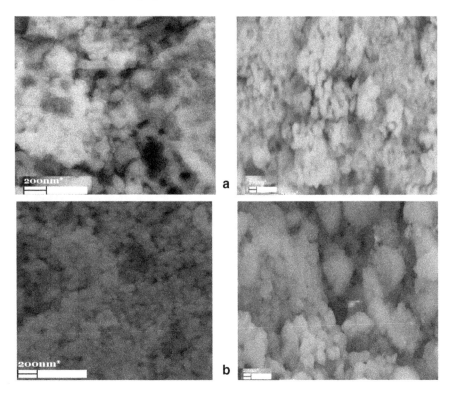

Fig. 50.3 SEM images for $La_{1-x}Sr_xAlO_{3-\delta}$ (x = 0; 0.1): **a** prepared with TA, **b** prepared with EDTA

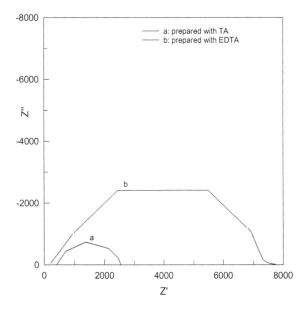

Fig. 50.4 Impedance spectra at 700 °C for $La_{0.95}Sr_{0.05}AlO_{3-\delta}$

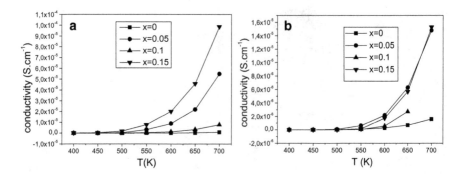

Fig. 50.5 Effect of temperature and x rate on conductivity for $La_{1-x}Sr_xAlO_{3-\delta}$: **a** prepared by TA, **b** prepared by EDTA

linked to the creation of oxygen deficits on the one hand and to the free path on the other hand. We also note that the values of conductivity are greater in the case of the phases prepared in the presence of tartaric acid; this change is related to the variation of the primitive cell volume and to the existence of the insulating secondary phase ($LaSrAl_3O_7$) in the case of the phases prepared in the presence of EDTA.

50.4 Conclusion

This work shows the effect of organic complexing agents on the structural, morphological and electrical properties of the perovskite type ionic conductor $La_{1-x}Sr_xAlO_{3-\delta}$ (x = 0, 0.05, 0.1, 0.15) that can be used as solid electrolyte in SOFCs.

The preparation of our samples was made by the modified Pechini method. Under the same operating conditions, we used two different types of complexing agents: ethylene diamine tetraacetic acid and tartaric acid.

XRD analysis shows that the compounds prepared by the two complexing agents crystallized in a perovskite structure with a rhombohedral system. The doped phases prepared by EDTA have a secondary phase $LaSrAl_3O_7$ for x greater than 0.05, which is absent in the compounds prepared by tartaric acid. The refinement of the cell parameters, has shown that the type of the complexing agent and the doping lead to a slight variation of the cell parameters as well as its volume.

References

1. S. Giddey, S.P.S. Badwal, A. Kulkarni, C.A. Munnings, comprehensive review of direct carbon fuel cell technology. Prog. Energy Combust. Sci. **38**, 360–399 (2012)
2. S.P.S. Badwal, S. Giddey, C. Munnings, A. Kulkarni, Review of progress in high temperature solid oxide fuel cells. J. Aust. Ceram. Soc. **50**, 23–37 (2014)
3. X. Zhang, S.H. Chan, G. Li, H.K. Ho, J. Li, Z. Feng, A review of integration strategies for solid oxide fuel cells. J. Power Sources **195**, 685–702 (2010)
4. S.P.S. Badwal, F.T. Ciacchi, Oxygen–ion conducting electrolyte materials for solid oxide fuel cells. Ionics **6**, 1–21 (2000)
5. T. Ishihara, H. Matsuda, Y. Takita, Doped $LaGaO_3$ perovskite type oxide as a new oxide ionic conductor. J. Am. Chem. Soc. **116**, 380–3803 (1994)
6. O. Bohnke, C. Bohnke, J.L. Fourquet, Mechanism of ionic conduction and electrochemical intercalation of lithium into the perovskite lanthanum lithium titanate. Solid State Ionics **91**(1–2), 21–31 (1996)
7. T. Ishihara, H. Matsuda, Y. Takita, Effects of rare earth cations doped for La site on the oxide ionic conductivity of $LaGaO_3$-based perovskite type oxide. Solid State Ionics **79**, 147–151 (1995)
8. M.P. Pechini, U.S. Patent No. 3.330.697 (1967)
9. F. Bouremmad, A. Benabbas, H. Bouridah, K. Rida, S. Shawuti, M.A. Gülgün, Structural, morphological and electrical properties of $La_{1-x}Sr_xAlO_{3-\delta}$ (x = 0, 0.1, 0.15) synthesized by the Pechini method. Acta Chim. Slov. **59**(4), 927–933 (2012)
10. S.D. Neelapala, N.K. Yaranal, H. Dasari, Synthesis of $La_{1-x}Sr_xAlO_3$ perovskites by reverse strike co-precipitation method and its soot oxidation activity. Iran. J. Chem. Chem. Eng. **38**(4), 69–77 (2018)

Chapter 51
Production of Bio-Oil for Chemical Valorization by Flash Pyrolysis of Lignocellulosic Biomass in an Entrained Bed Reactor

Imane Ouarzki, Aissa Ould Dris, and Mourad Hazi

Abstract Flash pyrolysis is used for chemical valorization of lignocellulosic biomass in an entrained bed reactor for the bio-oil production from wood particles of 350 μm diameter. The experimental conditions were selected by predicting the spatio-temporal temperature profile of the biomass particles. These calculations are used to compare different tests carried out under different operating conditions. The experiments were implemented in the temperature range between 400 and 550 °C and at different residence times (1.07, 1.64 and 2.74 s). The quality of the bio-oil depends essentially on the heating rate, temperature and the effective residence time of the particles at gas temperature. Long residence time slows down the heating rate of the particles and extends the effective residence time of the condensable gases. The secondary reactions intensify the cracking of the condensable vapors and induce the production of CO and CH_4 from furan decomposition and primary lignin degradation products.

Keywords Pyrolysis · Lignocellulosic biomass · Entrained bed reactor · Value added chemicals

51.1 Introduction

The three macro-components of lignocellulosic biomass: cellulose, hemicellulose and lignin can constitute a source of raw material for the chemical industry [1]. Their decomposition during flash pyrolysis conditions (temperature between 450 and 550 °C and particles residence time <2 s) show more than 70% of bio-oil [2]. This fraction contain about 300 value added chemicals with different physic-chemical proprieties witch make their isolation difficult [1]. The decomposition of

---

I. Ouarzki (✉)
University of Boumerdes, Rue de l'indépendance, Boumerdes, Algeria
e-mail: ouarzkimane@yahoo.fr

A. Ould Dris · M. Hazi
Laboratoire transformations intégrées de la matière renouvelable (TIMR), Compiègne, France

cellulose and hemicelluloses during flash pyrolysis led to a bio-oil fraction rich in furfural and hydroxyaldehyde. Lignin decomposes to form monomeric phenols, cresol, syringuol and guaiacol [3]. Other studies focusing on the kinetic pyrolysis for converting biomass into bio-oil are mostly based on the coupled time conservation equations including kinetics of biomass decomposition [4]. The most applied model considers the conversion of the biomass into char, condensable gas (bio-oil) and gas [5]. In this work; we will focus on the heating of biomass particles and the mechanism of formation of the major chemical compounds present in the bio-oil fraction.

51.2 Materiel and Methods

The experiments were carried out in an entrained bed reactor. Firstly, a model for predicting the heating particles along the reactor is developed in order to determine the spatio-temporal temperature profile of the biomass particles. This model is used to compare the tests under different operating conditions of the nitrogen flow and the reactor temperature.

51.2.1 Experimental Set-Up

The installation (Fig. 51.1) consists of biomass feed system with a vibrating rail allowing the continuous supply of biomass at 1 g/min driven by N_2 flow.

51.2.2 Transport and Heating of Wood Particles in the Entrained Bed Flow Reactor

The equation of the heating and transport of the biomass particles.

The heating is assured by both preheated nitrogen gas flow and radiation from the metal wall of the reactor. The residence time of the particles at the reactor temperature is determined by the resolution of the coupled equations of momentum and energy balance applied to the isolated particle. The calculation does not consider neither the pyrolytic process which intervenes at a certain temperature nor a possible contraction of the particle volume.

$$\frac{1}{6}\pi d^3 \rho_s c_p \frac{dt_p}{dt} = F\varepsilon\sigma\left(T_w^4 - T_p^4\right)\pi d^2 + h\left(T_f - T_p\right)\pi d^2 \quad (51.1)$$

The Eq. (51.1) assumes that: the number of Biot is sufficiently low to consider the temperature uniform at each instant in the particle. The particles are spherical and the flow is sufficiently diluted to be considered as isolated particles. Calculations are of

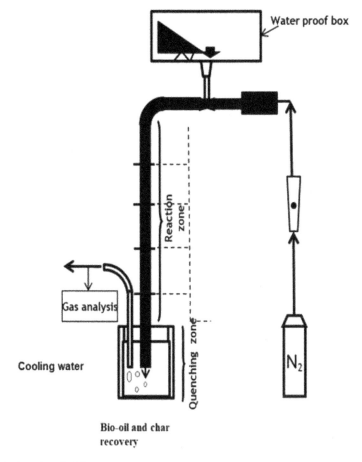

Fig. 51.1 Entrained bed flow reactor

interest only for particle temperature because it is not possible to evaluate the thermal losses along the reactor. The coefficient of the heat transfer (h) by convection of the fluid towards the particle is determined from the Nusselt number which is given by the correlation of Ranz and Marshall

$$Nu = 2 + 0.62\, Re_p^{0.5} Pr^{0.33} \tag{51.2}$$

Re_p: Particle Reynolds:

$$Re_p = \frac{\rho_f(U_f - U_p)}{\mu} \tag{51.3}$$

ρ_p: Density of the particle, ρ_f: Density of the fluid U_f: Fluid velocity calculated from the gas state equation as a function of the local temperature in the reactor, U_p:

Particle velocity, μ: Viscosity of nitrogen given as a function of temperature by the relation.

$$\frac{\mu(T)}{\mu(700\,K)} = \left(\frac{T}{700}\right)^{0.64} \tag{51.4}$$

The momentum equation applied to the particle takes into account only the drag force:

$$\frac{1}{6}\pi d^3 \rho_p \frac{DU_P}{Dt} = C_d \rho_f (U_f - U_p) \frac{2d^2}{8} \tag{51.5}$$

The apparent weight of the particle is neglected and the equation then becomes valid in the horizontal and in the vertical section of the reactor. The drag coefficient Cd is calculated according to the values of the Reynolds number. The resolution of coupled equations of balance of momentum and energy applied to the particle, gives the position and the temperature of the particles in the reactor as a function of the time. The resolution of the equations system was carried out by the first tangent method with: $D_p(m) = 0.000315$, $\rho_p(Kg/m^3) = 850$, $C_p(J/Kg) = 2500$, $\lambda(W/m\,K) = 0, 36$.

51.2.3 Production of Bio-oil Pyrolysis in the Entrained Bed Reactor

The effect of pyrolysis conditions was studied at the temperature range between 450 and 550 °C and particle residence time (1.07, 164 and 2.74 s) for the pine wood particles with 350 μm size. The calculations presented previously allow the determination of the temperature profile of the particles along the reactor as well as the time during which the particles are actually brought to the pyrolysis temperature.

51.3 Results and Discussions

51.3.1 Influence of Operating Conditions on the Yield of Flash Pyrolysis Products

Effect of Temperature and the Residence Time of Biomass Particles
The effective residence time of the particles at the gas temperature depends on the nitrogen injection rate (Fig. 51.2). Figure 51.3 shows a maximum yield of 48% of bio-oil at 450 °C for a total particle residence time of 1.07 s. When the residence time increases from 1.64 to 2.74 s, a decrease in the yield of the bio-oil is registered.

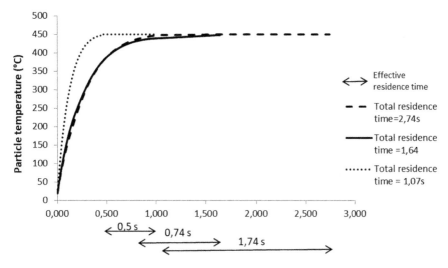

Fig. 51.2 Evolution of the particle temperature in the entrained bed reactor at T = 450 °C

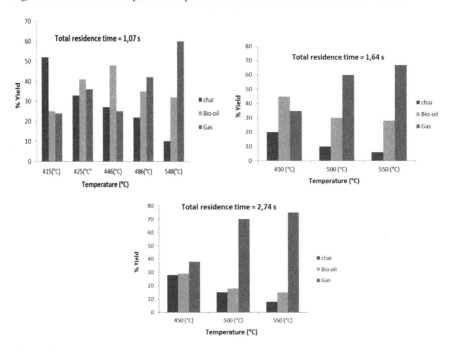

Fig. 51.3 Yields of the pyrolysis products of biomass as a function of temperature at different residence times

Fig. 51.4 Gas pyrolysis produced at 450 °C at different residence times

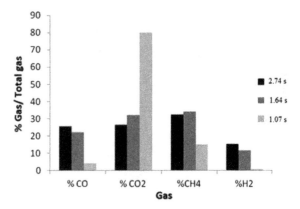

The yield of the char decreases with the temperature and the residence time of the biomass particles. These results are due to the high heating rate associated with the high nitrogen injection, and the effective residence time of the particles at the gas temperature. The heterogeneous reactions between the vapors and the char, contribute to the increase of the incondensable gases yield and the decrease of the char yield. With a total residence time of 2.74 s, the particles stay at the temperature of the gas for 1.74 s, which favors the secondary decomposition reactions.

51.3.2 Gases Analysis

When the particles remain in the reactor for 1.07 s, the gases are essentially composed of CO_2 (Fig. 51.4), which is explained by the primary decomposition of the biomass due to the breakdown of cellulose and lignin to produce respectively furan derivatives and lignin pyrolysis products. In case of total residence time of 1.64 and 2.74 s, the effective residence time of the particles at the gas temperature increases. This leaves the time for the secondary decomposition reactions of the condensable vapors to produce essentially CO, CH_4 and H_2.

51.3.3 Oil Analysis

Chromatographic peaks identification was performed using the Xcalibur software database. The results of the analysis are presented in the Table 51.1. Under 450 °C and at short residence time Fig. 51.5a, the bio-oil essentially consists of 3,4,5 trimethoxybenzaldehyde derivatives resulting from the decomposition of lignin, and products of the decomposition of holocelluloses like furfural, furfuryl alcohol, 2,5-dimethylfuran and 5-furfuraldehyde. Indeed, at 450 °C and residence time equal to 1.07 s, the primary decomposition of lignin by thermo-oxidation occurs

Table 51.1 GC/Ms analysis of bio-oils produced from pine wood pyrolysis

Retention time (min)	Compound
1.95	Furfural
2.14	Furfuryl alcohol
2.18	2,5-dimethylfuran
2.22	5-furfuraldehyde
2.33	1,4 benzenediol
2.42	Phenol
2.54	2-acetyl-5-methylfuran
2.86	Corylon
2.99	Cyclohexane-1,4-dione
3.58	4-methoxyphenol
3.72	2-methoxyphenol
4.63	p-xylene
5.42	2 methoxy-4methylphenol
6.89	2,4-dimethoxytoluee
8.75	3-methylcetechol
10.61	2,6-dimethoxyphenol
13.28	Vanillin
15.78	Syringaldehyde
27.09	3,4,5-trimethoxybenzaldehyde
28.78	Diphenylmethane
36.36	ND
38.36	ND
40.39	ND

to produce tri-methoxybenzaldehyde according to the mechanism proposed by Nonier et al. 2006 [6]. GC/MS analysis of the bio-oils shows that syringaldehyde, 3,4,5-trimethoxybenzaldehyde and 2-methoxy-4-methylphenol are affected by the rise of particles residence time Fig. 51.5b which cause their decomposition to produce phenolic monomers and polycyclic aromatic compounds accompanied by the production of CH_4, CO, and H_2 as is shown in Fig. 51.4.

51.4 Conclusion

The results of this study demonstrate that the yield and the composition of the bio-oil depend essentially on the actual temperature of the particle and the effective residence time of the condensable vapors in the reactor. The increase of the temperature of the gas favors the conversion of the biomass into incondensable gases. Short residence

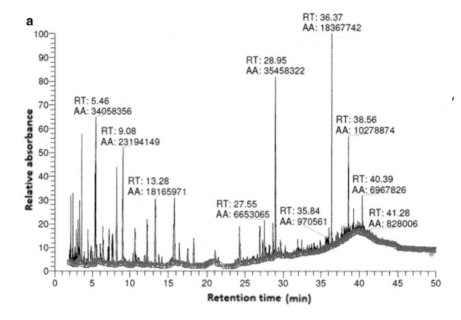
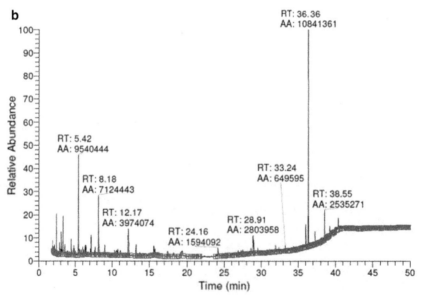

Fig. 51.5 GC/MS analysis of bio-oil pyrolysis at 450b °C at **a** 1.07 s and **b** 2.74 s

time of the particles at the gas temperature improves the yield of the bio-oil by limiting the secondary decomposition reactions. Markers of the thermo-decomposition of biomass, such as furan derivatives from cellulose and hemicellulose and 3,4,5 trimethoxybenzaldehyde from lignin, decompose as function of the temperature and the effective residence time to form incondensable gases containing mainly CO, CH_4 and H_2.

References

1. A. Kumar, L. Varmaa, T. Singh, S. Ravi, M. Prasenjil, Pyrolysis of wood sawdust: effects of process parameters on products yield and characterization of products. Waste Manage **89**, 224–235 (2019)
2. I. Ouarzki, M. Hazi, D. Luart, C. Len, A. Ould-dris, Comparison of direct and staged pyrolysis of the ligno-cellulosic biomass with the aim of the production of high added value chemicals from bio-oil. J. Mater. Environ. Sci. **7**(3), 1008–1022 (2016)
3. Y. Xiao, F. Zewu, H. Duoduo, Z. Yuying, L. Rui, W. Yulong, Unveiling the pyrolysis mechanisms of cellulose: experimental and theoretical studies. Renew. Energy **147**, 1120–1130 (2020)
4. A. Sharma, V. Pareeka, S. Wang, Z. Zhang, H. Yang, D. Zhang, A phenomenological model of the mechanisms of lignocellulosic biomass pyrolysis processes. Comput. Chem. Eng. **60**, 231–241 (2014)
5. C. Di Blasi, Modeling chemical and physical processes of wood and biomass pyrolysis. Prog. Energy Combust. Sci. **34**, 47–90 (2008)
6. M.F. Nonier, N. Vivas, A. De Gaulejac, C. Absalon, Ph Soulie, E. Fouquet, Pyrolysis—Gas chromatography/mass spectrometry of *Quercus* sp.: wood application to structural elucidation of macromolecules and aromatic profiles of different species. J. Anal. Appl. Pyrol. **75**, 181–193 (2006)

Chapter 52
Suitable Sites for Wind Hydrogen Production Based on GIS-MCDM Method in Algeria

Djilali Messaoudi, Noureddine Settou, Belkhir Negrou, Belkhir Settou, Charafeddine Mokhtara, and Chetouane Mohammed Amine

Abstract Hydrogen production driven by renewable energy sources (RES) represents an attractive energy solution to global warming. This paper deals with site selection problems for wind hydrogen production and aims to propose a structural procedure for determining the suitable sites. The study area is Algeria. The methodology focuses on the combined use of geographic information systems (GIS) and multi-criteria decision making (MCDM), aiming to provide a decision tool for wind hydrogen production sites. The first stage excludes sites that are infeasible for wind hydrogen production, based on land use, water bodies, water ways, roads, railways, power lines, and also their buffer around these zones. The second stage weighting criteria will be chosen according to the objective to be reached, in this case they will be distance to roads, to railways, to power lines, hydrogen demand, wind hydrogen potential, digital elevation models (DEMs), and slope. Through the use of MCDM the criteria mentioned will be weighted in order to evaluate potential sites to produce hydrogen from wind energy. Analysis and calculation of the weights of these criteria will be conducted using SWING weighting method. The overlaid result map showed that 23.59% (561,836 km^2) of the study area is promising and suitable for deploying wind hydrogen farms while the most suitable areas to be in the southwest of the Algeria. It has been found that suitable lands are following the pattern of the wind hydrogen potential. The integration of the GIS with MCDM methods is a powerful tool to deal with a geographical information data and vast area as well as manipulate criteria importance towards introducing the best sites for wind hydrogen production.

Keywords Hydrogen energy · Geographical information system (GIS) · MCDM method

---

D. Messaoudi (✉) · N. Settou · B. Negrou · B. Settou · C. Mokhtara · C. M. Amine
Lab. Promotion et valorisation des ressources sahariennes (VPRS), Université Kasdi Merbah Ouargla, Route de Ghardaïa, BP 511, 30000 Ouargla, Algeria
e-mail: messaoudi.djilali30@gmail.com

D. Messaoudi · B. Settou · C. Mokhtara · C. M. Amine
Department of Mechanical Engineering, Faculty of Applied Sciences, Univ. Kasdi Merbah, Ouargla, Algeria

© The Editor(s) (if applicable) and The Author(s), under exclusive license to Springer Nature Singapore Pte Ltd. 2021
A. Khellaf (ed.), *Advances in Renewable Hydrogen and Other Sustainable Energy Carriers*, Springer Proceedings in Energy,
https://doi.org/10.1007/978-981-15-6595-3_52

52.1 Introduction

An increasing need of a human for energy resources has always been the fundamental issues in the life of a human being. The population explosion is the main factor that leading to increase in energy demand in the world. According to BP statistical review, more than 80 percent of today's energy demand is produced using fossil fuels [1]. However, fossil fuel reserves are limited and their usage has significant environmental effects. The burning of fossil fuels such as coal, oil and gas releases greenhouse gases such as carbon dioxide and methane into Earth's atmosphere and oceans. According to the International Energy Agency, the world CO_2 emissions from fuel combustion rose from 15,458 in 1973 to 32,294 Mt of CO_2 in 2015 [2]. The transport sector is the fastest growing contributor to climate emissions, which is for approximately 23% of total energy-related CO_2 emissions (6.7 $GtCO_2$) [3]. Almost all (95%) of the world's transportation energy comes from petroleum-based fuels, largely gasoline and diesel. Such problems strongly encourage the research, development and demonstrations of clean energy resources, energy carriers, and in the case of transportation. In recent study, hydrogen is one of the energy carriers that can replace fossil fuels [4]. Hydrogen may be produced from renewable energy sources through a variety of pathways and methods [5]. Hydrogen as an energy vector, together with electrolyzer and fuel cell technologies can provide a technical solution to this challenge. Moreover, the use of hydrogen for a clean transportation fuel will increase the need of renewable hydrogen producing [6]. Among renewable energy sources, wind power has had a faster growth than other renewable sources, because the use of wind turbine leads distributed generation system to a system with variable production in addition to the environmental and economic capabilities in production of clean and sustainable energy [7]. Algeria has enormous potential of renewable energy namely solar, wind, geothermal and biomass and hoping to increase its renewable energy status by producing as much electricity from green sources as it currently produces from its natural gas power plants by 2020 [8]. The selection of an installation site from other sites is MCDM problem containing many criteria [9]. MCDM offers as set of procedures and techniques for structuring decision problems, and designing, evaluating and prioritizing decision alternatives. GIS are designed to store, manage, analyze and visualize geospatial data required by decision-making processes [10]. In recent years GIS integrated with MCDM have been widely used as a decision support system (DSS) to assist in locating suitable sites for wind farms [11]. Several studies have been done on the wind hydrogen production in Algeria. Hamane and Belhamel [12] gave an estimation of hydrogen production from wind power in the south of Algeria. They considered two aspects of the system. Estimation of the wind power produced by three types of wind turbines generators and the energy required for the electrolysis process. Their results indicate that the hydrogen production strongly depends on the wind speed and its frequency distribution. Douak and Settou [13] gave a simplified methodology to evaluate the hydrogen production from the wind energy available at each site using different wind turbine. Their results indicated that the hydrogen production strongly depends on the site and the selected wind turbine. Messaoudi et al. [14] developed a methodology for site selection of hydrogen refueling stations with on-site hydrogen production from wind energy sources. The authors examined a case study for the region of Adrar in Algeria and they followed an analytical hierarchy process belonging to the multi-criteria decision making tools in order to identify which of the available petrol stations can be reverted to H_2 refueling stations. Out of 24 conventional petrol stations in total, 15 were investigated and only 4 sites were found suitable for wind powered onsite H_2 stations. The rest 11 however, could be modified as off-site hydrogen stations in the retail refueling market. In this paper, a GIS-based methodology for evaluating alternative locations of wind hydrogen production sites is developed by using MCDM. The main objective is to produce a decision support system which can assist authorities and decision makers to identify priority sites for wind hydrogen production in Algeria.

52.2 Study Area

Algeria is situated in northern Africa, bordered to the east by Tunisia and Libya, to the southeast by Niger, to the southwest by Mali, and to the West by Mauritania, occidental Sahara, and Morocco as shown in Fig. 52.1. Algeria is the largest country in Africa with a total surface of 2,381,741 km² and a population of 41.2 Millions of

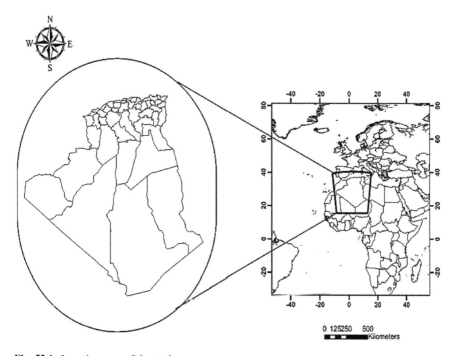

Fig. 52.1 Location map of the study area

inhabitants. The climate is transitional between maritime in the north region and semi-arid to arid middle and south. The mean annual precipitation varies from 500 mm in the north to 150 mm in the south. The average annual temperature is about 12 °C [15]. Algeria has promising wind energy potential of about 35 TWh/year. Almost half of the country experience significant wind speed [16].

52.3 Methodology

To select suitable sites for wind hydrogen production, many criteria must be considered according to goal that will be achieved. Based on several literatures and case studies concerning wind farm site selection, seven criteria were selected for evaluating of suitable sites for wind hydrogen production in Algeria [11, 17] (Fig. 52.2).

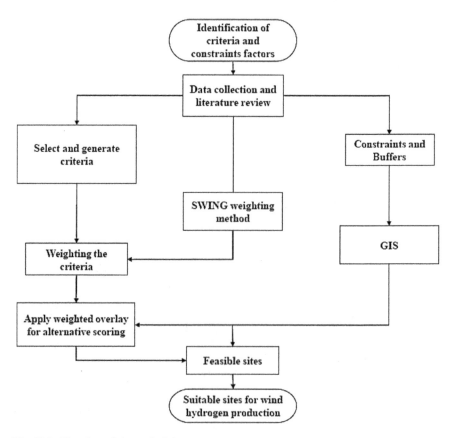

Fig. 52.2 Flowchart of the methodology

The selection of the most suitable sites to produce wind hydrogen is a very complex issue. It requires consideration of multiple alternative solutions and evaluation criteria. Multi-Criteria Decision Making (MCDM) have been used by many researchers and decision makers in different fields [11]. SWING weighting method is one of the MCDM methods [18].

52.4 Results and Discussion

The weights of the criteria are carried out in excel spreadsheet.

Table 52.1 presents the weight of criteria with respect to the goal obtained by SWING weighting method. The wind hydrogen potential is the important criteria with a value of 20.51% and the less important criteria is distance to railways with 8.25%.

The final locations were defined by overlaying the results of the restrictive and the different weighted criteria, which is allowed to classify the study area on a scale between 0 and 8.063 where the pixel that have a high value corresponds to the most suitable area for wind renewable hydrogen production as shown in Fig. 52.3.

Table 52.2 presents the final suitability results were divided into four discrete categories: very low suitable areas, low suitable areas, moderate suitable areas, and high suitable areas which is used to define the degree to which site is suitable for wind hydrogen production according to the associated criteria and excluding all restrictions.

The results indicate that 0.11%, of the study area has very low suitability, 54.51% has low suitability, 23.53% has moderate suitability and 0.06% has high suitability for a wind renewable hydrogen production. The other 21.79% of the study area is not suitable for wind hydrogen production as shown in Fig. 52.4.

According to the LSI analysis, the result showed that many of the highly suitable locations are located in the southwest region as illustrated in Fig. 52.4. The low suitable areas are located Southeast to High Plains, mainly due to lower wind energy in that region.

Table 52.1 Criteria weights

Criteria	Distance to roads	Distance to railways	Distance to power lines	Hydrogen demand	Wind hydrogen potential	DEMs	Slope
Weight (%)	16.87	8.25	16.38	10.92	20.51	14.74	14.32

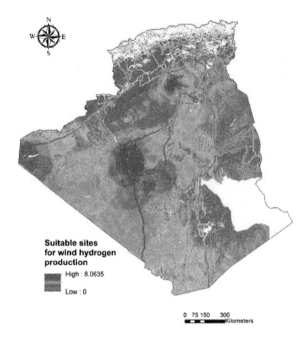

Fig. 52.3 The most suitable sites for wind hydrogen production

Table 52.2 Suitability index

Suitability	Unsuitable	Very low suitable	Low suitable	Moderate suitable	High suitable
Value score	0	0–2	2.01–4	4.01–6	6.01–8.063

52.5 Conclusion

This paper presents an application of combining MCDM with GIS for sites selection of wind renewable hydrogen production in Algeria. The objective of the study was to find suitable sites for wind hydrogen production taking into account a number of different criteria. SWING weighting method was utilized to assign the relative weights of the evaluation criteria, while GIS established the spatial dimension of constraints and evaluation criteria and elaborated them in order to produce the overall suitability map. The MCDM methodology integrated with GIS is a powerful tool for the effective evaluation of the wind hydrogen production site selection.

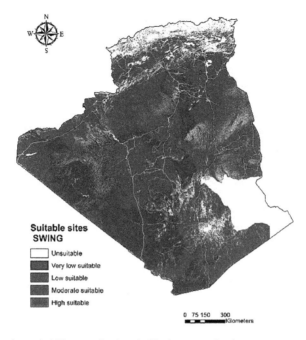

Fig. 52.4 The sites suitability map for the wind hydrogen production

References

1. *British Petroleum Statistical Review of World Energy*. https://www.bp.com/worldenergystatistics. Accessed 01 Mar 2017
2. *International Energy Agency Key World Energy Statistics*. https://webstore.iea.org/statistics-data. Accessed 01 Mar 2019
3. R. Sims, R. Schaeffer, B. Soares Moreira Cesar Borba, R. Schaeffer, F. Creutzig, X. CruzNúñez, D. Dimitriu, M. Figueroa Meza, L. Fulton, S. Kobayashi, O. Lah, A. McKinnon, P. Newman, M. Ouyang, J. Schauer, D. Sperling, G. Tiwari, Y. Sokona, E. Farahani, S. Kadner, K. Seyboth, A. Adler, I. Baum, S. Brunner, P. Eickemeier, B. Kriemann, J. Savolainen, S. Schlömer, C. von Stechow, T. Zwickel, J. Minx, Transport 599–670 (2014) (IPCC)
4. M. Balat, Potential importance of hydrogen as a future solution to environmental and transportation problems. Int. J. Hydrogen Energy **33**(15), 4013–4029 (2008)
5. I. Dincer, C. Acar, Review and evaluation of hydrogen production methods for better sustainability. Int. J. Hydrogen Energy **40**(34), 11094–11111 (2014)
6. M. Belhamel, B. Benyoucef, M. Hamane, L. Aiche-hamane, M. Belhamel, B. Benyoucef, Study of hydrogen production from wind power in algeria study of hydrogen production from wind power in Algeria, in *18th World Hydrogen Energy Conference* (Essen, 2010), pp. 16–21
7. A. Mostafaeipour, Economic evaluation of small wind turbine utilization in Kerman, Iran. Energy Convers. Manage. **73**, 214–225 (2013)
8. A.B. Stambouli, Z. Khiat, S. Flazi, Y. Kitamura, A review on the renewable energy development in Algeria: current perspective, energy scenario and sustainability issues. Renew. Sustain. Energy Rev. **16**(7), 4445–4460 (2012)
9. S. Chakhar, V. Mousseau, Spatial Multicriteria Decision Making, vol. 20, no. 11 (Springer, 2008), p. 1392

10. P.A. Longley, M.F. Goodchild, D.J. Maguire, D.W. Rhind, Geographical information systems and science. Comput. Oper. Res. **29**, 536 (2002)
11. K. Baris, A. Bahadir, A. Ulucan, M. Umur, A GIS-based multiple criteria decision analysis approach for wind power plant site selection. Utilities Policy (2015)
12. M. Hamane, M. Belhamel, Estimation of hydrogen production from different wind turbine sizes in the south of Algeria. Revue des Energies Renouvelables 129–134 (2007)
13. M. Douak, N. Settou, Estimation of hydrogen production using wind energy in Algeria. Energy Procedia **74**, 981–990 (2015)
14. D. Messaoudi, N. Settou, B. Negrou, S. Rahmouni, B. Settou, I. Mayou, Site selection methodology for the wind-powered hydrogen refueling station based on AHP-GIS in Adrar, Algeria. Energy Procedia **162**, 67–76 (2019)
15. A.B. Stambouli, An overview of different energy sources in Algeria. Jordan Eng. 1–15 (2009)
16. Ecomena.orgrenewables-algeria. www.ecomena.orgrenewables-algeria. Accessed 01 Mar 2019
17. D. Messaoudi, N. Settou, B. Negrou, B. Settou, GIS based multi-criteria decision making for solar hydrogen production sites selection in Algeria. Int. J. Hydrogen Energy **44**(60), 31808–31831 (2019)
18. J.J. Wang, Y.Y. Jing, C.F. Zhang, J.H. Zhao, Review on multi-criteria decision analysis aid in sustainable energy decision-making. Renew. Sustain. Energy Rev. **13**(9), 2263–2278 (2009)

Chapter 53
Liquefaction of Hydrogen: Comparison Between Conventional and Magnetic Refrigeration Systems

Mustapha Belkadi and Arezki Smaili

Abstract Hydrogen, under atmospheric pressure, becomes liquid at 20.3 K. The liquefaction is carried out by extracting of 4914 kJ/kg of heat. This liquefaction requires the use of some high level cryogenic technology whether to liquefy it or to keep it in the liquid state. In general, three processes are applied: Claude cycle, Brayton cycle and liquefaction by magnetocaloric effects (Magnetic Refrigeration). The present work aims at performing comparison between conventional liquefaction and magnetic liquefaction systems. It deals with a comparison between performances and energy consumption evaluated for the two systems at similar operating conditions. Precooled Claude liquefaction cycle has been considered for the first one. Here, energy and material balances has been performed by use of Aspen Hysys simulator. However, the second one is based on a multistage Active Magnetic Regenerator (AMR) cycles operating with real magnetic materials. A new simulation method has been proposed to use Aspen Hysys simulator for thermal analysis of the AMR liquefier.

Keywords Hydrogen liquefaction · Active magnetic refrigeration · Magnetocaloric effect

53.1 Introduction

Hydrogen, under atmospheric pressure, becomes liquid at 20.3 K. The liquefaction is carried out by extracting of 4914 kJ/kg of heat (divided between sensible heat, latent heat and conversion heat from n-H_2 to p-H_2 [1]). This liquefaction requires the use of some high level cryogenic technology whether to liquefy it or to keep it in

---

M. Belkadi (✉) · A. Smaili
Laboratoire de Génie Mécanique et Développement, Ecole Nationale Polytechnique, B. P. 182, 16200 El-Harrach Algiers, Algeria
e-mail: mustapha.belkadi@sonatrach.dz

M. Belkadi
Sonatrach/Direction Centrale recherche et Développement, Avenue du 1er novembre, 35000 Boumerdes, Algeria

© The Editor(s) (if applicable) and The Author(s), under exclusive license to Springer Nature Singapore Pte Ltd. 2021
A. Khellaf (ed.), *Advances in Renewable Hydrogen and Other Sustainable Energy Carriers*, Springer Proceedings in Energy,
https://doi.org/10.1007/978-981-15-6595-3_53

the liquid state. In general, three processes are applied: Claude cycle, Brayton cycle and liquefaction by magnetocaloric effects (Magnetic Refrigeration) [2].

The use of magnetic refrigeration to liquefy hydrogen is based on the magnetocaloric effect (MCE) which occurs in some materials when they are subjected to external magnetic field changes. The MCE is defined as the change of material temperature when applying or removing the magnetic field (magnetization/demagnetization process). In fact, if a magnetic material is placed in a magnetic field, there is usually an increase in its temperature. Conversely, demagnetization process has a cooling effect on it, as for the compression and expansion of gas [3]. Liquefaction could be carried out by cooling the gas through a thermomagnetic cycle, known as Active Magnetic Regenerative cycles (AMR cycles). An AMR cycle consists essentially of a regenerator bed (magnetic material) which is subjected to cyclic changes in magnetic-field intensity, alternating between zero-field and the maximum field. Temperature span induced by the magnetization and demagnetization process is amplified by forcing a working fluid to move alternatively through the regenerator bed between two heat sources (hot and cold reservoirs). Thus, a large temperature span can be obtained [4].

The present work aims at performing comparison between conventional liquefaction and magnetic liquefaction systems. It deals with a comparison between performances and energy consumption evaluated for the two systems at similar operating conditions.

53.2 Hydrogen Liquefaction Process

53.2.1 Conventional Liquefaction Cycles

In conventional systems, liquefaction of hydrogen at 20.3 K can be obtained by the combined effect of cooling and adiabatic expansion of gas after it has been previously compressed. In general, two process are applied [1]:

- Claude cycle: it consists of a pre-cooling by liquid nitrogen followed by the use of hydrogen as refrigerant, after being expanded through two turbines and a Joule-Thomson valve (JT valve) as illustrated in Fig. 53.1. The use of JT valve avoids liquid formation at the end of the cold turbine.
- Brayton cycle: it uses helium, often mixed with neon, as an external refrigerant.

53.2.2 Magnetic Liquefaction Cycle

Application of Magnetic Refrigeration to hydrogen liquefaction consists in performing a thermal contact between an AMR regenerator cycle and the gas to

53 Liquefaction of Hydrogen: Comparison Between Conventional ...

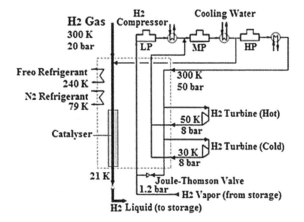

Fig. 53.1 Claude liquefaction cycle. Adapted from [1]

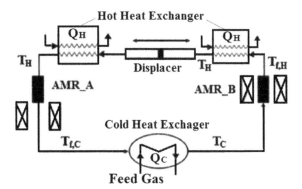

Fig. 53.2 Description of an AMR cycle

be liquefied. As illustrated in Fig. 53.2, the AMR cycle considered in the present paper is mainly constituted of the following components [5]:

- Two Regenerator beds working alternately (when the first regenerator bed is magnetized, the second one is demagnetized) to provide a continuous production of cooling power (Q_C).
- Reciprocating displacer which is able to move the fluid alternatively;
- Magnetic field source able to magnetize and demagnetize the regenerator bed;
- Hot heat exchanger able to reject the thermal energy into the hot reservoir;
- Cold heat exchanger able to receive the thermal energy from the cold reservoir.

A complete AMR cycle consists of two isentropic steps and two isofield steps [6]:

1. Adiabatic demagnetization step: with no flow, the bed is demagnetized adiabatically.
2. Cold Blow at zero fields: the fluid is then forced by the displacer to move from the hot to the cold reservoirs. Passing through the cold heat exchanger, the fluid absorbs heat at a rate Q_C, which represents the cooling power of the regenerator.
3. Adiabatic magnetization step: the bed is magnetized adiabatically with no flow.

4. Hot Blow at applied field: the fluid is then forced from the cold to the hot ends. Passing through the hot exchanger, the fluid rejects heat at a rate Q_H.

Furthermore, the use of one AMR cycle is not suitable to provide the large temperature span required for hydrogen liquefaction. Thus, applying magnetic refrigeration to the liquefaction of hydrogen is performed by assembling a number of AMR cycles operating in series (cascade cycles). The cooling power produced by one stage is used to cool hydrogen entering the stage and absorbs heat rejected by the lower stage. The required cooling power of each stage is regulated by adjusting the volume of magnetic material.

53.3 Thermal Analysis Procedure

For conventional liquefaction, Claude liquefaction cycle has been considered. Here, classical simulation method has been performed by use of Aspen Hysys simulator which is a powerful software that was developed by AspenTech to simulate gas processing plants, oil refineries and petrochemical plants. The simulation consists of building a Process Flow Diagram (PFD) of the liquefaction system, performing material and energy balances and calculating heat transfer and energy consumption. The cycle efficiency is then computed by introducing the coefficient of performance (COP):

$$COP = \frac{Q_{Liq}}{W_C}$$

where Q_{Liq} is the total heat absorbed from hydrogen and W_C is the total work consumed.

For magnetic liquefaction system, the AMR regenerator and the reciprocating displacer are not included in Hysys palette of process units; even the alternative flow cannot be simulated. Hence, a new method has been developed to simulate an AMR cycle with two regenerator beds working alternately, as shown in Fig. 53.2. The method consists of modeling the AMR system, once the steady state is reached, by an ordinary system where the magnetized regenerator is idealized as a heater, whereas the demagnetized one is idealized as a cooler (Fig. 53.3). The carrier fluid flows continuously through the cycle, absorbing form the heater an amount of heat equal to the heat absorbed from the regenerator bed during the hot blow (Q_{HB}). The carrier fluid rejects through the cooler an amount of heat equal to the heat exchanged with the regenerator bed during the cold blow (Q_{CB}) [5].

The exchanged heats, Q_{HB} and Q_{CB}, are calculated through the resolution of an AMR heat transfer numerical model developed separately [5]. The model is based on the fundamental equations of energy conservation. The MCE characterization is performed according to the Molecular Field Theory (MFT) [6]. The COP of the AMR cycle is calculated as follow:

Fig. 53.3 AMR cycle simulation

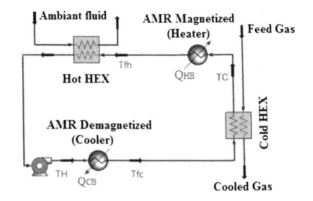

$$COP = \frac{Q_{Liq}}{Q_{Rej} - Q_{Liq} + W_{in}}$$

where Q_{Rej} is the heat rejected into the environment and W_{in} is the total work input used to move the carrier fluids through the regenerator beds.

53.4 Results and Discussion

To perform comparison between AMR liquefaction cycle and conventional cycles; Precooled Claude cycle has been simulated by Aspen Hysys for different pressure of hydrogen feed gas. Figure 53.4 illustrates the evolution of the COP and energy consumption (W_C) as function of the feed gas pressure (P_{FG}). Results show the increase of the COP with the pressure. This rise becomes insignificant from the pressure of 30 bar. At 20 bar the total energy required to liquefy 1 kg/h is 13 kW. The corresponding COP is 0.105 (10.5%). This efficiency is not so far compared with published values. Pre-cooled Claude efficiency, e.g., as published by Krasae-in [7], is situated between 6.2 and 8.8%.

Fig. 53.4 COP evolution of Claude cycle as function of feed gas pressure

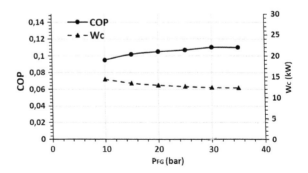

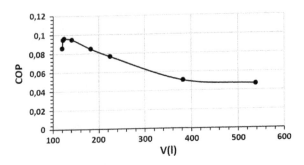

Fig. 53.5 COP of AMR cycle as function of magnetic material volume

Table 53.1 Comparison between conventional system and AMR system

	P (bar)	COP	W_C (kW)
Claude cycle	20	0.105	13
AMR system	1	0.096	0.053

Using Gadolinium (Gd), Terbium (Tb), Dysprosium (Dy) and Holmium (HoN) [8] as magnetic materials in the regenerator beds of an AMR liquefier composed of 6 stages, for which, the total volume (V) and the corresponding COP are calculated. Results obtained (COP as function of magnetic material volume) are illustrated in Fig. 53.5. The most efficient volume is 124 L. The corresponding COP is 0.096 (9.6%). Here, the total work input used to move the carrier fluid through the regenerator beds is insignificant (0.053 kW to liquefy 1 kg of H_2 supplied at atmospheric pressure).

To highlight the difference between the two systems, efficiency and energy consumption, calculated according to the thermal analysis procedure presented in Sect. 53.3, are illustrated in Table 53.1.

53.5 Conclusion

The present work deals with a comparison between conventional liquefaction system (precooled Claude cycle) and AMR liquefaction cycle. In conventional systems:

- Liquefaction can be obtained by the combined effect of cooling and adiabatic expansion of gas after it has been previously compressed. This liquefaction requires the use of large amounts of energy through cooling loops and at the recompression of gas (13 kW to liquefy 1 kg/h supplied at 20 bar).
- Low liquefaction efficiency (theoretical work is 10.5% of real work consumed).

By use Gadolinium (Gd), Terbium (Tb), Dysprosium (Dy) and Holmium (HoN) [8] as magnetic materials in the AMR considered:

- The total work input used to move the carrier fluid through the regenerator beds is insignificant (0.053 kW to liquefy 1 kg of H_2 supplied at atmospheric pressure).
- Low efficiency found with selected magnetic materials. This efficiency can be enhanced by using other magnetic materials having large EMC.

Acknowledgements The support from Directorate-General for Scientific Research and Technological Development (DGSRDT) of Algerian government in the form of research grand to Prof. A. Smaili is gratefully acknowledged

References

1. J. Gallarda, Liquéfaction de l'hydrogène. Technique de l'Ingénieur, j3603 (2001)
2. P. Malbrunot, Alleau, Th.: Mémonto de l'hydrogène, Fiche 4.3 (2019)
3. A. Lebouc, F. Allab, J.M. Fournier, J.P. Yonnet, Réfrigération magnétique. Technique de l'Ingénieur, RE 28 (2005)
4. C. Aprea, A. Maiorino, A flexible numerical model to study an active magnetic refrigerator for near room temperature applications. Appl. Energy **87**, 2690–2698 (2010)
5. M. Belkadi, A. Smaili, Thermal analysis of a multistage active magnetic regenerator cycle for hydrogen liquefaction. Int. J. Hydrogen Energy **43**, 3499–3511 (2018)
6. A. Smaili, S. Aït Ali, R. Chahine, Performance predictions of a first stage magnetic hydrogen liquefier. Int. J. Hydrogen Energy **36**, 4169–4177 (2011)
7. S. Krasae-in, J.H. Stang, P. Neksa, Development of large-scale hydrogen liquefaction processes from 1898 to 2009. Int. J. Hydrogen Energy **35**, 4524–4533 (2010)
8. T. Nakagawa, T.A. Yamamoto, T. Numazawa, Research on a magnetic refrigeration cycle for hydrogen liquefaction. Cryocoolers **14**, 645–653 (2007)

Chapter 54
Numerical Study of Heat and Mass Transfer During Absorption of H_2 in a LaNi$_5$ Annular Disc Reactor Crossed by a Tubular Heat Exchanger

Abdelaziz Bammoune, Samir Laouedj, and Bachir Dadda

Abstract In this work, we studied numerically the two dimensional phenomena of heat and mass transfer during the exothermic chemical reaction of hydrogen absorption. These phenomena take place within a thin disk drive filled with a metallic powder of LaNi$_5$ and crossed by cooling and injection tubes of hydrogen gas. A cooling jacket, surrounding its peripheral surface has been added. The commercial ANSYS FLUENT R17.2 CFD simulation code has been used to perform a series of numerical investigations of the arrangement effect of the incorporated cooling tubes, on the heat and mass transfers during the storage process of hydrogen. For the selected optimal configuration, we illustrated the temperature contours for the four configurations proposed at different times and the time average of temperature bed as well as the fraction hydrogen absorbed are assessed in the history curves. Our simulation results have been validated by experimental results found in the literature, and a good agreement has been noticed, which confirms the reliability of our simulation procedure.

Keywords H_2 absorption · Cooling tubes · Numerical simulation · Metal hydride

54.1 Introduction

Faced with the energy transition, research efforts on new clean and sustainable energy alternatives are faced with multiple challenges, notably that which depends on energy storage. Speaking of the automotive sector of the future, the emergence of hydrogen as fuel requires very strict storage criteria in terms of safety and technology, as set by the DOE energy department in the United States. Besides this requirement, the

---

A. Bammoune (✉) · S. Laouedj
Laboratory of Materials and Reactive Systems, Faculty of Technology, University Djillali Liabes, Sidi Bel-Abbés, Algeria
e-mail: abdelazizbammoune@gmail.com

B. Dadda
Applied Research Unit for Renewable Energies, URAER, Renewable Energy Development Center, CDER, 47133 Ghardaïa, Algeria

technique of solid hydrogen storage from physicochemical phenomena of adsorption/absorption by different materials is more advantageous compared to other techniques (under pressure, cryogenic). However, this process has two major obstacles around which research is focused, namely; optimizing thermal management of the huge amount of heat released/supplied during storage/destocking and improving its low thermal conductivities. In order to study the phenomena of mass and heat transfers within the several geometric configurations of metal hydride reactors. Jemni et al. [1] have developed a two-dimensional mathematical model describing the different equations governing the phenomena of thermal and mass transfer within a cylindrical metal hydride $LaNi_5$ reactor. The authors validated the ETL hypothesis (local thermal equilibrium between the solid-gas phases) as well as the neglect of the convective effect at the solid-gas interface inside the porous medium. After that, Askri et al. [2, 3] confirmed the neglect of the thermal effect of radiation on the storage process and studied the effect of the impact of the volume expansion of the absorbent medium during storage. The development of reactors equipped with fluid heat exchangers has known several geometric configurations [4]. Yang et al. [5, 6] studied analytically and numerically the behavior of heat and mass transfers with non-LTE (Local Thermal Equilibrium), by two parameters of heat and mass transport, in order to describe the storage process within the reactor for different designs: (a) TR tubular reactor, (b) DR disc reactor, (c) ADR annular disc reactor. Boukhari et al. [7] have shown, from their two-dimensional study of H_2 absorption in an ADR $LaNi_5$ reactor, that the addition of cooling tubes and/or the increase in their ratio significantly, increases the transfer of heat and mass and reduce the absorption time by around 60% for an ADR configuration with 6 tubes incorporated compared to other reactors of the same ADR type with fewer cooling tubes. In this work, we have chosen an ADR reactor design proposed by Yang et al. [5] based on the study of Boukhari et al. [7]. However, we have modified the way of incorporating the tubular heat exchanger, in which the coolant flows. The purpose of this study is to determine the effect of tube arrangement on the behavior of heat and mass transfers during the absorption of H2. In this regard, we have used the commercial ANSYS Fluent 17.2 CFD simulation code. In addition, in order to well personalize our thermochemical phenomenon, we have introduced UDFs (User-Defined-Function).

54.2 Mathematical Model

54.2.1 Assumptions and Governing Equations

The two-dimensional mathematical model describing the behavior of heat and mass exchanges is based on the model developed by Jemni et al. [1], the hypotheses on which the equations are based, are as follows:

1. The thermal equilibrium between the gas and the solid is considered, and the heat transfer by radiation is negligible.

2. The gas phase is ideal, from a thermodynamic point of view.
3. The temperature and pressure of hydrogen are maintained during its passage by the injection filters.
4. The porosity is uniform and its variation is negligible.
5. The equilibrium pressure of the gas in the absorbent bed follows the Van't Hoff equation.
6. The thermophysical properties of the gas and solid phases are constant.

Equation of continuity for hydrogen gas:

$$\varepsilon \frac{\partial \rho_g}{\partial t} + \frac{\partial}{\partial x}(\rho_g u) + \frac{\partial}{\partial y}(\rho_g v) = -\dot{m} \qquad (1)$$

Mass conservation equation for the LaNi$_5$ metal alloy:

$$(1-\varepsilon)\frac{\partial \rho_s}{\partial t} = \dot{m} \qquad (2)$$

Density of hydrogen gas:

$$\rho_g = \frac{PM_{H_2}}{R_g T} \qquad (3)$$

Energy equation:

$$(\rho C_p)_{eff} \frac{\partial T}{\partial t} = \frac{\partial}{\partial x}\left(\lambda_{eff}\frac{\partial T}{\partial x}\right) + \frac{\partial}{\partial y}\left(\lambda_{eff}\frac{\partial T}{\partial t}\right) - (\rho_g C_{pg} u)\frac{\partial T}{\partial x}$$
$$- (\rho_g(\rho_g C_{pg} v))\frac{\partial T}{\partial y} - \dot{m}(\Delta H - T(C_{pg} - C_{ps})) \qquad (4)$$

The effective heat capacity:

$$(\rho C_p)_{eff} = \varepsilon \rho_g C_{pg} + (1-\varepsilon)\rho_s C_{ps} \qquad (5)$$

Effective thermal conductivity:

$$\lambda_{eff} = \varepsilon \lambda_g + (1-\varepsilon)\lambda_s \qquad (6)$$

Reaction kinetics:

$$\dot{m} = -C_a \exp\left(-\frac{E_a}{R_g T}\right)\ln\left(\frac{P}{P_{eq}}\right)(\rho_{sat} - \rho_s) \qquad (7)$$

Equilibrium pressure (Van't Hoff equation):

$$\ln\left(\frac{P_{eq}}{P_0}\right) = \frac{\Delta H}{R_g T} - \frac{\Delta S}{R_g} \tag{8}$$

54.2.2 Initial and Boundary Conditions

At $t = 0$: $P(x, y, 0) = P_0$, $T(x, y, 0) = T_0$, $\rho(x, y, 0) = \rho_{s0}$

At $t > 0$: the injection of H2 through the filters is maintained at constant temperature and pressure $P(x, y, t) = P_0$, $T(x, y, t) = T_0$

The incorporated tubes and the peripheral envelope exchange heat by convection with the porous medium (cooling of the hydride bed) according to the following formula: (Table 54.1)

$$\lambda_{eff} s \frac{\partial}{\partial \vec{n}} T(x, y, t) = hs(T - T_{ref}) \tag{9}$$

Table 54.1 The different constants and operational parameters

Constants and operational parameters	Values
Radius of the annular reactor unit R (mm)	300
Radius of central H$_2$ injection tube r_{ce} (mm)	17.5
Radius of interior H$_2$ injection tubes r_{in} (mm)	6
Injection pressure of H$_2$ P_0 (bars)	12
Injection temperature of H$_2$ T_0 (K)	291
Temperature of surrounding fluid and tubes T_{ref} (K)	291
Heat transfer coefficient h (W/m^{2}K)	1652
Porosity of the metal ε	0.5
H$_2$ gas thermal conductivity λ_g (W/mK)	0.24
Metal thermal conductivity λ_s (W/mK)	2.4
Density of empty/saturated H$_2$ metal ρ_0/ρ_{st} (kg/m^3)	8400/8521
Enthalpy of formation ΔH (J/kg)	31,000

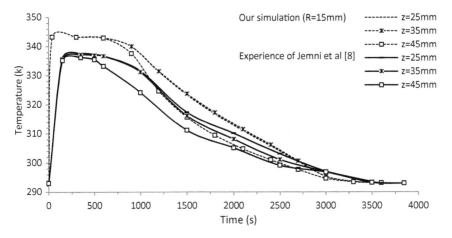

Fig. 54.1 Comparison of the calculated temperatures with the experimental measurements of Jemni et al. [8] in three points of the reactor

54.3 Results and Discussion

54.3.1 Validation

In order to validate our simulation procedure, we compare the obtained numerical results by ANSYS Fluent with the experiment of Jemni et al. [8]. From Fig. 54.1, we notice a good agreement between our numerical results and the experimental temperature measurements for the three points during the absorption of H_2 in a cylindrical metal hydride $LaNi_5$ reactor.

54.3.2 Arrangement Effect of the Cooling Tubes

In this study, we proposed four configurations of a thin $LaNi_5$ disk drive, with a radius of 300 mm, crossed by a central H_2 injection tube with a radius of 17.5 mm. Other injection tubes with a radius of 6 mm are distributed inside the disc. Each configuration has 45° angular symmetry, with different distribution and different numbers of incorporated cooling tubes. To study this effect we kept the same perimeter of the cooling (17.58 cm) and injection (8.91 cm) tubes for the four configurations.

From Fig. 54.2 which presents the distribution of the absorbent bed temperature during the different indicated periods of time, we note that there is a sensitivity of the evolution of the bed temperature to the different configurations, despite the fact that they have the same cooling and hydrogen injection tube perimeter. We also observe that the reactor in configuration 4 cools down quickly compared to the others,

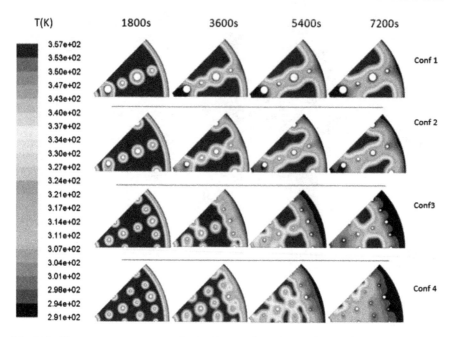

Fig. 54.2 The temperature contours for the four configurations at times (1800, 3600, 5400 and 7200 s)

because it is clear that its cold zones (in white) are very wide compared to the other configurations.

Figure 54.3 shows the time variation of the average temperature of the absorbent bed for the different configurations. We notice a sudden increase in temperature

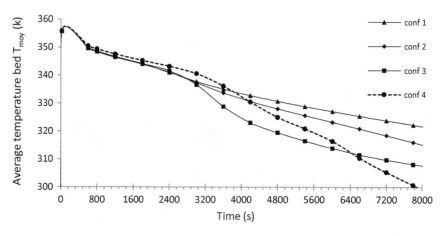

Fig. 54.3 Profiles of the time variation of the average bed temperature for the four proposed configurations

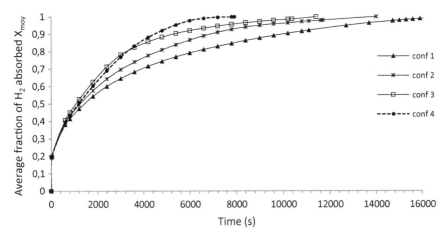

Fig. 54.4 Time evolution of the average fraction of H_2 absorbed for the four proposed configurations

during the first seconds, due to the strong release of heat during the exothermic reaction of absorption of H2. Then, the profiles will gradually decrease by the cooling effect, but we distinguish a rapid cooling of the fourth configuration compared to the others. This confirms our previous interpretation.

Figure 54.4 illustrates the evolution over time of the average fraction of the bed for the proposed configurations. This parameter is essential for distinguishing the quantitative behavior of the chemical process from H_2 storage. From the profiles of the average fraction, we notice that configuration 4, which carries more tubes, reached saturation more quickly with a reduction of approximately 50% compared to configuration 1. This remark means that the more the distribution of the cooling tubes incorporated on the porous medium is homogenized, the less the time required to reach saturation.

54.4 Conclusion

In this work, a two-dimensional numerical study of the effect of arrangement of cooling tubes, on the H_2 storage process in an ADR reactor in $LaNi_5$ has been presented. The obtained results show that, when the arrangement of the cooling tubes is homogenized, the cooling of the bed is accelerated and consequently, the saturation time is minimized with very high rates and without having to modify the volumetric capacities or the mass of the storage system. Thus, we manage to solve the problem set by Boukhari et al. [7], of the inability to dissipate heat from areas which are far from convective walls (region of the heart), because of the low thermal conductivity absorbent bed.

References

1. A. Jemni, S. Ben Nasrallah, Study of two-dimensional heat and mass transfer during absorption in a metal-hydrogen reactor. Int. J. Hydrogen Energy **20**(1), 43–52 (1995)
2. F. Askri, A. Jemni, S. Ben Nasrallah, Study of two-dimensional and dynamic heat and mass transfer in a metal–hydrogen reactor. Int. J. Hydrogen Energy **28**(5), 537–557 (2003)
3. F. Askri, A. Jemni, S. Ben Nasrallah, Dynamic behavior of metal–hydrogen reactor during hydriding process. Int. J. Hydrogen Energy **29**(6), 635–647 (2004)
4. M. Afzal, R. Mane, P. Sharma, Heat transfer techniques in metal hydride hydrogen storage: a review. Int. J. Hydrogen Energy **42**(52), 30661–30682 (2017)
5. F. Yang, X. Meng, J. Deng, Y. Wang, Z. Zhang, Identifying heat and mass transfer characteristics of metal hydride reactor during adsorption—parameter analysis and numerical study. Int. J. Hydrogen Energy **33**, 1014–1022 (2007)
6. F. Yang, X. Meng, J. Deng, Y. Wang, Z. Zhang, Identifying heat and mass transfer characteristics of metal hydride reactor during adsorption: improved formulation about parameter analysis. Int. J. Hydrogen Energy **34**(4), 1852–1861 (2009)
7. A. Boukhari, R. Bessaïh, Numerical heat and mass transfer investigation of hydrogen absorption in an annulus-disc reactor. Int. J. Hydrogen Energy **40**(39), 13708–13717 (2015)
8. A. Jemni, S. Ben Nasrallah, J. Lamloumi, Experimental and theoretical study of a metal–hydrogen reactor. Int. J. Hydrogen Energy **24**(7), 631–644 (1999)

Chapter 55
Photocatalytic Hydrogen Production on 5% CuO/ZnO Hetero-Junction

Meriem Haddad, Akila Belhadi, and Mohamed Trari

Abstract The hetero-junction 5% CuO/ZnO elaborated by co-precipitation is studied by thermal analysis, X-ray diffraction, ATR spectroscopy, and optical reflectance to assess its performance for the H^2 photoproduction. The specific surface area averages ~7 m^2 g^{-1} while the crystallite size varies from 20 to 50 nm. The diffuse reflectance spectra shows an indirect transition at 3.13 eV for ZnO and a direct transition at 1.60 eV for CuO. The capacitance-potential plot of the sensitizer CuO exhibits p-type conduction, due copper deficiency, with a flat band potential of 0.075 V_{SCE}. ZnO works as an electron bridge; its conduction band of ZnO (-0.68 V_{SCE}) is cathodically localized with respect to that of CuO and $CuOH_2O/H_2$ level (~-0.5 V_{SCE}), yielding H_2 evolution under visible light illumination. H_2 liberation rate of 340 μmol h^{-1} (g catalyst)$^{-1}$ and a quantum yield of 0.38% were obtained under full light (27 mW cm^{-2}) for a catalyst dose of 0.25 mg/mL and pH ~ 7 in presence of SO_3^{2-} as reducing agent; H_2 liberation rate of 340 μmol h^{-1} (g catalyst)$^{-1}$ and a quantum yield of 0.38% are determined.

Keywords Hetero-junction CuO/ZnO · Photocatalyst · Co-precipitation · Hydrogen · Visible light

---

M. Haddad · A. Belhadi (✉)
Laboratory of Natural Gas, Faculty of Chemistry, USTHB, BP 32, 16111 Algiers, Algeria
e-mail: sarakila@yahoo.fr

M. Haddad
e-mail: haddadmeriem16@gmail.com

M. Trari
Laboratory of Storage and Valorization of Renewable, Faculty of Chemistry, USTHB, BP 32, 16111 Algiers, Algeria

© The Editor(s) (if applicable) and The Author(s), under exclusive license to Springer Nature Singapore Pte Ltd. 2021
A. Khellaf (ed.), *Advances in Renewable Hydrogen and Other Sustainable Energy Carriers*, Springer Proceedings in Energy,
https://doi.org/10.1007/978-981-15-6595-3_55

55.1 Introduction

The energy of the sun liberates is gigantic but is restricted because of the day/night alternation. Hydrogen is adequate [1, 2] and can be obtained from water and solar energy on low cost oxides semiconductors [3, 4].

In this regard, we have synthesized the hetero-junction CuO/ZnO by co-precipitation in the goal to augment the active surface area and to reduce the length the electrons cross to attain the solid/liquid interface. The present study aims at synthesizing by co-precipitation and characterizing the hetero-junctions 5% CuO/ZnO and their utilization for the H_2 formation upon visible illumination. They were characterized by X-ray diffraction, thermal analysis, ATR and UV-Vis spectroscopies and photoelectrochemistry.

The hetero-junctions were successfully tested for the H2 formation upon visible illumination and some physical parameters were optimized.

55.2 Experimental

55.2.1 Samples Preparation

The hetero-junction 5% CuO/ZnO is prepared by co-precipitation method. $Cu(NO_3)_2$, $3H_2O$ and $Zn(NO_3)_2$, $6H_2O$ with desired molar ratios were mixed in $NaHCO_3$ solution, then NaOH solution was added until reaching pH ~ 10. The solution was magnetically stirred at 80 °C for 2 h. The precipitates were filtered and thoroughly washed with distilled water and dried at 100 °C. The powders were finally calcined at 350 °C for 4 h.

For the photo electrochemical (PEC) study, the CuO powder was pressed into bars (thickness = 13 mm) and sintered at 600 °C to have good mechanical properties.

A double wall reactor linked to a thermostat permitting the elimination the infrared heating effects was used for the H_2 liberation, the temperature was maintained at its optimal value of 50 (±1 °C). A dose of catalyst powder is dispersed in 200 mL of neutral solution electrolyte Na_2SO_4 (0.1 M) and magnetically agitated. Three tungsten lamps are positioned around the reactor, the solution was bubbled with N_2 during 35 min, then the light was switched on. Hydrogen was characterized by gas chromatography; the amount was evaluated by volumetry in homemade setup.

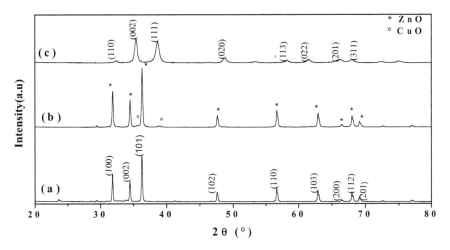

Fig. 55.1 XRD patterns of **a** ZnO, **b** 5% CuO/ZnO, **c** CuO calcined at 350 °C

55.3 Results and Discussion

55.3.1 Structural Properties

The patterns clearly indicate mixed phases. However, although the low temperature synthesis, the XRD lines (Fig. 55.1) are fairly narrow indicating a medium crystallinity of CuO (Tenorite JCPDS card No.: 05-0661) and ZnO (Wurtzite JCPDS card No.: 36-1451); no secondary impurity is detected. The crystallite size (D) of the hetero-junctions is between 20 and 49 nm, it was evaluated from the full width at half maximum indicating that the synthesized oxides possess nano-morphological aspect.

55.3.2 Optical and Photoelectrochemical Properties

The relationship between the absorption coefficient (α) and photon energy ($h\nu$) is expressed by the relation [5]:

$$(\alpha h\nu)^n = \text{Const} \times (h\nu - E_g) \tag{1}$$

where n = 2 and 0.5 for direct and indirect transition respectively. Figure 55.2 represents the diffuse reflectance spectrum and the plot $(\alpha h\nu)^2$ as a function of $h\nu$ for CuO; the transition is evaluated from the intersection of the linear plot with the $h\nu$-axis. The E_g value is found to be 1.60 eV for CuO (direct transition).

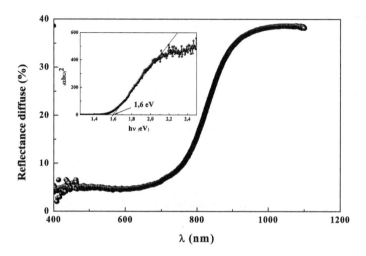

Fig. 55.2 The direct optical transitions of CuO at room temperature

It would be imperative to undertake an electrochemical characterization to be close to the working conditions, Fig. 55.3 gives the intensity-potential characteristic I(E) of CuO plotted under illumination in neutral solution (Na_2SO_4 0.1 M, pH ~ 7). The peak (O1, ~ 0.2 V) corresponds to the couple Cu^{2+}/Cu^+, a value close to that given in the literature, indicating electrons localization in the rock salt structure CuO; the decreased current below ~ −0.8 V is attributed the H_2 evolution as supported by gas bubbling on the electrode.

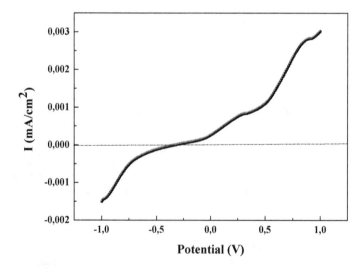

Fig. 55.3 The I(V) curve of CuO under visible light illumination in Na_2SO_4 electrolyte

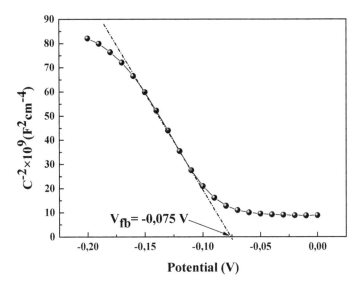

Fig. 55.4 The Mott-Schottky characteristic of CuO plotted in Na_2SO_4 solution (pH ~ 7) at 10 kHz

The flat band potential (E_{fb}) permits to localize the electronic bands with respect to redox levels in solution; it is obtained by plotting the capacitance (C) versus applied potential (E):

$$C^{-2} = \pm(0.5S^2 \times e\varepsilon\varepsilon_o N_A)^{-1}\{E - E_{fb}\} \qquad (2)$$

The negative slope indicates p-type conduction of CuO; the potential Efb (−0.075 V) and the holes density NA (8.7×10^{18} cm^{-3}) are determined from the intercept with the abscissa axis and the slope of the linear plot (C^{-2}–E) respectively (Fig. 55.4). The potential E_{fb} comes from the high ionization energy of copper (7.72 eV) and gives the position of the conduction band (CB: −1.425 V = e E_{fb} + E_a/e − E_g/e), such value is more cathodic than the potential of the H_2 evolution (~−0.8 V).

55.4 Photocatalysis

Hydrogen is cleanly produced from water under visible light; however CuO alone produces a low H_2 volume because of the large difference between CuO-CB and H_2 level. Accordingly, ZnO is used to accelerate the electrons transfer because its conduction band is a midway between CuO-CB and H_2 level. An optimal percentage of 5% CuO/ZnO (Fig. 55.5) is found with a formation rate of 340 µmol min^{-1} (g catalyst)$^{-1}$ due to the synergetic, higher amounts block the light penetration.

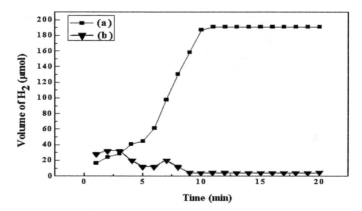

Fig. 55.5 Hydrogen evolution on **a** 5% CuO/ZnO hetero-junction and **b** CuO under visible light (Na_2SO_4 solution (pH ~ 7) in presence of SO_3^{2-}

55.5 Conclusion

The co-precipitation method at low temperature was used to prepare the heterojunction p-5% CuO/ZnO to preclude the formation of impurity phases and to get small crystallites size. The oxide showed photocatalytic properties and the photoelectrochemical characterization indicated p type behavior. The origin of the gap of CuO comes from the crystal field splitting of Cu^{2+}: $3d$ orbital, octahedrally coordinated.

References

1. A. Belhadi, I. Nadjem, S. Zaidat, A. Boudjemaa, M. Trari, Hydrogen evolution under visible light over the heterojunction p-CuO/n-ZnO prepared by impregnation method. Int. J. Energy Res. **39**(14), 1909–19016 (2015)
2. K. He, L. Guo, NiS modified CdS pyramids with stacking fault structures: highly efficient and stable photocatalysts for hydrogen production from water. Int. J. Hydrogen Energy **42**(38), 23995–24005 (2017)
3. Y. Gyeong Kim, W.K. Jo, Photodeposited-metal/CdS/ZnO heterostructures for solar photocatalytic hydrogen production under different conditions. Int. J. Hydrogen Energy **42**(16), 11356–11363 (2017)
4. X.J. Zheng, Y.J. Wei, L.F. Wei, B. Xie, M.B. Wei, Photocatalytic H_2 production from acetic acid solution over CuO/SnO_2 nanocomposites under UV irradiation. Int. J. Hydrogen Energy **35**(21), 11709–11718 (2010)
5. J. Zeng, H. Wang, Y.C. Zhang, M.K. Zhu, H. Yan, Hydrothermal synthesis and photocatalytic properties of pyrochlore $La_2Sn_2O_7$ nanocubes. J. Phys. Chem. C **111**(32), 11879–11887 (2007)

Chapter 56
Field Enhancement Factor Around Hydrogen-Negative Index Metamaterial Waveguide

Houria Hamouche and Mohammed M. Shabat

Abstract The improvement in the electromagnetic field energy density around planar surfaces of hydrogen-absorbing metals is necessary for the development of effective hydrogen storage technologies. We explore in this research the possibility to substitute the planar metal by a planar metamaterial in order to enhance the field enhancement factor. Metamaterials are artificial materials with, in a certain frequency region, simultaneously negative effective electric permittivity and negative effective magnetic permeability. A comparison of the field enhancement factor is undertaken for structures with a planar metamaterial and a planar metal with, respectively, negative and positive real part of refractive indices. This research takes into account the reflectivity at the metamaterial hydrogen interface. This reflectivity is determined by the Transfer Matrix Method. The dependence of the field enhancement factor on the metamaterial thickness is also discussed. The computational results show an increasing of the field enhancement factor of the planar metamaterial compared to the field enhancement factor of the planar metal.

Keywords Planar · Hydrogen · Metal · Metamaterial · Field enhancement factor · Reflectivity · Transfer matrix method

56.1 Introduction

Hydrogen with its safety characteristics appears to be one of the important energy carriers for the future [1]. Nevertheless, the harnessing of the energy stored in hydrogen to produce electrical energy is greatly hindered by the lack of efficient storage methods [2]. Basic hydrogen storage methods include storage in gaseous

---

H. Hamouche (✉)
Centre de Développement des Energies Renouvelables, CDER, BP 62 Route de l'Observatoire, Bouzaréah, 16340 Algiers, Algeria
e-mail: hamouche_houri@yahoo.fr

M. M. Shabat
Department of Physics, The Islamic University of Gaza, P. O. Box 108, Gaza, Gaza Strip, Palestine

© The Editor(s) (if applicable) and The Author(s), under exclusive license to Springer Nature Singapore Pte Ltd. 2021
A. Khellaf (ed.), *Advances in Renewable Hydrogen and Other Sustainable Energy Carriers*, Springer Proceedings in Energy,
https://doi.org/10.1007/978-981-15-6595-3_56

form, storage in liquid forms or media and storage in metal hydrides [1, 3]. The reduction of the adsorption of hydrogen is mainly caused by the weak interaction between hydrogen and the surface of solid materials [4]. To enhance the adsorption energy, a lot of attention has been paid to the use of the transition metals as absorbers for hydrogen energy storage [5]. In recent years, research interest is directed to metallic nanoparticles. Indeed, a remarkably enhanced capacity and speed of hydrogen storage have been obtained in palladium nanoparticles covered with the metal–organic framework [6]. Lately, the degrees of field enhancements for hydrogen-absorbing transition metals, Pd, Ti, and Ni have been investigated [7]. Inspired by the work of Fukuoka and Tanabe [7], we explore in this research the possibility to substitute the planar metal by a planar metamaterial in order to enhance the field enhancement factor. In this paper, we are starting with a brief description of the electromagnetic parameters of the proposed hydrogen-metamaterial waveguide, the optical field energy enhancement at the interface between hydrogen and the metamaterial layer is studied in term of the field enhancement factor. This research takes into account the reflectivity at the metamaterial hydrogen interface. This reflectivity is determined by the Transfer Matrix Method. The dependence of the field enhancement factor on the metamaterial thickness is also discussed. Finally, a comparison of the field enhancement factor is undertaken for structures with a planar metamaterial and a planar metal with, respectively, negative and positive real part of refractive indices.

56.2 Electromagnetic Parameters of the Proposed Waveguide

This work attempts to replace the planar metal in the structure investigated in references [7, 8] by a planar metamaterial then the proposed structure is identical to the structure investigated in references [7, 8]. Metamaterials are artificial materials with, in a certain frequency region, simultaneously negative effective electric permittivity $\varepsilon(\omega)$ and negative effective magnetic permeability $\mu(\omega)$ [9–11]. In this study, the Drude model is used to express the metamaterial's relative electric permittivity $\varepsilon(\omega)$

$$\varepsilon(\omega) = \varepsilon_{lattice} - \frac{\omega_p^2}{\omega^2 + i.\gamma.\omega} \quad (1)$$

where $\varepsilon_{lattice}$ is the lattice permittivity. Values of $\varepsilon_{lattice}$, ω_p and γ used in the program are extracted from [12]: $\varepsilon_{lattice} = 1.5$, $\omega_p = 73000 \, cm^{-1}$ and $\gamma = 140 \, cm^{-1}$. The metamaterial's relative magnetic permeability is assumed to be -1 and the refractive index of hydrogen is taken equal to 1 [7]. The complex relative electric permittivity and the complex refractive index of metamaterial are presented in Figs. 56.1 and 56.2, respectively.

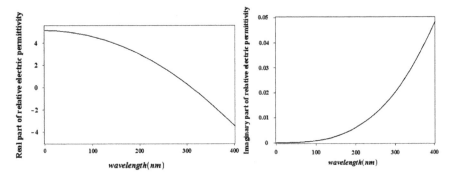

Fig. 56.1 Complex relative electric permittivity versus wavelength

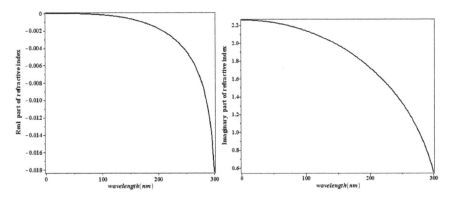

Fig. 56.2 Complex refractive index of metamaterial versus wavelength

56.3 Field Enhancement Factor

The application of an external electric field to the structure is one of means to enhance the interaction between hydrogen and the surface/interface of the metamaterial and consequently hydrogen energy storage. This concept has been introduced earlier in reference [4] where the authors have showed the improvement of the hydrogen storage properties of polarizable substrates by an applied electric field. In this work, an electric field is assumed to be incident on the structure and calculations are made in order to determine the field enhancement factor. The field enhancement factor is defined as the intensity ratios for fields around the object to those in the absence of the object, or the original incident fields [7]. The field enhancement factor η is expressed by

$$\eta = \frac{\lambda \left(|q_1|^2 + |k|^2 \right) \cos \theta (1 - R)}{\varepsilon_1^{1/2} k'' \operatorname{Re} \left\{ \frac{k \left(\varepsilon_1 q_1' + \varepsilon_2 q_2' \right)}{\varepsilon_2 q_1' q_2'} \right\}} \tag{2}$$

where λ is the wavelength of the field; R is the reflectivity at the metamaterial interface; q_1 and q_2 are the complex wave vectors perpendicular to the structure of the hydrogen and the metamaterial, respectively; k is the complex wave vectors parallel to the structure. The wave vectors of a nonmagnetic metal with a relative magnetic permeability equal to 1 are calculated by

$$q_j = \frac{1}{\lambda}\left(\frac{-\varepsilon_j^2}{\varepsilon_1 + \varepsilon_2}\right)^{1/2} \quad (j = 1, 2) \tag{3}$$

$$k = \frac{1}{\lambda}\left(\frac{\varepsilon_1 \varepsilon_2}{\varepsilon_1 + \varepsilon_2}\right)^{1/2} \tag{4}$$

The wave vectors of a metamaterial with a relative magnetic permeability equal to -1 are calculated by

$$q_j = \frac{1}{\lambda}\left(\frac{\varepsilon_j^2}{\varepsilon_2 - \varepsilon_1}\right)^{1/2} \quad (j = 1, 2) \tag{5}$$

$$k = \frac{1}{\lambda}\left(\frac{\varepsilon_1 \varepsilon_2}{\varepsilon_2 - \varepsilon_1}\right)^{1/2} \tag{6}$$

The real and imaginary parts of complex quantities are indicated by primes and double primes, respectively. The reflectivity required for the calculation of the field enhancement factor is performed using Transfer Matrix Method. A detailed description of the Transfer Matrix Method has been given in our previous papers [9, 13]. The expressions of the reflectivity and the field enhancement factor are simulated with MAPLE software.

56.4 Numerical Simulation Results

By a numerical computation of Eq. (2), we have derived the field enhancement factors for structures with different thicknesses of the metamaterial layer, taking into account the reflectivity. Figure 56.3 displays the reflectivity for normal incidence as a function of wavelength. The highest values of reflectivity are obtained in the range from approximately 50 nm to 275 nm for metamaterial layer thickness of 50 and 75 nm. It is also noticed that the reflectivity increases with the increase of metamaterial layer thickness in the range of wavelength from approximately 50 nm to 300 nm.

The field enhancement factors have been plotted in Fig. 56.4. As seen from this figure, the field enhancement factor attains higher values for 10 nm metamaterial layer thickness (d_{LHM}). In contrary to the reflectivity, the field enhancement factor decreases with the increase of metamaterial layer thickness for layers of 10, 15, 25, 50 and 75 nm, in the range of wavelength from approximately 50 nm to 300 nm.

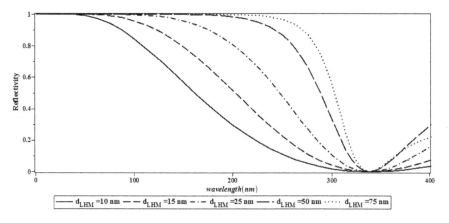

Fig. 56.3 Reflectivity at normal incidence versus wavelength

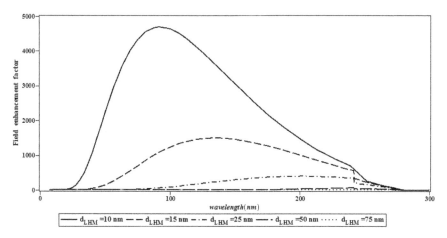

Fig. 56.4 Field enhancement factors versus wavelength around hydrogen-metamaterial waveguide

56.5 Comparison of the Field Enhancement Factor for Structures with Metamaterial Layer and Metal Layer

This section shows that the field enhancement factor for the proposed structure is higher than for structure with metal layer. For this comparison, the metal relative electric permittivity is taken the same as the metamaterial relative electric permittivity (Eq. 1) and the metal's relative magnetic permeability is assumed to be 1. Assuming a layer thickness of 10 nm, Fig. 56.5 clearly illustrates a quite large difference between the field enhancement factors for two structures in the range of wavelength from approximately 25 nm to 250 nm.

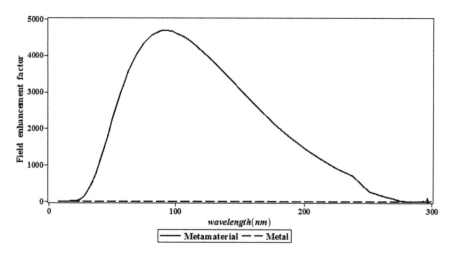

Fig. 56.5 Field enhancement factors around hydrogen-metamaterial waveguide and hydrogen-metal waveguide

56.6 Conclusion

A straightforward hydrogen-metamaterial waveguide has been investigated to enhance the field enhancement factor. The increase of this factor signifies the increase of the hydrogen energy storage. In this investigation, the dependence of the field enhancement factor on the metamaterial thickness was analyzed and a comparison of the field enhancement factor for a metamaterial and a metal with opposite real part of refractive indices was given. It can be concluded that the field enhancement factor is affected by varying the thickness. An increasing of the field enhancement factor of the planar metamaterial structure was also observed compared to the field enhancement factor of the planar metal structure.

References

1. C.J. Winter, J. Nitsch, *Hydrogen as an Energy Carrier: Technologies, Systems, Economy*, 1st edn. (Springer Verlag, Berlin, New York, 1988)
2. R. Mohtadi, S. Orimo, The renaissance of hydrides as energy materials. Nat. Rev. Mater. **2**(16091), 1–115 (2016)
3. Tzimas, E., Filiou, C., Peteves, S.D., Veyret, J.B.: Hydrogen storage: state-of-the-art and future perspective. Institute for Energy, JRC IE, PR& Communication, Européen Commission, Netherlands (2003)
4. J. Zhou, Q. Wang, Q. Sun, P. Jena, X.S. Chen, Electric field enhanced hydrogen storage on polarizable materials substrates. Proc. Natl. Acad. Sci. **107**(7), 2801–2806 (2010)
5. T. Tanaka, M. Keita, D.E. Azofeifa, Theory of hydrogen absorption in metal hydrides. Phys. Rev. B **24**(4), 1771–1776 (1981)

6. Li, G., Kobayashi, H., Taylor, J.M., Ikeda. R., Kubota, Y., Kato, K., Takata, M., Yamamoto, T., Toh, S., Matsumura, S., Kitagawa, H.: Hydrogen storage in Pd nanocrystals covered with a metal–organic framework. Nat. Mater. **13**, 802–806 (2014)
7. N. Fukuoka, K. Tanabe, Large plasmonic field enhancement on hydrogen-absorbing transition metals at lower frequencies: implications for hydrogen storage, sensing, and nuclear fusion. J. Appl. Phys. **126**(023102), 1–7 (2019)
8. K. Tanabe, Plasmonic field enhancement on planar metal surfaces for condensed matter nuclear fusion. J. Condens. Matter Nucl. Sci. **27**, 152–157 (2018)
9. H. Hamouche, M.M. Shabat, D.M. Schaadt, Multilayer solar cell waveguide structures containing metamaterials. Superlattices Microstruct. **101**, 633–640 (2017)
10. P. Markos, C.M. Soukoulis, *Wave Propagation from Electrons to Photonic Crystals and Left-Handed Materials* (Princeton University Press, United States of America, 2008)
11. M.F. Ubeid, M.M. Shabat, M.O. Sid-Ahmed, Effect of negative permittivity and permeability on the transmission of electromagnetic waves through a structure containing left-handed material. Nat. Sci. **3**(4), 328–333 (2011)
12. Tanner, D.: Optical effects in solids. University of Florida, Lecture notes (2013)
13. Hamouche, H., Shabat, M.M.: Artificial metamaterials for high efficiency silicon solar cells. In: 3rd International Symposium on Materials and Sustainable Development, pp. 105–115. Springer, Algeria (2018)

Chapter 57
Effect of Silicates and Carbonates on the Structure of Nickel-Containing Hydrotalcite-Derived Materials

Baya Djebarri, Nadia Aider, Fouzia Touahra, Ferroudja Bali, Juan Paul Holgado, and Djamila Halliche

Abstract The NiAl and NiAlSi were prepared by coprecipitation method at constant basic pH, calcined at 800 °C, and characterized by ICP analysis, XRD, FTIR, SEM, BET method and (TPR). The CH_4/CO_2 reaction was carried out in a fixed-bed tubular reactor at 750 °C. A reactant mixture CH_4 and CO_2 were mixed at a ratio of 1 diluted in He (10:10:80). The XRD patterns of the "as prepared" materials exhibit the characteristic diffractions of hydrotalcite-like layered double hydroxide materials, although were not clearly evidenced in the Si containing samples. Nevertheless, after calcination this hydroltacite-like structure is destroyed. We have studied the reaction of dry reforming of methane by carbon dioxide in presence of various catalysts at 750 °C. In the case of NiAlSi, the CH_4 conversions were close to 70%, with a H_2/CO ratio close to 0.9. Nevertheless, the NiAl sample presented a very variable conversion and a high carbon formation, what resulted in a blocking of the reactor at the end of the experience and a decrease in the H_2/CO ratio, which indicates that probably the CO_2 is consumed in parallel in the reverse water gas shift reaction (RWGS). We have shown that Ni/Al/Si mixed oxides obtained through the thermal decomposition of layered double hydroxide-type precursors are promising catalysts in the reforming of methane with CO_2.

Keywords CH_4 · CO_2 · Nickel

B. Djebarri (✉)
Department of Chemistry, Faculty of Sciences, University of M'hamed Bougara, Independence Avenue, 35000 Boumerdès, Algeria
e-mail: djebaribaya@gmail.com

F. Touahra
Centre de Recherche Scientifique et Technique En Analyses Physico-Chimiques (CRAPC), BP 384-Bou-Ismail-RP42004, Tipaza, Algeria

N. Aider · F. Bali · D. Halliche
Laboratory of Natural Gas Chemistry, Faculty of Chemistry (USTHB), BP 32, 16111 Algiers, Algeria

J. P. Holgado
Departamento de Quimica Inorganica, Instituto de Ciencia de Materiales de Sevilla, CSIC-University of Seville, Avda. Americo Vespucio, 49, 41092 Seville, Spain

© The Editor(s) (if applicable) and The Author(s), under exclusive license to Springer Nature Singapore Pte Ltd. 2021
A. Khellaf (ed.), *Advances in Renewable Hydrogen and Other Sustainable Energy Carriers*, Springer Proceedings in Energy,
https://doi.org/10.1007/978-981-15-6595-3_57

57.1 Introduction

CO_2 reforming of methane (dry reforming) shows a growing interest from both industrial and environmental viewpoint. From an environmental perspective, CO_2 and CH_4 are undesirable greenhouse gases and both are consumed by the proposed reaction. From the industrial point of view, the reaction allows to transform these invaluable gases into synthesis gas with a low H_2/CO ratio, adequate for hydroformylation and carbonylation reactions as well as for methanol and Fischer–Tropsch synthesis. The dry reforming reaction has been studied over numerous supported metal catalysts including Ni-based and noble metal-based catalysts. The latter are more active but, considering the high cost and the limited availability of noble metals, Ni based-catalysts are more interesting for industrial application. The main drawback of a conventional Ni-based catalyst is its poor stability caused, mainly, by a high coking-rate [1, 2]. As consequence, very intensive research efforts are currently being carried out to improve both performance and lifetime of Ni-based catalysts. In this context, basic supports. [3, 4] and/or highly dispersed Ni catalysts have been used to lower the rate of coke deposition. The main purpose of this paper is to study the effect of anions incorporated in interlamellar of hydrotacites such as carbonates or silicates precursors for CO_2 reforming of methane reaction.

57.2 Experimental

57.2.1 Catalysts Preparation

The NiAl carbonates containing materials and NiAlSi silicate-containing samples were prepared by coprecipitation method in the following steps: An aqueous solution containing an appropriate amount of $Ni(NO_3)_2 6H_2O$ and $Al(NO_3)_3 9H_2O$, with the total metal ion concentration of 1 M was added dropwise into an equal volume of 0.5 M of Na_2CO_3 solution or 0.5 M silicates (sodium silicate solution) with vigorous stirring at room temperature. The PH maintained at 10 by simultaneous addition of NaOH solution. The resulting suspension was then maintained at 70 °C for 18 h under stirring. The products were filtered and washed several times with water and then dried overnight at 80 °C. The catalysts were obtained after calcination at 800 °C for 6 h. The samples were denoted NiAl and NiAlSi.

57.2.2 Samples Characterization

Chemical composition of the samples was determined by inductive coupling plasma-atomic emission spectroscopy (ICP-AES). Analyses were performed with a Horiba Jobin-Yvon, Ultima2 spectrometer.

Samples were analyzed by isothermal nitrogen adsorption using standard BET and BJH methods for the specific surface area and pore size distribution calculations, respectively. Measurements were carried out on a Micromeritics ASAP 2010 system. Prior to each experiment, samples were degassed at 150 °C under vacuum for 2 h.

XRD powder analyses were carried out using a Siemens D-501 equipment, with a Bragg- Brentano configuration, using CuKα (λ=1.5418Å). Spectra were collected at room temperature in the range of $2\theta = 5$–$80°$, with a step size of $0.05°$ and an acquisition time of 1 s for each point.

FTIR spectra were recorded using the KBr pellet technique on PERKIN-Elmer spectrometer in the range 4000–400 cm^{-1} using a resolution of 4 cm^{-1}.

The calcined samples were analyzed by temperature programmed reduction (TPR) experiments which were done on a H_2 (5%)/Ar mixture (50 ml/min) from room temperature up to 900 °C, with a heating rate of 10 °C/min. A thermal conductivity detector (TCD), previously calibrated using CuO, were used to evaluate the H_2 consumption. Experimental conditions were adjusted in order to avoid peak coalescence [5].

57.2.3 Catalytic Activity

Catalytic Activity Tests were carried out in a quartz tubular down-flow reactor. Dimensions of the fix-bed reactor were 9 mm inner diameter and 16 cm long. Catalyst was placed in the middle of the reactor between two pompons of quartz wool. Reaction temperature was measured in the middle of the reactor by means of a K-type thermocouple. Feedstock gases were fed to reactor by independent mass flow controllers (Bronkhorst). The CH_4 and CO_2 reactants were mixed diluted in He (CH_4:CO_2: He = 10:10:80 in vol.). Sample was put in contact with the reaction mixture, and heated from room temperature up to 750 °C at 1 °C/min rate, holding the samples at 750 °C during 8 h. Previous to reaction conditions, all catalysts were treated "in situ" with a H_2 (5%)/Ar mixture (50 ml/min) during 1 h at the same temperature of reaction (750 °C). Finally, reaction was cooled down to room temperature in the same reaction mixture. In all the cases, the experiments have been carried out using a gas hourly space velocity (GHSV) of 300.000 L/kg•h (20 mg of sample). Analysis of reactants and products were done by a Gas Chromatograph (Varian CP-3800) with a Thermal Conductivity Sensor and a column (Porapak Q).

57.3 Results and Discussion

57.3.1 Characterization

Table 57.1 summarizes the chemical composition and textural properties of Ni/Al/Si samples.

Elemental chemical analysis data of all calcined samples indicated that the molar metal ratio in the solids was very close to the value in the parent solutions, i.e. $M^{2+}/M^{3+}=2$. This means that the precipitation step was carried out effectively.

Values of specific surface areas were calculated following the B.E.T method. The incorporation of silicates into the structure of NiAl sample leads a significant development of specific surface area it decrease from 114 to 31 m^2/g.

The PXRD patterns of NiAl and NiAlSi samples before calcination was shown in Fig. 57.1. The diffractogram of NiAl exhibit sharp and symmetrical reflexions for (003, (006), (110) and (113) plans characteristic of a well crystallised of hydrotalcites according to the literature [6]. The insertion of silicates, in the interlayer leads to solid with a lower crystalline degree [7]. However, some carbonate anions, coming from CO_2 in the atmosphere, could also be in the interlayer region as contaminant species because of the high affinity of hydrotalcites to them. This poor crystallinity is due to the formation of polysilicate where SiO_4 units condense, moreover a grafting process of silicate layers onto brucite-like layers with a diffusion of Al^{3+} ions from the layers occurred during preparation [8]. After calcinations at 800 °C the hydrotalcite structure

Table 57.1 Chemical composition and surface areas of catalysts

Samples (m^2/g)	M^{2+}/M^{3+}	Surface areas
NiAl	1.94	114
NiAlSi	1.94	31

Fig. 57.1 X-ray diffraction of fresh samples

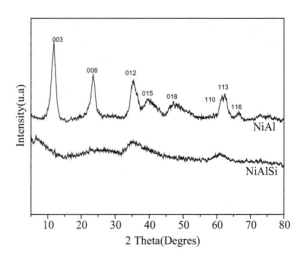

is destroyed. The XRD patterns Fig. 57.2 of NiAl sample show formation of NiO. The aluminum oxides Al_2O_3 are in the amorphous state, therefore not detected by XRD [4].

The FT-IR spectra of the LDH precursors are shown in Fig. 57.3. They shown the characteristic bands of hydrotalcite-like compounds [9]. The broad band at 3400 cm^{-1} is attributed to the υ (OH) vibration mode of lamellar OH groups and water molecules, either inserted in the interlamellar space or physisorbed on the surface. Furthermore, at around 1640 cm^{-1} the bending mode of water molecules is also observed. The shoulder at around 2900 cm^{-1} is recorded, whereas the shoulder cannot be observed in silicate spectra, this shoulder is attributed to OH groups bonded to carbonates species in the interlayer region [10] (Fig. 57.4).

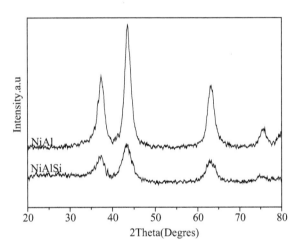

Fig. 57.2 X-ray diffraction of calcined samples

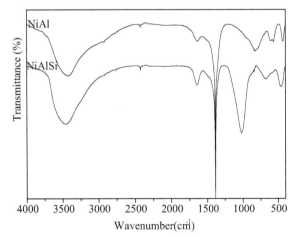

Fig. 57.3 FT-IR spectra of fresh samples

Fig. 57.4 FT-IR spectra of calcined samples

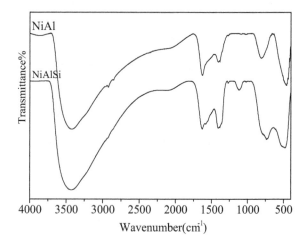

However, at around 1365 cm^{-1} the band due to the $\upsilon 3$ (CO_3^{-2}). Vibration mode is recorded for both samples, thus suggesting contamination by carbonate species, because of the high affinity of hydrotalcite to these species.

The results of H_2-TPR measurements of the calcined samples showed in Fig. 57.5 NiAlSi sample give rise to one reduction peak at lower temperature ≤600 °C which can due to the reduction of Nickel with less interaction with the support. In the case of NiAl sample which show a two peaks of reduction, the peak at 500 °C characteristic of the reduction of NiO oxide particles (free) well dispersed and without interaction with the matrix the highest one about 700 °C is probably do to reduction of nickel with strong interaction with the matrix. In addition, a minor shoulder can be seen at 490 °C, the temperature of reduction of this minor shoulder could be attributed to reduction of NiO weakly interacting with the support

Fig. 57.5 Temperature programmed reduction profiles

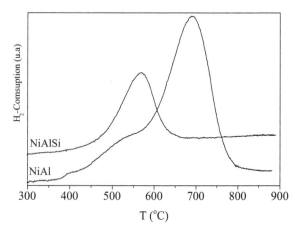

57.3.2 Catalytic Activity

The stability of the catalysts was examined at temperature reaction of 750 °C and a 1:1 CH_4/CO_2 feed ratio, after in-situ reduction at 750 °C for 1 h. In the case of NiAl, CH_4 and CO_2 conversion are respectively 65, 85%. The CH_4 conversion of NiAl lower than conversion of CO_2. The unequivalence of CO_2 and CH_4 conversions reveals the presence of secondary reaction under the operating conditions, which indicates that probably the CO_2 is consumed in parallel in the reverse water gas shift reaction (RWGS) [11]:

$$CO_2 + H_2 \Rightarrow CO + 2H_2O \tag{57.1}$$

Much coke have observed in this catalyst obstruction of the reactor. In the case of NiAlSi sample, the CH_4 conversion is 70% with a H_2/CO ratio close to 0.9.

57.4 Conclusion

NiAl and NiAlSi catalysts were prepared by coprecipitation method. They were characterized by ICP method, XRD, FTIR, TPR and BET methods. The data obtained from chemical analysis of the calcined catalysts confirmed that the n (M^{2+})/n (M^{3+}) ratio is close to the intended value of 2, 0. The XRD patterns of the materials exhibit the characteristic diffractions of hydrotalcite-like layered double hydroxide materials, although were not clearly evidenced in the Si containing samples. The catalysts present catalytic performances at 750 °C that vary in the following sequence: NiAlSi > NiAl.

References

1. Touahra, F., Sehailia, M., Ketir, W., Bachari, K., Chebout, R., Trari, M., Cherifi, O. Halliche, D.: Effect of the Ni/Al ratio of hydrotalcite-type catalysts on their performance in the methane dry reforming process. Appl. Petrochem. Res. **6**, 1–13 (2016)
2. Abdelsadek, Z., Sehailia, M., Halliche, D., Gonzalez-Delacruz, V., Holgado, J., Bachari, K., Caballero, A, Cherifi, O.: In-situ hydrogasification/regeneration of NiAl-hydrotalcite derived catalyst in the reaction of CO_2 reforming of methane: A versatile approach to catalyst recycling. J. CO_2 Util. **14**, 98–105 (2016)
3. N. Aider, F. Touahra, F. Bali, B. Djebarri, D. Lerari, K. Bachari, D. Halliche, Improvement of catalytic stability and carbon resistance in the process of CO_2 reforming of methane by CoAl and CoFe hydrotalcite-derived catalysts. Int. J. Hydrog. Energy **43**(17), 8256–8266 (2018)
4. B. Djebarri, V. Gonzalez-Delacruz, D. Halliche, K. Bachari, A. Saadi, A. Caballero, J.P. Holgado, O. Cherifi, Promoting effect of Ce and Mg cations in Ni/Al catalysts prepared from hydrotalcites for the dry reforming of methane. React. Kinet., Mech. Catal. **111**, 259–275 (2014)

5. P. Malet, A. Caballero, The selection of experimental conditions in temperature-programmed reduction experiments. J. Chem. Soc., Faraday **84**, 2369–2375 (1988)
6. A.H. Iglesias, O.P. Ferreira, D.X. Gouveia, A.G.S. Filho, J.A.C. de Paiva, J.M. Filho, O.L. Alves, Structural and thermal properties of Co–Cu–Fe hydrotalcite-like compounds. J. Solid State Chem. **178**, 142–152 (2005)
7. S. Albertazzi, F. Basile, P. Benito, P. Del Gallo, G. Fornasari, D. Gary, V. Rosetti, A. Vaccari, Effect of silicates on the structure of Ni-containing catalysts obtains from hydrotalcite-type precursors. Catal. Today **128**, 258–263 (2007)
8. C. Depege, F.Z. El Metoui, C. Forano, A. de Roy, J. Dupuis, J.P. Besse, Polymerisation of silicates in layered double hydroxide. Chemestry Mater. **8**, 952–960 (1996)
9. F. Touahra, M. Sehailia, D. Halliche, K. Bachari, A. Saadi, O. Cherifi, (MnO/Mn3O4)-NiAl nanoparticles as smart carbon resistant catalysts for the production of syngas by means of CO2 reforming of methane: Advocating the role of concurrent carbothermic redox looping in the elimination of coke. Int. J. Hydrog. Energy **41**(46), 21140–21156 (2016)
10. Kruissink, E. C., van Reijden, L. L., Ross, J. R. H.: Coprecipitated nickel–alumina catalysts for methanation at high temperature. Part 1—Chemical composition and structure of the precipitates. J. Chem. Soc., Faraday Trans. 1: Phys. Chem. Con-Densed Phases. **77**, 649–663 (1981)
11. Benrabaa, R., Boukhlouf. H., Lo¨fberg, A., Rubbens, A. R., Vannier, A., Bordes-Richard, E., Barama, A.: Nickel ferrite spinel as catalyst precursorin the dry reformingof methane: Synthesis, characterization and catalytic properties. J. Nat. Gas Chem. **21**, 595–604 (2012)

Chapter 58
On the Effect of the Inlet Hydrogen Amount on Hydrocarbons Distribution Produced via Fischer-Tropsch Synthesis

Abdelmalek Bellal and Lemnouer Chibane

Abstract Hydrogen has a significant importance through its benefit for sustainable energy producing that can respond to the worldwide demand and address environmental problems. It can be used as an alternative resource for the generation of clean and renewable fuels via Fischer-Tropsch synthesis (FT). In this context, our investigation focuses on studying the impact of injected amount of hydrogen on reactor outlet for assuring readily control and manage of the distribution of the products. For this purpose, a simulation study by using a mathematical model of FT synthesis carried out in a packed bed reactor was established under various values of molar flow rate of hydrogen at constant carbon monoxide flow rate. It was found that increasing the inlet hydrogen to carbon monoxide molar ratio was able to shift the distribution of hydrocarbons towards the production of paraffins. Either, the optimal reactor performances were also could be achieved at high injection of hydrogen.

Keywords Hydrogen · Sustainable energy · Fischer-Tropsch synthesis · Hydrocarbons distribution

Nomenclature

A	Cross-section (m^2)
d_p	Diameter of catalyst particle (m)
D_c	Diameter of cooling tube (m)
F_ξ	Molar flow rate of hydrocarbon ξ (mol s^{-1})
k_i	Reaction rate constant
K_i	Reaction equilibrium constant
L	Reactor length (m)
l	Dimensionless reactor length

A. Bellal (✉) · L. Chibane
Laboratory of Chemical Processes Engineering, Department of Processes Engineering, Faculty of Technology, Ferhat Abbas University of Setif 1, 19000 Setif, Algeria
e-mail: bellalabdelmalek.science@gmail.com

O/P Olefin over paraffin ratio
R_j Rate of reaction j (mol kg^{-1} s^{-1})
U_{sh} Heat transfer coefficient shell-gases (Wm^{-2} K^{-1})
ρ Catalyst density (kg m^{-3})

58.1 Introduction

According to global growth in energy needs, the development of new resources that can reduce the dependence on conventional forms of power to satisfying the imposed environmental requirements has become increasingly investigated [1]. In line with the suggested research in this field, H_2 has attracted intense attention as an alternative solution to produce clean and renewable fuels [2]. The synthesis gases issues from biomass represent the appropriate feedstock for the synthesis of eco-friendly hydrocarbons (HC). In fact, the produced diesel can guarantee low emission of pollutants such as hydrogen sulfide and CO_2 and can also offer good ignition properties attributed to the high cetane number. Over the years, the FT process has known plenty of actual intensifications in the objective of enhancing the reaction efficiency and thus reaching high fuel quality. In light of the foregoing research, achieving the desired HC composition represents one of the most challenging issues and requires a deep optimization of reactor parameters that can affect FTS, including temperature, pressure and catalyst composition [3, 4]. It was reported [5] that low values of gas hourly space velocity can favor the production of heavier paraffins while unsubstantial levels of pressure will adverse the production of paraffins and swift the formation rate of olefins. In spite of this remarkable variation in hydrocarbons distribution, the effect of such parameters is complex and thoroughly restricted under a narrow range of investigations, in which implicate delicate control of the process. Usually, H_2/CO molar ratio is the most adequate parameter that can offer resilient and readily orientation of hydrocarbons distribution as the amount of the injected H_2 in the reactor inlet has a direct influence on FT kinetic, high or low surface recovery can shift FT reaction towards the formation of paraffins or olefins according to the energetic potential of hydrogenation. Hence, the inlet H_2/CO ratio plays an important role in the control of HC selectivity and their yields [3]. For the purpose of better understanding the effect of the H_2/CO ratio on products distributions, a reactor simulation was performed to analyze the evolution of HC selectivity and the HC yields along the reactor.

58.2 Reactor Model and Reactions Kinetic

In the present work, FT synthesis was studied in a fixed bed reactor packed with catalyst particles. This kind of reactor is commonly used on industrial scale, due to his simplicity and its high capacity of production. However, other types of reactors such as slurry and fluidized bed reactors are highly recommended for achieving high syngas conversions but their application is limited due to several drawbacks including catalyst attrition and the hard separation of products and catalyst particles [6].

FT process was carried out over Fe–Cu–K catalyst. The reaction was conducted under specific levels of temperature (533 K) and pressure (20 bar) that are restricted with respect to the validity of the kinetic model [7]. Other selected parameters are taken from previous experimental works [7, 8]. Some known assumptions are considered in order to simplify the construction of the nonisothermal mathematical model. Different model equations are given as follow:

Mass balance:

$$\frac{dF_\xi}{dl} = \frac{\rho AL}{F_{CO}^0} \upsilon_{\xi j} R_j \qquad (3)$$

Selectivity:

$$S_\xi(\%) = \frac{F_\xi}{\sum F_\xi} \times 100\% \qquad (4)$$

Yields:

$$Y_\xi(\%) = \frac{F_\xi}{F_{CO}^0} \times 100\% \qquad (5)$$

Momentum balance:

$$\frac{dP_T}{dl} = -\frac{L\upsilon}{d_p}\left(\frac{1-\varepsilon}{\varepsilon^3}\right)\left(\frac{150(1-\varepsilon)\mu}{d_p} + 1.75\rho_g\upsilon\right) \qquad (6)$$

Heat balance:

$$\frac{dT}{dl} = \frac{\rho AL}{F_T C_{pg}} \sum_{j=i}^{B} R_j(-\Delta H_{R_j}) + \frac{L\pi D_C}{F_T C_{pg}} U_{sh}(T_{sh} - T) \qquad (7)$$

A developed kinetic model for iron catalyst was used in this investigation [8]. This kinetic approach is widely accepted for iron catalyst under a processing temperature superior to 350 K. Different reaction rates are given as follow:

$$R_{CH_4} = k_{5M} P_{H_2} \alpha_1 / DEN \qquad (8)$$

$$R_{C_n H_{2n+2}} = k_5 P_{H_2} \prod_{k=1}^{n} \alpha_k / DEN \qquad (9)$$

$$R_{C_nH_{2n}} = k_6(1-\beta_n)\prod_{k=1}^{n}\alpha_k/DEN \qquad (10)$$

$$\alpha_1 = \frac{k_1 P_{CO}}{k_1 P_{CO} + k_{5M} P_{H_2}} \qquad (11)$$

$$\alpha_A = \frac{k_1 P_{CO}}{k_1 P_{CO} + k_5 P_{H_2} + k_6} \qquad (12)$$

$$\alpha_n = \frac{k_1 P_{CO}}{k_1 P_{CO} + k_5 P_{H_2} + k_6(1-\beta_n)} \qquad (13)$$

$$\beta_n = \frac{k_{-6}}{k_6}\alpha_A^{n-1}\frac{k_1 P_{CO}}{k_1 P_{CO}+k_5 P_{H_2}} + \frac{k_{-6}}{k_1 P_{CO}+k_5 P_{H_2}+k_6}\sum_{a=2}^{n}\alpha_A^{n-2}P_{C_{(n-a+2)}H_{2(n-a+2)}} \qquad (14)$$

$$DEN = 1 + \left(1 + \frac{1}{K_2 K_3 K_4}\frac{P_{H_2O}}{P_{H_2}^2} + \frac{1}{K_3 K_4 P_{H_2}} + \frac{1}{K_4}\right)\sum_{f=1}^{n}\left(\prod_{k=1}^{f}\alpha_k\right) \qquad (15)$$

$$R_{WGS} = \frac{k_v\left(\frac{P_{CO}P_{H_2O}}{P_{H_2}^{0.5}} - \frac{P_{CO_2}P_{H_2}^{0.5}}{K_{WGS}}\right)}{1 + K_v\frac{P_{CO}P_{H_2O}}{P_{H_2}^{0.5}}} \qquad (16)$$

The equilibrium constant of water gas shift reaction (WGSR) can be calculated by the following equation [8], and all kinetic parameters are taken from [7].

$$K_{WGS} = \frac{5078.0045}{T} - 5.8972089 + 13.958689 \times 10^{-4}T - 27.592844 \times 10^{-8}T^2 \qquad (17)$$

58.3 Results and Discussion

In first, we outline the evolution of HC selectivity at different inlet H_2/CO molar ratios (H_2/CO) as displayed in Fig. 58.1a–c. The used range of H_2/CO was chosen according to the reaction conditions conducted over an appropriate catalyst [7]. At constant molar flow rate of CO (0.1 mol s^{-1}), the inlet syngas composition can be set by changing the molar flow rate of H_2 with respect to the proposed range of H_2/CO ratio. The obtained results show that the increase in the injected amount of H_2 across the reactor section has a remarkable effect on reactor performance.

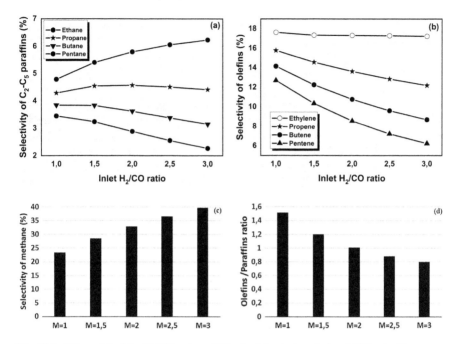

Fig. 58.1 Effect of the inlet H_2/CO ratio on HC selectivity (**a, b** and **c**) and O/P ratio (**d**)

A significant behaviour was attributed to CH_4 and C_2H_6, in which their selectivity were increased from 23.37 to 39.61 and 4.79 to 6.23% respectively, by raising the inlet H_2/CO from 1 to 3. It is noteworthy that the obtained selectivity of CH_4 is much higher than that of reported elsewhere [9, 10] under similar range of reaction conditions. It also appears that the production of propane can be enhanced upon the variation of H_2/CO ratio from 1 to 1.5 but a persistent increase in this critical metric will cause a slight decline in propane selectivity. This can be explained by the fact that the excessive insertion of H_2 on the catalyst surface will cause faster desorption of HC to smaller chain length rather than a continuous propagation for the formation of heavier paraffin. Furthermore, implying that H_2/CO molar ratio has a negative effect on the production of HC with high molecular weight, the total selectivity of butane and pentane was decreased from 7.3 to 5.42% by increasing the molar ratio from 1 to 3. In this case, the obtained results under our investigated conditions show a good agreement with that obtained previously [3, 11]. A similar behavior was reported for olefins selectivity. The production of olefins was highly influenced by the molar ratio change as it is shown in Fig. 58.1b. Moreover, the total selectivity of olefins with a chain length between C_3 and C_5 was decreased from 42.62 to 27.1% by increasing the H_2/CO ratio from 1 to 3. It is noteworthy that the selectivity of ethylene was nearly constant considering the intense increase in the molar ratio due to the high hydrogenation degree of ethylene even at low H_2 concentrations. from the obtained results, it is evident that the H_2/CO molar ratios is the adequate requirement for achieving a light mixture of HC including CH_4 and C_2H_6, whereas lower molar ratios are desirable for high production of olefins and long-chain paraffin. These opposite trends in HC selectivity justify the decrease in olefins to paraffins ratio.

As illustrated in Fig. 58.1d, the O/P ratio tends to reach the minimum value that corresponds to 0.8 at the highest molar ratio due to the great desorption rate of paraffins compared to that of olefins. Also, it can be caused by the secondary hydrogenation that takes place to convert olefins to saturated HC. Considering HC chain length variation, the decline in O/P ratio was the same for all range of HC, which indicate that olefins readsorption for paraffins formation by hydrogenation reactions is independent on carbon number. These results agree well with the proposed kinetic model for FT reactions and with the results that are reported in literature [10], in which exhibits an experimental investigation of H_2/CO ratio effect on O/P ratio. Hence, H_2/CO molar ratio is the key parameter to control and manage reactor performances.

FT reaction efficiency can be further examined by analyzing the evolution of products yields according to the inlet H_2/CO molar ratio. The results presented in Figs. 58.2 and 58.3 show that the increase in the molar ratio can lead to a positive

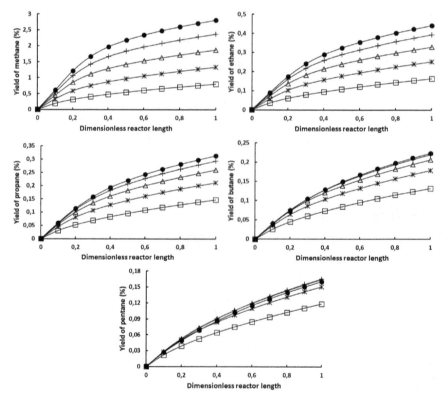

Fig. 58.2 Evolution of paraffins yields along the reactor at different values of inlet H_2/CO ratio: 1 (–□–), 1.5 (–∗–), 2 (–▲–), 2.5 (–+–) and 3 (–●–)

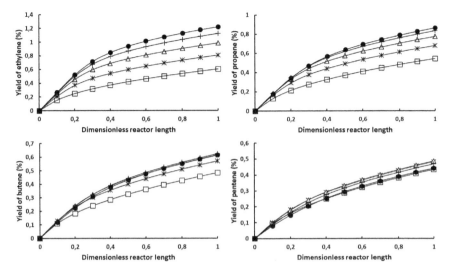

Fig. 58.3 Evolution of olefins yields along the reactor at different values of inlet H_2/CO ratio: 1 (-☐-), 1.5 (-✱-), 2 (-▲-), 2.5 (-+-) and 3 (-●-)

effect on the yield of paraffins and C_2–C_4 olefins, while a nonlinear effect was observed for pentene, in which exhibit a primary increase and then a decrease in the production yield upon molar ratio change. This is typically expected due to the high reported values of hydrogen conversion at high H_2/CO molar ratios [12]. Also, it can be seen that H_2/CO molar ratio incrementation from 1 to 1.5 can conduct to a major improvement in the HC yields, whereas the enhancement in HC yields are being less remarkable at higher molar ratios. Noteworthy, light paraffins are more sensitive to H_2/CO molar ratio change. For instance, raising the molar ratio from 1 to 3 was conducted to a valuable increase in CH_4 yield from 0.8% to 2.8. These results justify the observed variation of CH_4 selectivity according to H_2/CO molar ratio. At least, it can be said that the incorporation of a high initial amount of H_2 at the reactor inlet can shift the HC distribution by decreasing the selectivity of olefins and favoring the formation of light paraffins and also it can be a preferential parameter for increasing the HC yields.

58.4 Conclusion

Understanding the behaviour of reactor performances under the predicted composition of syngas and its effect on HC distribution is highly required for imposing the right conditions that can guarantee a flexible orientation of syngas conversion to the formation of the desired mixture of HC. Based on the obtained results, it can be concluded that high H_2/CO ratio favors the production of light-saturated HC including methane (39.61%) and ethane (6.23%) and disfavors the overall formation of olefins, as suggested by a strong decrease in olefins to paraffins ratio.

References

1. S.R. Sharvini, Z.Z. Noor, C.S. Chong, L.C. Stringer, R.O. Yusuf, Energy consumption trends and their linkages with renewable energy policies in East and Southeast Asian countries: challenges and opportunities. Sustain. Environ. Res. **28**(6), 257–266 (2018)
2. S. Yang, S. Lee, S.C. Kang, S.J. Han, K.-W. Jun, K.-Y. Lee, Y.T. Kim, Linear α-olefin production with Na-promoted Fe–Zn catalysts via Fischer-Tropsch synthesis. RSC Adv. **9**(25), 14176–14187 (2019)
3. N. Moazami, M.L. Wyszynski, K. Rahbar, A. Tsolakis, Parametric study and multiobjective optimization of fixed-bed Fischer-Tropsch (FT) reactor: the improvement of FT synthesis product formation and synthetic conversion. Ind. Eng. Chem. Res. **56**(34), 9446–9466 (2017)
4. A.A. Mirzaei, S. Vahid, M. Feyzi, Fischer-Tropsch synthesis over iron manganese catalysts: effect of preparation and operating conditions on catalyst performance. Adv. Phys. Chem. **2009**, 1–12 (2009)
5. J. Gorimbo, A. Muleja, X. Liu, D. Hildebrandt, Fischer-Tropsch synthesis: product distribution, operating conditions, iron catalyst deactivation and catalyst speciation. Int. J. Ind. Chem. **9**(4), 317–333 (2018)
6. M.P. Rohde, *In-Situ H_2O Removal via Hydrophilic Membranes During Fischer-Tropsch and Other Fuel-Related Synthesis Reactions* (KIT Scientific Publishing, Karlsruhe, 2011)
7. Y.-N. Wang, W.-P. Ma, Y.-J. Lu, J. Yang, Y.-Y. Xu, H.-W. Xiang, Y.-W. Li, Y.-L. Zhao, B.-J. Zhang, Kinetics modelling of Fischer-Tropsch synthesis over an industrial Fe–Cu–K catalyst. Fuel **82**(2), 195–213 (2003)
8. D. Alihellal, L. Chibane, Simulation study of the effect of water removal from Fischer-Tropsch products on the process performance using a hydrophilic membrane reactor. ReacKinet Mech. Cat. **117**(2), 605–621 (2016)
9. Y.H. Kim, D.-Y. Hwang, S.H. Song, S.B. Lee, E.D. Park, M.-J. Park, Kinetic parameter estimation of the Fischer-Tropsch synthesis reaction on K/Fe-Cu-Al catalysts. Korean J. Chem. Eng. **26**(6), 1591–1600 (2009)
10. B. Todic, L. Nowicki, N. Nikacevic, D.B. Bukur, Fischer-Tropsch synthesis product selectivity over an industrial iron-based catalyst: effect of process conditions. Catal. Today **261**, 28–39 (2016)
11. Raje, A.P., Davis, B.H.: Fischer-Tropsch synthesis over iron-based catalysts in a slurry reactor. Reaction rates, selectivities and implications for improving hydrocarbon productivity. Catal. Today **36**(3), 335–345 (1997)
12. T. Olewski, B. Todic, L. Nowicki, N. Nikacevic, D.B. Bukur, Hydrocarbon selectivity models for iron-based Fischer-Tropsch catalyst. Chem. Eng. Res. Des. **95**, 1–11 (2015)

Chapter 59
Predictive Current Control in Grid-Connected Fuel Cell–Photovoltaic Based Hybrid System for Power Quality Improvement

M. R. Bengourina, L. Hassaine, and M. Bendjebbar

Abstract Due to the use of nonlinear loads, the current quality at the point of common coupling (PCC) is degraded. To improve this, a three-phase double-stage multifunctional grid-connected inverter interfaced with a fuel cell–photovoltaic based hybrid system is proposed with embedded active filter function. The studied system consists of a fuel cell stack, a photovoltaic array, a dc-dc boost converter and a three-phase voltage source inverter connected to the utility at the PCC. The objective of the proposed system is to supply the nonlinear currents by the fuel cell–photovoltaic inverter, and thus, the grid current will become sinusoidal with low harmonics. The control of system is performed via P&O MPPT and predictive current control, which uses a discrete-time model of the system to predict the future value of the filter current for different voltage vectors. Simulation in MATLAB is carried out to verify the operation and the control principle. The results are obtained for different operating conditions to prove the effectiveness of the entire system.

Keywords Grid · Inverter · Fuel cell · Photovoltaic · PCC · Harmonic · P&O MPPT

59.1 Introduction

For the future, renewable energy technologies are seen as the most appropriate solution and need to be further developed in order to capture most of the energy production. Several distributed Generation (DG) systems are expected to have a significant impact on the energy market in the near future. These DG systems include, but are not limited to: photovoltaics (PV), wind and microturbine. In addition, several energy

---

M. R. Bengourina (✉) · L. Hassaine
Centre de Développement des Energies Renouvelables (CDER), BP. 62 Route de l'Observatoire, Bouzareah 16340, Algiers, Algeria
e-mail: bengourina.rida@yahoo.fr

M. Bendjebbar
Diagnostic Group, LDEE Laboratory, Electrical Engineering Faculty, University of Science and Technology of Oran MB, BP 1505, El-Mnaouer, Bir El Djir 31000, Oran, Algeria

© The Editor(s) (if applicable) and The Author(s), under exclusive license to Springer Nature Singapore Pte Ltd. 2021
A. Khellaf (ed.), *Advances in Renewable Hydrogen and Other Sustainable Energy Carriers*, Springer Proceedings in Energy,
https://doi.org/10.1007/978-981-15-6595-3_59

storage systems such as Fuel cell are under consideration for DG systems, as they have been expected for many years to increase their presence in applications over a wide range of power ratings [1]. Fuel cells are similar to PV systems in that they produce DC power. Power conditioning systems, including inverters and DC-DC converters, are often required in order to supply normal customer load demand or send electricity into the grid [2]. Nowadays, the integration of small scale DG systems at distribution level is growing extensively [3]. The DG systems such as photovoltaic (PV) and fuel-cell are interfaced with the utility at the Point of Common Coupling (PCC) by means of power electronic converters [4]. The large use of power electronics-based loads connected to the utility causes serious Power Quality (PQ) problems, by generating harmonics and reactive current in the utility [5]. To improve this, the harmonic filter needs to be installed, which incurs more additional costs. When renewable inverters are connected to the PCC, the harmonics can be compensated by enabling the renewable systems as active power filters. Considerable research has been done on the active power filter and its control technique [6].

In order to resolve the PQ problems at PCC in DG system, a multifunctional grid-connected inverter interfaced with a fuel cell–photovoltaic based hybrid system is proposed with using the basic working principles of a predictive current control scheme. Whereas the control is used to transfer the power from the fuel cell–photovoltaic system, it also assures the compensation of the harmonic currents, reactive power and unbalanced current. The overall control system is built in the Matlab/Simulink environment. Then, the simulations results are provided to validate the correctness of the adopted control system.

59.2 Proposed Hybrid System

The structure of the grid-connected Fuel Cell–Photovoltaic Based Hybrid System is shown in Fig. 59.1. The system consists of a photovoltaic array and fuel cell

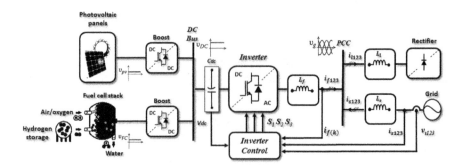

Fig. 59.1 Proposed grid-interactive fuel cell–photovoltaic based hybrid system

Fig. 59.2 Equivalent circuit of a PV cell

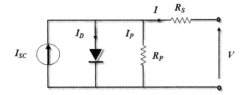

connected through a boost converter to a three-phase inverter that is connected to a grid through a simple filter and nonlinear load.

59.2.1 Introduction Photovoltaic Module Modeling

The PV module is the unity of base for the construction of a PV array. We have used the PV module model of a single diode [4] (Fig. 59.2). This model is the most commonly used due to good results. The current-voltage relationship is given by:

$$I = I_{SC} - I_D - \frac{V + I \cdot R_S}{R_P} \tag{59.1}$$

$$I = I_{SC} - I_0 \left(\exp\left[\frac{q}{m \cdot k \cdot T_c}(V + I \cdot R_S)\right] - 1 \right) - \frac{V + I \cdot R_S}{R_P} \tag{59.2}$$

59.2.2 Fuel Cell Stack

The FC stack is modelled as a controlled voltage source E in series with a constant resistance R_{fc} (internal resistance of the FC stack) as shown in Fig. 59.3. The controlled voltage source is represented as:

$$E = E_{oc} - NA \ln\left(\frac{i_{fc}}{i_0}\right) \tag{59.3}$$

And FC stack output voltage as:

$$V_{fc} = E - R_{fc} i_{fc} \tag{59.4}$$

where E_{oc} is the open circuit voltage, N is the number of cells in the stack, A is the Tafel slope, i_0 is the exchange current resulting from the continual backward and forward flow of electrons from and to the electrolyte at no load, i_{fc} is the FC stack current and V_{fc} is the FC stack voltage [7].

Fig. 59.3 Fuel cell stack model

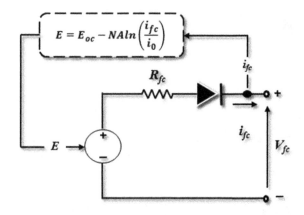

59.3 Proposed Control

In order to control the first conversion stage, a Maximum Power Point Tracking (MPPT) Perturb and Observe (P&O) algorithm is proposed to track permanently the optimum point through an adequate tuning of a boost converter (Fig. 59.4). On the other hand, the second stage control is performed via a predictive current control strategy by using a discrete model of SAPF to predict the behavior of the filter current. Then, via a cost function, an optimal switching state is selected [8]. Reference current is extracted from the dc link voltage loop devoted to maintaining dc link voltage at its desired value through a PI controller. This control strategy can be summarized in the following steps [9]:

- Define a cost function g.
- Build a model of the converter and its possible switching states.
- Build a SAPF model for prediction (Fig. 59.5).

Fig. 59.4 P&O MPPT

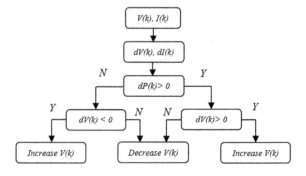

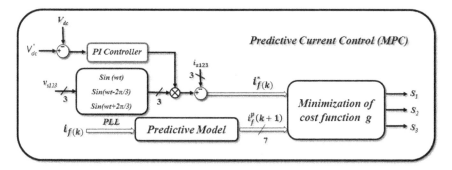

Fig. 59.5 Predictive current control strategy

59.3.1 Predictive Current Control Algorithm

1. The value of the reference current $i_f^*(k)$ is obtained (from DC link voltage algorithm) and the current of SAPF $i_f(k)$ is measured.
2. The model of the system is used to predict the value of the SAPF current in the next sampling interval $i_f^p(k+1)$ for each of the different voltage vectors.

$$v_f - v_c = (L_f + L_s)\frac{di_f}{dt}$$

3. The cost function g evaluates the error between the reference and predicted currents in the next sampling interval for each voltage vector.

$$g = i_f^p(k+1) = i_f(k) + \frac{T_s}{L_f}(v_f(k) - v_s(k))$$

4. The voltage that minimizes the current error is selected and the corresponding switching state signals are generated.

59.4 Simulation Results

In order to prove the robustness and the performance of the proposed control, the system depicted in Fig. 59.1 has been simulated using the SimPowerSystem of Matlab/Simulink. Parameters used in simulation are as in Tables 59.1 and 59.2 (Figs. 59.6, 59.7 and 59.8).

The nonlinear loads which appears in Fig. 59.1 causes a harmonics disturbance in the utility. This effect can totally deform the shape of the source currents. Figure 59.6 presents the load current of the first phase and its harmonic spectrum. The THDi for this load is 27.88%. This THDi is calculated for the first forty harmonics according

Table 59.1 Parameters of DC-DC boost converter

DC-DC boost converter				
Parameter	f switching	Output voltage	Inductor L	Capacitor C
Values	5 kHz	400 V	10 mH	100 µF

Table 59.2 Parameters of SAPF

SAPF	
Parameter	Values
Source voltage	125 V
Source impedance R_S, L_S	0.1 Ω, 0.1 mH
Filter impedance R_f, L_f	0.1 Ω, 2.6 mH
Line impedance R_l, L_l	0.1 Ω, 0.5 mH
DC-link capacitor C	2.2 µH
Diode rectifier load R_L, L_L	20 Ω, 2 mH

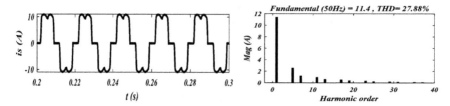

Fig. 59.6 Supply current and current harmonic spectrum

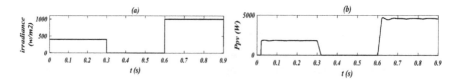

Fig. 59.7 **a** Solar radiation. **b** Power PV array

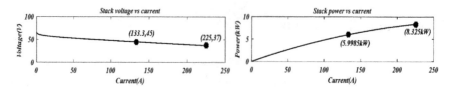

Fig. 59.8 FC stack V-I and P-I characteristic

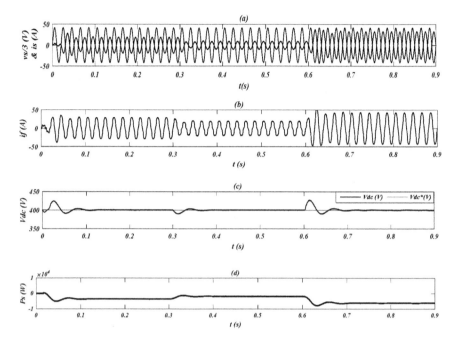

Fig. 59.9 Simulation results: **a** source voltage and source current of a-phase, **b** filter current, **c** DC-link voltage Vdc, **d** active power

to the "CEI standard". The purpose system is to reduce this THDi to less than 5%, as required by the CEI standard.

Figure 59.9a, b show source and filter currents after filtering with fuel cell-PV array. The source currents become perfectly sinusoidal compensated by the filter current at unity power factor. We can observed the opposite phase between voltage and current source and the negative sign of the utility active power (Ps), meaning the filter current has information of harmonic and fuel cell-PV array currents to ensure elimination of harmonic and injection of current to the load. The dc link voltage returns to its reference value in few milliseconds with PI controller. The active filter decreases the total harmonic distortion (THD) from 27.88 to 1.15% with 400 W/m^2, 2.14% with 0 W/m^2, 0.81% with 1000 W/m^2 (Fig. 59.10). Proposed predictive current control confirms its effectiveness and robustness.

59.5 Conclusion

In this paper, the validity of the proposed predictive current control in Grid-connected Fuel Cell–Photovoltaic Based Hybrid System is investigated. The proposed system injects solar and fuel cell power to the grid and compensates reactive power and harmonic currents existing at PCC caused by the non-linear load. Implementation

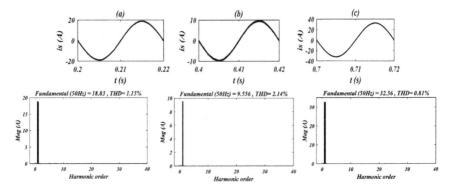

Fig. 59.10 Supply current and current harmonic spectrum for fuel cell with PV: **a** 400 W/m^2, **b** 0 W/m^2, **c** 1000 W/m^2

of predictive control is performed on a Matlab/Simulink^{TM} environment. Simulated results have proved that the proposed predictive control algorithm is good in term of THD. The predictive current control algorithm is a stable and robust solution.

References

1. Y. Yan, K. Kim, F. Blaabjerg, A. Sangwongwanich, *Advances in Grid-Connected Photovoltaic Power Conversion Systems*. Woodhead Publishing Series in Energy (Elsevier, Kidlington, United Kingdom, 2019)
2. W. Kramer, S. Chakraborty, B. Kroposki, H. Thomas, *Advanced power electronic interfaces for distributed energy systems, part 1: systems and topologies* [Technical Report NREL/TP-581-42672] (2008)
3. B. Boualem, J. Gaubert, C. Abdelmadjid, H. Mabrouk, Predictive current control in multi-functional grid connected inverter interfaced by PV system. Sol. Energy **139**, 130–141 (2016)
4. M.R. Bengourina, M. Rahli, S. Saadi, L. Hassaine, PSO based direct power control for a multi-functional grid connected photovoltaic system. Int. J. Power Electron. Drive Syst. **9**(2), 610–621 (2018)
5. P. Sandipan, N. Modem, R.M. Soumya, K. Nand, Power quality improvement in grid-connected photovoltaic-fuel cell based hybrid system using robust maximum power point tracking controller. Electr. Power Compon. Syst. **43**(20), 2235–2250 (2015)
6. B. Singh, K. Al-Haddad, A. Chandra, A review of active filters for power quality improvement. IEEE Trans. Ind. Electron. **46**(5), 960–971 (1999)
7. M. Gitanjali, S. Singh, R. Patidar, Active, reactive and harmonic compensation control of grid interfaced fuel cell system. Int. J. Emerg. Electr. Power Syst. **13**(4) (2012)
8. A. Pablo, L. Moran, M. Rivera, J. Dixon, J. Rodríguez, Improved active power filter performance for renewable power generation systems. IEEE Trans. Power Electron. **29**(2), 687–694 (2014)
9. J. Rodríguez, J. Pontt, C. Silva, P. Correa, Predictive current control of a voltage source inverter. IEEE Trans. Ind. Electron. **50**(1), 495–503 (2008)

Chapter 60
Hydrogen Production by Photo Fermentation via *Rhodobacter sp.*

Sabah Menia, Ilyes Nouicer, and Hammou Tebibel

Abstract Efficiency of photo-hydrogen production is highly dependent on the culture conditions. Initial pH, temperature were optimized for maximal hydrogen production using response surface methodology with central composite design. Photo fermentative hydrogen production is influenced by several parameters, including pH and temperature. *Rhodobacter sp.* KKU-PS1 was isolated from the methane fermentation broth of an UASB reactor. This study presents the experimental results obtained from *Rhodobacter sp. KKU-PS1* cultures as a function of temperature and pH. For this purpose, a complete factorial design was used, for the temperatures of 26, 30 and 34 °C and the pH of 6.5; 7 and 7.5. *Rhodobacter sp. KKU-PS1* has been isolated from the fermentation of methane. Optimum conditions for maximizing cumulative hydrogen production (H_{max}) were an initial value of pH 7.0 and a temperature of 25 °C. The regression models revealed a maximum hydrogen production of 1623 ml at these conditions. KKU-PS1 can produce hydrogen from organic acids. By optimizing the pH and the temperature, a maximal production of hydrogen by this strain was obtained. Validation experiments at calculated optima confirmed these results.

Keywords Hydrogen · Photo fermentation · *Rhodobacter*

60.1 Introduction

At present, fossil fuels are considered the main source of energy supply. However, due to their increasing demand and limited availability, an alternative source of energy is needed. Hydrogen is one of the most promising fuels of the future because of its high-energy content (122 kJ g^{-1}), which is 2.72 times higher than that of gasoline. In addition, H_2 is a major energy carrier and can be used in fuel cells for power generation. However, hydrogen gas is not readily available in nature, as are fossil fuels and natural gas, but it can be produced from renewable materials, such as

---

S. Menia (✉) · I. Nouicer · H. Tebibel
Centre de Developpement des Energies Renouvelables, BP. 62 Route de l'Observatoire, Bouzareah 16340, Algiers, Algeria
e-mail: s.menia@cder.dz

biomass and water. Organic hydrogen production is ecological. In particular, this process can utilize a variety of organic substrates as carbon sources, including waste. Therefore, there is a dual benefit: waste reduction and energy production.

The combination of photosynthetic bacteria with fermentative bacteria can increase the amount of hydrogen produced by the system, and at the same time, there is a reduction in residual organic matter. The photo fermentation culture is very promising because of the purity of the biogas produced. However, the potential of different substrates must be considered, as in the case when dark fermentation is to be coupled to the photo fermentation process, where acetate, propionate and butyrate are generated. In addition, it may be interesting to evaluate the use of a consortium instead of a pure culture because the sterile conditions are useless, which makes the operation of a reactor easier to control. In addition, a microbial consortium may be more robust against inhibitors that may be produced by variations in the substrate introduced into the system [1].

In this study, hydrogen is produced by photo fermentation via the bacteria *Rhodobacter sp.* The aim of bio-hydrogen production by photo fermentation is to facilitate waste treatment and reduce greenhouse gas emissions.

60.2 Material and Method

The Response Surface Methodology (RSM) with the Central Composite Design (CCD) was applied to determine the major and interactive effects of two independent factors, namely, initial pH (6.5–7.5) and temperature incubation (26–34 °C). The initial pH and temperature ranges were chosen following a search in the literature. Previous reports indicated that an initial pH of 7 was optimal for cell growth and hydrogen production while hydrogen can not be produced at an initial pH of 5 [2]. Therefore, the initial pH tested in this study was in the range of 6.5–7.5. There were several temperature ranges for photo hydrogen production. For example, the temperature range of hydrogen production by *Rhodobacter sp.* was between 31–36 °C [2] and 27.5–32.5 °C for *Rp. palustris CQK 01* [3]. Thus, in order to cover the optimal temperature possible for hydrogen production by strain *KKU-PS1*, the fermentation at 26–34 °C was studied. Cumulative maximum hydrogen production (H_{max}) was chosen as the desirable response in batch culture (Table 60.1).

A quadratic model has been used to optimize key environmental factors.

$$Y_i = \beta_0 + \Sigma \beta_i X_i + \Sigma \beta_{ii} X_{ii}^2 + \Sigma \beta_{ij} X_i X_j. \tag{60.1}$$

where Y_i is the predicted response (H_{max}), β_0 is a constant, β_i is the linear coefficient, β_{ii} is the square coefficient, β_{ij} is the interaction coefficient, and X_i is the variable. The response variables (H_{max}) were adjusted using a predictive quadratic polynomial equation to coelate the response variable to the independent variables. The test conditions were designed with experimental data. A differentiation calculation was

Table 60.1 Central composite plan

pH	Temperature (°C)	H_{max} (ml/L)
7	30	1122.4
6.5	26	1232.5
7.5	34	714
6.5	34	916.5
7.5	26	1048
7.5	30	690.5

then used to predict the optimal values of the different factors maximizing H_{max} of the new isolate in the batch fermentation.

60.3 Results and Discussion

The effects of initial pH (X_1) and incubation temperature (X_2) on H_{max} were studied. The regression model is shown in Eq. (60.2).

$$H_{max} = -51{,}750.62 + 18{,}642.40 X_1 - 739.25 X_2 \\ - 1340.60 X_{11}^2 + 11.91 X_{22}^2 - 2.25 X_1 X_2 \qquad (60.2)$$

The results indicated that an increase in initial pH and incubation temperature resulted in a substantial decrease in H_{max}. A maximum value of 1623 ml H_2/L at an initial pH of 7 and an incubation temperature of 25 °C was observed. pH is a very important factor that influences the metabolism. The active site and the characteristics of the biochemical reaction of the enzyme nitrogenase, the enzyme responsible for the production of hydrogen, have been affected by its ionic form at different pH values in the culture medium [4].

Temperature has an important function of moving metabolic pathways towards photo-hydrogen production. It affects cell synthesis, the rate of hydrogen production and the conversion efficiency of substrates of photo-hydrogen-producing bacteria. The optimal initial pH of 7.0 in our study was consistent with other reports [5]. The optimal temperature in our study was slightly lower than other reports, where the optimal temperature range of *Rhodobacter sp.* was between 30 and 32 °C [6]. In contrast, Wang et al. found that the optimum temperature for *Rp. palustris CQK 01* was between 27.5 and 32.5 °C [3] (Fig. 60.1 and Table 60.2).

Fig. 60.1 Effect of pH and temperature on hydrogen production

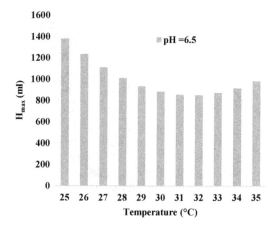
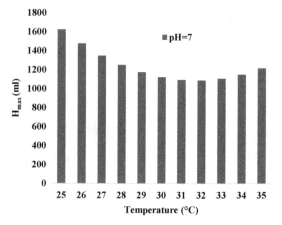
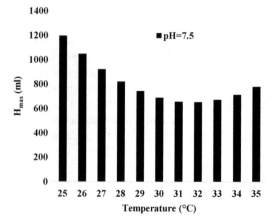

Table 60.2 Confirmation of photo-hydrogen production experiences

Condition	pH	Temperature (°C)	H_{max} (ml/L)
Optimum	7	25	1623
Bottom	6.5	26	1232
Center	7	30	1122
Top	7.5	34	714

60.4 Conclusion

KKU-PS1 strain effectively used ten carbon sources for hydrogen production. Malic acid was a preferred carbon source for hydrogen production. The initial pH and incubation temperature had a significant effect on H_{max}, but there were no interaction effects between the initial pH and the incubation temperature. Maximization of H_{max} occurred at the optimal initial pH of 7 at an incubation temperature of 25 °C. Under these conditions, a maximum value of H_{max} of 1623 ml H_2/L was obtained.

References

1. R. Cardena, G. Moreno, I. Valdez-Vazquez, G. Buitron, Optimization of volatile fatty acids concentration for photo fermentative hydrogen production by a consortium. Int. J. Hydrogen Energy **40**, 17212–17223 (2015)
2. N. Basak, A.K. Jana, D. Das, Optimization of molecular hydrogen production by *Rhodobacter sphaeroides* O.U.001 in the annular photo bioreactor using response surface methodology. Int. J. Hydrogen Energy **39**, 11889 (2014)
3. Y.Z. Wang, Q. Liao, X. Zhu, X. Tian, C. Zhang, Characteristics of hydrogen production and substrate consumption of *Rhodopseudomonas palustris* CQK 01 in an immobilized-cell photobioreactor. Bioresour. Technol. **101**, 4034 (2010)
4. T. Assawamongkholsiri, A. Reungsang, Photo-fermentation hydrogen production of *Rhodobacter* sp. KKU-PS1 isolated from an UASB reactor. Electron. J. Biotechnol. **18**, 221–230 (2015)
5. M.S. Kim, D.H. Kim, J. Cha, Culture conditions affecting H_2 production by phototrophic bacterium *Rhodobacter sphaeroides* KD131. Int. J. Hydrogen Energy **37**, 14055 (2012)
6. A. Pandey, N. Srivastava, P. Sinha, Optimization of hydrogen production by *Rhodobacter sphaeroides* NMBL-01. Biomass Bioenergy **37**, 251–256 (2012)

Chapter 61
Estimation of Hydrogen Production in Three Cities in the North of Algeria

Ilyes Nouicer, M. R. Yaiche, Sabah Menia, Fares Meziane, and Nourdine Kabouche

Abstract Great increasing interest in both commercial community and research on the production of hydrogen in the energy sector as an energy carrier. In addition, protecting our earth from global warming and the depletion of crude oils and gasses are the main objectives for our energy strategies worldwide. The Westinghouse Corporation proposed a new method known as the Hybrid Sulphur (HyS) cycle, which forms part of the so-called thermo-chemical cycles. Two sub-reactions in this system are used to complete the functionality of this process: thermochemical and electrochemical reactions. In the second sub-reaction, several parameters can affect the electrochemical reaction and electrolysis efficiency such as: cell temperature, membrane thickness and catalyst loading … etc. This study will focus on the influence solar irradiation and ambient temperature on hydrogen production in three cities in the north of Algeria. According to the study, it was found that the power consumption to produce a flow 8 Nm^3/h is 31 kWh for Algiers and 32 kWh for Constantine and Oran.

Keywords Solar hydrogen production · SDE · Photovoltaic system

61.1 Introduction

Searching for energy security, the depletion of crude oils and gasses and protecting our earth from global warming are the main objectives for our energy strategies worldwide. The best solution is to search for new sources of energy that can be sustainable and environment friendliness. Renewable energies can play an important role to resolve this problem especially hydrogen. However, hydrogen doesn't exist directly in nature, the best way is to produce it. Three ways available used to produce hydrogen: fossil fuels, nuclear power, and renewable energy sources [1]. But, three

---

I. Nouicer (✉) · M. R. Yaiche · S. Menia · F. Meziane · N. Kabouche
Centre de Développement des Energies Renouvelables (CDER), B.P. 62, Route de l'Observatoire, Bouzaréah 16340, Algiers, Algeria
e-mail: i.nouicer@cder.dz; ilyes.nouicer@gmail.com

© The Editor(s) (if applicable) and The Author(s), under exclusive license to Springer Nature Singapore Pte Ltd. 2021
A. Khellaf (ed.), *Advances in Renewable Hydrogen and Other Sustainable Energy Carriers*, Springer Proceedings in Energy,
https://doi.org/10.1007/978-981-15-6595-3_61

Table 61.1 Description of the three studied sites located in the north of Algeria

City number	City	Town	Latitude (°)	Longitude (°)	Altitude (m)	Slope angle (°)
25	Constantine	Airport	36.28	6.62	690	36
16	Algiers	Algiers port	36.79	3.07	12	36
31	Oran	Arzew	35.82	−0.27	4	35

criteria must be followed for the hydrogen production to be profitable, environmentally and economically viable: Competitiveness, Optimum energy efficiency and Environmental considerations [2]. The main approach is going to be green methods for clean hydrogen production such as the Westinghouse Process [3, 4]. This study focused on the influence of solar irradiation on tilted surface and ambient temperature on hydrogen production in three cities in the north of Algeria using SDE electrolyzer in Westinghouse Process.

61.2 Proposed PV/SDE System Components

61.2.1 Presentation of Liu and Jordan Model

The estimation of global solar radiation on horizontal and inclined surface using *Liu and Jordan* model [5]. The generalized equation of Liu and Jordan is given by the following equation:

$$\text{SRI} = \text{SR}_h R_b + H_d\left(\frac{1+\cos(\beta)}{2}\right) + \rho \text{SRH}\left(\frac{1-\cos(\beta)}{2}\right) \quad (61.1)$$

SRI Global solar radiation on inclined surface.
SRH Global solar radiation on horizontal surface.
SR_h Direct solar radiation on a horizontal surface.
H_d Diffusion solar radiation on a horizontal surface.
R_b Tilt factor.
β Slope angle (Table 61.1).
ρ Albedo of the underlying surface (unit less).

61.2.2 Proposed PV/SDE System Components

A schematic representation of PV/SDE systems based on SO_2 depolarized electrolyzer (SDE) is given in Fig. 61.1. It includes:

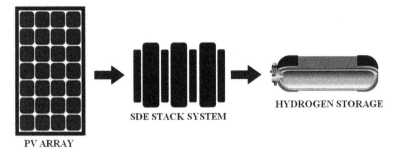

Fig. 61.1 Schematic diagram of the solar hydrogen production system

1. PV array to convert solar radiation into electricity (Model type: KC200GT (54 cells connected in series), conversion efficiency: 16%).
2. An MPPT (Maximum Power Point Tracker) to optimize the generated power from PV.
3. All the energy produced from PV was transferred to the SDE electrolyzer with direct coupling through MPPT.
4. SDE stack system based on SO_2 Depolarized Electrolyzer (Membrane N212, membrane surface: 1 cm^2, membrane pressure differential $\Delta P = 6$ bars, catalyst loading: 1.5 mg/cm^2).
5. The produced H_2 from electrolyzer was stored in storage tanks.

61.3 Results and Discussions

61.3.1 Modeling of Photovoltaic Modules

This study was conducted for three place (Table 61.1) in the north of Algeria: Constantine, Algiers and Oran on tilted PV module. The location of the system design where the study will be realized in three different sites (Fig. 61.2) in the north (east, center and west) of Algeria.

The electrical power available from a photovoltaic device can be modeled with the well known equivalent circuit shown in Fig. 61.3. This circuit includes a series resistance, a diode in parallel with a shunt resistance and current generator. The validation model was presented in previous work [3].

The daily hydrogen requirement can be produced from the change in power generated by the photovoltaic modules. Figure 61.4 shows the monthly energy production from horizontal and tilted PV array in three places in Algeria.

Several parameters must be known in order to determine the power generated from PV modules to the SDE electrolyzer. We can quote here: the light current I_{Ph}, the diode reverse saturation current I_0, the series resistance R_s, the shunt resistance R_p, and the ideality factor a.

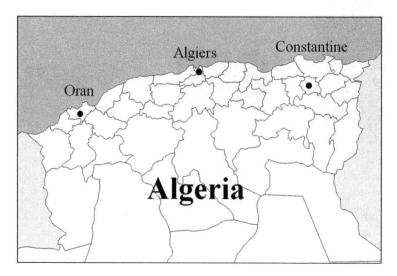

Fig. 61.2 The north of Algeria map and location of the three studied cities

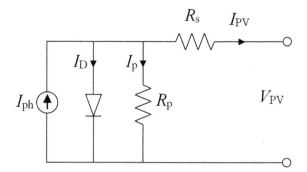

Fig. 61.3 Equivalent circuit representing the five-parameter model

61.3.2 The Daily Energy Production of PV Module for SDE Electrolyzer

The daily-calculated power generated us a function of global irradiation and ambient temperature for the three sites are represented in Fig. 61.4. However, all annual irradiations measurement has been taken from Ref. [6].

The lowest ambient temperature was obtained over the year is 5.4, 7.2, −2 °C and the highest value is 36, 38, 44 °C for Algiers, Constantine and Oran respectively.

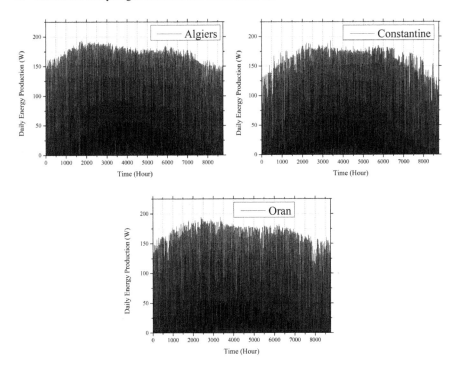

Fig. 61.4 Daily energy production from tilted PV array

61.3.3 Daily Hydrogen Production from SDE Electrolyzer in the Three Cities: Constantine, Algiers and Oran

According to the modeling results, the optimal hydrogen production from the SDE electrolyzer is represented in Table 61.2 (determined by Faraday's law). These total quantities of hydrogen production are given during the year. It has been conclude that the optimal hydrogen production is almost the same in the north of Algeria. Our results are lower than found in the work of Little et al. [7] (Fig. 61.5).

Table 61.2 Description of the three studied sites located in the north of Algeria

City number	City	Town	Cell temperature (°C)	SDE efficiency (%)	Optimal hydrogen production (kWh/Nm^3)
16	Algiers	Algiers port	80	60	3.193
25	Constantine	Airport	80	60	3.203
31	Oran	Arzew	80	60	3.206

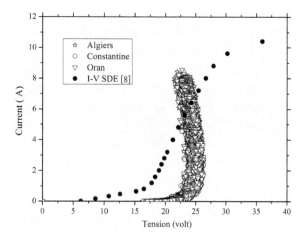

Fig. 61.5 PV array coupled to SDE electrolyzers with a 30 series/8 parallel cells configuration

The power consumption to produce a flow 8 Nm³/h of hydrogen is 34 kWh. In our case, it has been found that the power consumption to produce a flow 8 Nm³/h is 31 kWh for Algiers and 32 kWh for Constantine and Oran. Certainly, several parameters can affect the energy demand such as: types of sky, Sunshine duration, site location, solar radiation and temperature … etc.

61.4 Conclusion

Different components of a detailed modeling approach have been used to estimate hydrogen production in three sites. In the first part, a method of modeling and simulation of a photovoltaic module using a five-parameter model to predict the maximum power output as a function of any operating conditions (ambient temperature, solar radiation … etc.). A validation model with experimental data of a photovoltaic module manufacturer has been determined. The second part, the estimation equation using Faraday's law for hydrogen production of SO_2 depolarized electrolysis as a function of current consumed was presented. The coupling of these two parts allows us to estimate hydrogen production rate as a function of power consumption of PV systems in three sites. It has been found that the hydrogen production is important compared to other systems of hydrogen production, but the optimal hydrogen production is almost the same at each location.

References

1. F. Safari, I. Dincer, A review and comparative evaluation of thermochemical water splitting cycles for hydrogen production. Energy Convers. Manag. **205**, 112182 (2020)
2. F. Jomard, J.P. Feraud, J.P. Caire, Numerical modeling for preliminary design of the hydrogen production electrolyzer in the Westinghouse hybrid cycle. Int. J. Hydrogen Energy **33**, 1142–1152 (2008)
3. I. Nouicer, A. Khellaf, S. Menia, M.R. Yaiche, N. Kabouche, F. Meziane, Solar hydrogen production using direct coupling of SO_2 depolarized electrolyzer to a solar photovoltaic system. Int. J. Hydrogen Energy **44**(39), 22408–22418 (2019)
4. I. Dincer, Green methods for hydrogen production. Int. J. Hydrogen Energy **37**, 1954–1971 (2012)
5. M.R. Yaiche, A. Bouhanik, S.M.A. Bekkouche, A. Malek, T. Benouaz, Revised solar maps of Algeria based on sunshine duration. Energy Convers. Manag. **82**, 114–123 (2014)
6. L.K. Lawrie, D.B. Crawley, *Development of Global Typical Meteorological Years (TMYx)* (2019). Last accessed 2020/02/05
7. M. Little, M. Thomson, D. Infield, Electrical integration of renewable energy into stand-alone power supplies incorporating hydrogen storage. Int. J. Hydrogen Energy **32**, 1582–1588 (2007)
8. J.A. Staser, M.B. Gorensek, J.W. Weidner, Quantifying individual potential contributions of the hybrid sulfur electrolyzer. J. Electrochem. Soc. **157**(6), B952–B958 (2010)

Chapter 62
Comparative Study of Different PV Systems Configurations Combined with Alkaline and PEM Water Electrolyzers for Hydrogen Production

Nourdine Kabouche, Fares Meziane, Ilyes Nouicer, and Rafika Boudries

Abstract Algeria is endowed with one of the world's largest solar potential. Therefore, photovoltaic effect is one of the most common techniques used in exploiting this huge potential. Nevertheless, the amount of energy generated by solar PV panels is influenced by several factors. Indeed, PV cells technologies, site location and weather (solar radiation, temperature and wind speed) and system design (tracking and orientation) determine the output power of PV solar systems. This paper investigated PV-Hydrogen production by two water electrolyzers technologies, alkaline and PEM. Indeed, four different PV systems configurations, two fixed PV panels and two solar tracking system, have been considered. All these cases are studied using meteorological data of the site of Ghardaïa, in the south of Algeria. Thus, a photovoltaic system of about 1.6 kW supplies an electrolysis system of about 1.35 kW across an MPPT (Maximum Power Point Tracking) and buck DCDC converters. The annual hydrogen production shows the advantage of PEM technology in hydrogen production as well as that of sun trackers for PV systems.

Keywords Hydrogen production · Water electrolysis · Solar tracking

62.1 Introduction

The ceaseless progress in the use of renewable energy is due to the challenges facing the world such as climate change, environmental pollution and its impacts on public health and the unavoidable depletion of hydrocarbons. Renewable Energy entails sources that do not deplete, namely, solar, wind, biomass, geothermal, marines and hydropower. In 2018, the total capacity of renewable energy worldwide was about 2350 GW of which about 20.5% are Photovoltaic [1]. Photovoltaic conversion is one of the most important means for sunlight exploitation as well as the pathway to energy

---

N. Kabouche (✉) · F. Meziane · I. Nouicer · R. Boudries
Centre de Développement des Énergies Renouvelables, CDER, Bouzareah 16340, Algiers, Algeria
e-mail: n.kabouche@cder.dz

© The Editor(s) (if applicable) and The Author(s), under exclusive license to Springer Nature Singapore Pte Ltd. 2021
A. Khellaf (ed.), *Advances in Renewable Hydrogen and Other Sustainable Energy Carriers*, Springer Proceedings in Energy,
https://doi.org/10.1007/978-981-15-6595-3_62

transition. In fact, about 99% of the world's installed solar energy capacity is solar PV; the rest is mainly concentrated solar power (CSP). However, the exploitation of solar energy by PV panels remains insufficient due to several technical, political and economic problems.

The irregular availability of sunlight along the year and its unavailability during the night, are limiting the expansion of PV integration rate. Converting PV electricity into hydrogen through water electrolysis is therefore one among many intermittent energy storage solutions. In fact, hydrogen is a promising energy carrier because of its diverse sources of supply (fossil fuels, biomass, industrial crops, nuclear and renewable energies) and its numerous uses (energy storage, fuel cell cars, chemical and the petrochemical industry, injection into the natural gas network, etc.).

Thus, several studies have been conducted on hydrogen production from water electrolysis and PV systems in Algerian sites. Boudries and Dizene studied the hydrogen production potential in eight Algerian sites using three PV array configurations, it has been found that N/S tracking allow much more hydrogen production than E/W tracking. Thus, they predicted a reduction of 57.5% and 58.2% in the cost of hydrogen production in the case of an N/S tracking and a tilted plane, respectively [2, 3]. Tebibel and Labed studied hydrogen production intended for HCNG fuel mixture [4]. They found that PEM water electrolysis supplied by a grid connected PV system leads to the production of 15 g/m^2 day of hydrogen [5]. Ghribi et al. found that about 29 m^3/year of hydrogen could be produced by a PV module of 60 W and a PEM electrolyzer of 50 W in Tamanrasset, south of Algeria [6]. Laoun et al. studied the hydrogen production system using direct and optimal coupling of PV array and PEM water electrolyzer. They found that the largest solar to hydrogen efficiency was about 10.8% at 9 a.m. and 4 p.m., with the production rate reaching the maximum of 0.08 Nm^3/h at midday [7]. Saadi et al. studied hydrogen production using PV-PEM electrolyzer in the southeast of Algeria. Their results have revealed that a PV module of 190 W produced 2.39 m^3/day [8]. Mraoui et al. found that direct-coupling PV-PEM electrolyzer system, in Algiers site allowed a maximum hydrogen production rate of 0.033 ml/s hydrogen flow between noon and 1 p.m. [9].

In the current work, we have investigated hydrogen production via two electrolytic hydrogen productions technologies and different PV system configurations. The hydrogen production technologies considered are water alkaline and PEM electrolyzers. Alkaline technology is a mature technology that is widely used in the industry mainly for large systems. PEM (Proton Exchange Membrane) electrolysis, pricey but operate at high-current densities and can be used for small scales applications. Direct current (DC) produced by the PV source supplies the electrolyzers' cells through two DC–DC converters. Since solar tracking, orientation, and tilt of PV panels are among a number of available means used to increase PV systems output, we have considered four case studies with solar tracking and fixed PV modules.

62.2 Materials and Methods

The present study is carried out at a site located in Ghardaïa (Latitude 32.47 °N, Longitude 3.68 °E, 549 m), an arid region located in the north-central of Algeria. The proposed study is conducted using one-hour time resolution data of a typical meteorological year (TMY). In fact, 8760 data of the principal meteorological parameters, obtained from Meteonorm have been used. Thus, the studied site is characterized with an annual average temperature of about 22.5 °C (ranging from 2 to 46 °C), an average annual wind speed of about 3.6 m/s and an annual global horizontal irradiation of about 2.2 MWh/m².

Instead of reporting the large and unclear hourly data, we preferred to use the cumulative distribution function (CDF) that shows the percentage of data points less than or equal to a certain value (F(x) = P(X ≤ x)). From Figs. 62.1 and 62.2, we can

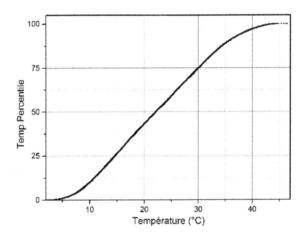

Fig. 62.1 Cumulative distribution function for the hourly air temperature

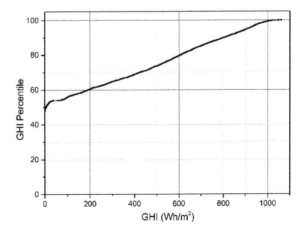

Fig. 62.2 Cumulative distribution function for the hourly GHI

Table 62.1 PV module and electrolyzer characteristics

	Cell number	Nominal power	Total power
PV module	07	228.7 W_p	1601.3 W_p
PEM electrolyzer	17	78.7 W	1337.9 W
ALK electrolyzer	01	1350 W	1350.0 W

notice that about 79% of time (6920 h), the temperature is between 10 and 35 °C, corresponding to a mild temperature. In addition, about 56% of time (4900 h), the global horizontal irradiation (GHI) take a zero value (night time, no solar radiation) or insignificant value (lower than 100 Wh/m^2). Therefore, these statistics show that this site is broadly acceptable for an optimal and efficient PV panel output.

The studied standalone system consists of a PV system that powers an electrolyzer system through two DCDC converters, an MPPT device and a buck converter. In fact, seven PV modules of about 1.6 kW have been considered to supply two types of electrolyzers PEM and alkaline with nominal power of about 1.35 kW each. The DCDC conversion block leads to a loss of power of about 16%, the power difference between the nominal power of the PV source and the load (electrolyzers). The characteristics of the different components of the system are reported in Table 62.1.

Taking into account PV tracking and orientation, four different cases are considered. Then, the SAM "System Advisor Model" software has been used to determine the energy produced by the four PV system configurations. The configurations difference is in the type of solar tracking and orientation. Thus, these four case studies are, namely:

- two axis solar tracking: PV modules rotate from north to south to follow the seasonal movement of the sun throughout the year, and from east in the morning to west in the evening to follow the daily movement of the sun in the sky.
- one axis solar tracking (east-west): PV modules are fixed at an angle to the horizontal (after several tests the optimized angle was found equal to 30.8°) and rotates around the inclined axis from east in the morning to west in the evening to follow the daily movement of the sun across the sky.
- an optimal annual tilt angle, found after tests, equal to 30.62°.
- a fixed tilt angle equal to the site latitude (32.47°), several studies have claimed that the optimal annual tilt angle is equal to the site latitude [10].

62.3 System Modeling

62.3.1 Electrolyzers

Hydrogen production via alkaline electrolysis technology is widely used than PEM. In fact, it is more economic, more mature, does not need a noble catalyst and its power range in MW. However, besides its fast response, PEM technology allows high current densities and a high purity of gases besides. Thus, each technology has a specific application.

In our case, the alkaline electrolyzer consists of a 0.25 m² cell with a nominal power of 1350 W. We have used the Ulleberg model to investigate the performance of this type of electrolyzer. The current-voltage characteristic of an alkaline electrolysis cell as a function of temperature, is given by [11]:

$$V = V_{rev} + \frac{r'}{A} I_E + S' \log\left[\frac{t'}{A} I_E + 1\right] \quad (62.1)$$

where r' is due to resistance in the electrolyte, S' and t' are due to the electrodes overvoltages (overpotentials), A is the active area of the cell.

Hydrogen production rate is given as a function of the current flowing through the electrolyzer cell, IE, and of the Faraday efficiency, η_F, by [11]:

$$\dot{n}_{H_2} = \frac{\eta_F N_C I_E}{n F} \quad (62.2)$$

The PEM electrolyzer consists of 17 cells of 27 cm² and a nominal power of 78.7 W each, the stack total power is 1337.9 W. A semi-empirical model of the hydrogen production rate as a function of the power supplied has been used [5]:

$$\dot{n}_{H_2} = -4.32 \cdot 10^{-5} P_{el}^2 + 2.39 \cdot 10^{-2} P_{el} - 1.43 \cdot 10^{-3} \quad (62.3)$$

62.3.2 PV Module

The photovoltaic module considered in this study is a multi-crystalline module (Hyundai Heavy Industries HiS-M228MG) with 60 cells in series and 1.617 m² of area. The remaining module parameters are shown in Table 62.2.

The DC output power of the PV system is estimated, using SAM (System Advisor Model) software. The CEC (California Energy Commission) performance model with module database has been selected; it is an implementation of the one diode equivalent circuit model and an extension of the five-parameter model described by

Table 62.2 PV module characteristics at reference conditions

Parameter	Value	Parameter	Value
I_{ref} (W/m²)	1000	V_{OC} (V)	36.8
T_{ref} (°C)	25	I_{SC} (A)	8.1
P_{mp} (W)	228.76	α (A/°C)	0.004
V_{mp} (V)	30.1	β (V/°C)	−0.118
I_{mp} (A)	7.6	T_{NOCT} (°C)	46.2

[12]:

$$I = I_L - I_0\left[\exp\left(\frac{V + IR_s}{a} - 1\right)\right] - \frac{V + IR_s}{R_{sh}} \quad (62.4)$$

Thus, the PV module output power is given by $P = IV$. Both the site temperature and the amount of solar irradiation received by PV modules influence this power.

62.4 Results and Discussion

Figure 62.3 represents the evolution over the year of the monthly PV produced power, for the four studied cases. First, it is shown that solar trackers allow the production of much more power than fixed PV systems, within a range from 16.9 to about 50%. Then, while a tilt angle equal to 30.62° leads to lightly more annual power, a tilt angle equal to the site latitude leads to more power on vulnerable months, from October to March. Finally, PV system with the 2-axis solar tracker allow the production of about 5.3% of annual power more than that with the 1-axis solar tracker.

Figures 62.4 and 62.5 represent the monthly hydrogen production rate, for the four studied cases and for the two electrolysis technologies. These curves follow perfectly those of monthly PV power on Fig. 62.3. It is shown that the PEM technology produces more hydrogen than alkaline technology; this production ranges from 5.06 to 8.4 kg for the first one and from 3.97 to 6.95 kg for the second. In fact, if we consider the monthly mean daily rate of hydrogen, the lowest results are obtained, in the four cases, for the month of December while the highest results are obtained

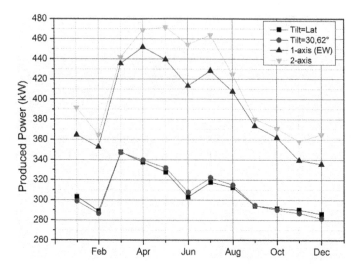

Fig. 62.3 Monthly PV produced power

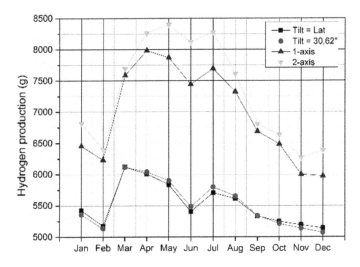

Fig. 62.4 Monthly hydrogen production by PEM technology

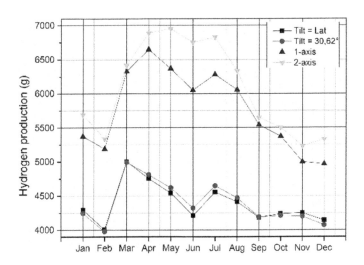

Fig. 62.5 Monthly hydrogen production by alkaline technology

for the month of April except for alkaline technology with fixed PV modules (tilt = latitude and tilt = 30.62°) the best month is March.

Figure 62.6 summarizes the annual hydrogen production rate, in kilograms, for the eight cases. The importance of solar tracking over fixed PV modules and of PEM electrolysis over alkaline technology are very clear. Indeed, while the power produced by PV modules inclined at 30.62° is 1.03% greater than that produced by PV modules inclined at site latitude only 0.36% and 0.11% amount of hydrogen is gained for, respectively, alkaline and PEM technologies. As for PV modules with solar tracking,

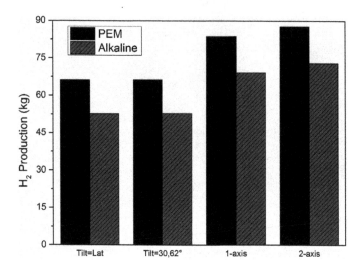

Fig. 62.6 Evolution of the annual hydrogen production

while 5.34% of more power is produced by the 2-axis system, 5.38% and 4.7% amount of hydrogen is gained for, respectively, alkaline and PEM technologies.

62.5 Conclusion

The sustained interest for hydrogen results from its diverse applications, the variety of processes and sources of its production as well as its importance in the future energy system, with low conventional resources and 100% renewable energy trend. Currently, the primary market for hydrogen is industrial applications such as petroleum refining and fertilizer production. However, it can be used in the transport field either in electric vehicles based on fuel cells or in internal combustion engines. In addition, an important integration rate of renewable energies in the electrical network could be reached through the storage of the excess of these intermittent energies as hydrogen.

Nowadays, hydrogen is produced mainly using the exhaustible and polluting hydrocarbons. While the combination of two abundant sources, water and the sun, allows the production of sustainable hydrogen. Indeed, alkaline or PEM water electrolysis technologies combined to PV panels produce a green and high purity hydrogen. Thus, the theoretical study of such a system has shown the advantage of PEM electrolysis over alkaline electrolysis, mature and available on a large scale, in term of hydrogen production. In fact, it has been found that PEM water electrolyzer produces between 20.25 and 25.86% more hydrogen than alkaline electrolyzer, this is due to higher current densities in PEM electrolyzer. Then, PV-Hydrogen system with sun trackers allow between 26.4 and 38.6% more hydrogen production than fixed PV systems.

References

1. IRENA, *Renewable Capacity Statistics 2019* (International Renewable Energy Agency (IRENA), Abu Dhabi, 2019). https://www.irena.org/publications/2019/Mar/Renewable-Capacity-Statistics-2019
2. R. Boudries, R. Dizene, Potentialities of hydrogen production in Algeria. Int. J. Hydrogen Energy **33**(17), 4476–4487 (2008)
3. R. Boudries, R. Dizene, Prospects of solar hydrogen production in the Adrar region. Renew. Energy **36**(11), 2872–2877 (2011)
4. H. Tebibel, S. Labed, Design and sizing of stand-alone photovoltaic hydrogen system for HCNG production. Int. J. Hydrogen Energy **39**(8), 3625–3636 (2014)
5. H. Tebibel, R. Medjebour, Comparative performance analysis of a grid connected PV system for hydrogen production using PEM water, methanol and hybrid sulfur electrolysis. Int. J. Hydrogen Energy **43**(6), 3482–3498 (2018)
6. D. Ghribi, A. Khelifa, S. Diaf, M. Belhamel, Study of hydrogen production system by using PV solar energy and PEM electrolyser in Algeria. Int. J. Hydrogen Energy **38**(20), 8480–8490 (2013)
7. B. Laoun, A. Khellaf, M.W. Naceur, A.M. Kannan, Modeling of solar photovoltaic-polymer electrolyte membrane electrolyzer direct coupling for hydrogen generation. Int. J. Hydrogen Energy **41**(24), 10120–10135 (2016)
8. A. Saadi, M. Becherif, H.S. Ramadan, Hydrogen production horizon using solar energy in Biskra, Algeria. Int. J. Hydrogen Energy **41**(47), 21899–21912 (2016)
9. A. Mraoui, B. Benyoucef, L. Hassaine, Experiment and simulation of electrolytic hydrogen production: case study of photovoltaic-electrolyzer direct connection. Int. J. Hydrogen Energy **43**(6), 3441–3450 (2018)
10. J.A. Duffie, W.A. Beckman, *Solar Engineering of Thermal Processes*, 4th edn. (Wiley, New Jersey, 2013)
11. Ø. Ulleberg, Modeling of advanced alkaline electrolyzers: a system simulation approach. Int. J. Hydrogen Energy **28**(1), 21–33 (2003)
12. W. De Soto, S.A. Klein, W.A. Beckman, Improvement and validation of a model for photovoltaic array performance. Sol. Energy **80**(1), 78–88 (2006)

Chapter 63
Wind Power System for Large-Scale Energy and Hydrogen Production in Hassi R'mel

Fares Meziane, F. Chellali, K. Mohammedi, Nourdine Kabouche, and Ilyes Nouicer

Abstract Wind energy is one of the most competitive renewable energy sources for providing energy to isolated area where the extension of the national grid is prohibitively expensive, or fossil fuel supply is difficult. The present work aims to evaluate the amount of energy and hydrogen produced from wind energy source over the site of Hassi R'mel, using hourly wind data (speed and direction) stretching from 2003 to 2017. As well as, estimating the cost of energy and hydrogen for large wind turbines. To this end, a model of wind-hydrogen system has been developed under Simulink environment, the main components are, a wind turbine to produce electricity from wind energy, an electrolyzer unit to produce hydrogen from the produced electricity by electrolysis technique, and a tank for hydrogen storage. The study pointed out, that the site of Hassi R'mel has an important wind potential with mean speed exceeding 6 m/s, and more profitable and adapted for wind energy conversion systems installation. The low cost of energy and hydrogen of 0.053 $/kWh, and 25.27 $/kg H_2 respectively, is obtained by the Nordex N100 wind turbine at 100 m.

Keywords Wind energy · Hydrogen · Cost · Hassi R'mel

63.1 Introduction

Wind energy is one of the most available renewable sources in Algeria [1], particularly the southern regions with significant wind speed exceeding 6 m/s at 10 m [2].

This important wind potential allows solving global energy problems and providing energy to its remote locations. Due to the intermittent nature of this energy

Table 63.1 Geographical coordinates of Hassi R'mel

Site	Longitude (°)	Latitude (°)	Altitude (m)	Period
Hassi R'mel	3.28 °E	32.93 °N	764	2003–2017

and mismatch between demand and offer, storing the excess of energy in the form of hydrogen proves to be a long-term sustainable solution for these remote areas.

However, wind-to-hydrogen based systems constitute an interesting option for energy and hydrogen production using wind energy [3–6]; it consists of supplying the excess electrical power produced by wind generator to an electrolyzer to produce clean and sustainable hydrogen through electrolysis. The cost is a very important task for energy and hydrogen production and their application for a given region [7, 8]; it allows not only to locate the most promising sites, but also to optimize the choice of wind energy systems. In Algeria, few studies dealing with the energy, hydrogen potentialities and feasibility were carried out [9–11]. Some works concerned wind hydrogen system and power supplying remote areas [12, 13].

The present work aims to evaluate the wind potential and estimate the cost of energy and hydrogen produced based on Alkaline water electrolysis technology over the site of Hassi R'mel, using data stretching from 2003 to 2017 provided by the National Office of Meteorology (NOM), and prospecting the unit cost of energy and hydrogen using large-scale wind turbines.

63.2 Site Description and Wind Data

The site of Hassi R'mel is situated in the north of Sahara of Algeria, between saharian and continental climates. Characterized by heavy and storms, it is poorly developed, and suffer from very poor communication means, and very limited connection to the national electricity grid. The hourly wind data stretching from 2003 to 2017 provided by the National Office of Meteorology were used. Table 63.1 shows the geographical coordinates of the site of Hassi R'mel and wind data.

63.3 System Description

The wind-hydrogen based system considered in this study is developed under Simulink environment. The system is mainly composed of a wind turbine (WT), an Alkaline electrolyzer (EL), and storage system (SS) (Fig. 63.1). It has been assumed that the power output of the wind turbine is used to provide the electrolyzer for hydrogen production. The technical data of the two selected wind turbines Nordex N54, and Nordex N100 with 1000 kW, and 2500 kW rated power respectively are presented in Table 63.2.

Fig. 63.1 Schematic diagram

Table 63.2 Wind turbines characteristics

Heading level	Nordex N54	Nordex N100
Hub height (m)	60	100
Rated power (kW)	1000	2500
Cut-in speed (m/s)	3.5	3
Rated speed (m/s)	14	13
Cut-out speed (m/s)	25	20
Rotor diameter (m)	54	100

Fig. 63.2 Electrolyzer cell voltage-current density curve

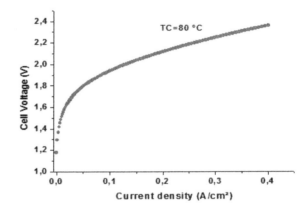

The cut-in speed for both wind turbines are 3 m/s, and 3.5 m/s, and the cut-out speeds are 25 m/s, and 20 m/s respectively.

The polarization characteristics of the electrolyzer depend mainly on voltage, current and temperature, the cell current-voltage characteristics for the temperature of 80 °C is plotted in Fig. 63.2.

63.4 Modeling

The Weibull distribution used to describe the wind speed distribution is defined by the following equation [13]:

$$f(V) = \left(\frac{k}{V}\right)\left(\frac{V}{C}\right)^{k-1} - \exp\left[-\left(\frac{V}{C}\right)^k\right] \tag{63.1}$$

where C is the scale parameter (m/s), k is the shape parameter (dimension-less), and V the mean wind speed (m/s). If the mean wind speed, and variance are known, the Weibull parameters c and k can be done as follows:

$$k = \left(\frac{\sigma}{\overline{V}}\right)^{-1.086} \quad (63.2)$$

where σ is the standard deviation.

$$C = \frac{\overline{V}}{\Gamma\left(1+\frac{1}{k}\right)} \text{ (m/s)} \quad (63.3)$$

where Γ is the gamma function.

The capacity factor based on Weibull parameters and wind turbine characteristics can be determined as follow [13]:

$$CF = \frac{\exp\left[-\left(\frac{V_{cin}}{C}\right)^k\right] - \exp\left[-\left(\frac{V_r}{C}\right)^k\right]}{\left(\frac{V_r}{C}\right)^k - \left(\frac{V_{cin}}{C}\right)^k} - \exp\left[-\left(\frac{V_{out}}{C}\right)^k\right] \quad (63.4)$$

V_{cin}, V_r, V_{out} are cut-in, rated, and cut-out speed (m/s).

Equation (63.5) gives the yearly energy produced by the wind turbine [13]:

$$E_{prod} = CF \times P_r \times 8760 \text{ (MWh/yr)} \quad (63.5)$$

The energy generated by the wind turbine is converted to hydrogen through electrolysis, the amount of hydrogen mass (M_{H2}) estimated is given as follows [13]:

$$M_{H2} = \frac{\eta_{el} \times E_{prod}}{LHV_{H2}} \text{ (kg H}_2\text{/yr)} \quad (63.6)$$

where η_{el} is the electrolyzer efficiency (%), LHV_{H2} is the lower heating value of hydrogen (MJ/kg).

The cost of energy (COE) depends on the yearly energy produced (E_{prod}), the operation maintenance and repair costs (C_{omr}), the initial investment cost (IC), the inflation rate (i) and interest rate (r) which were taken to be 15% and 12% respectively.

$$COE = \frac{1}{E_{prod}} \times \left(I_C + C_{omr} \times \left[\frac{1+i}{r-i}\right] \times \left[1 - \left(\frac{1+i}{1+r}\right)^t\right]\right)$$
$$- S \times \left(\frac{1+i}{1+r}\right)^t \right) \text{ (\$/kWh)} \quad (63.7)$$

The cost of hydrogen (COH) is calculated based on electrolyzer cost (C_{elec}), wind turbine cost (C_{wt}), and the amount of hydrogen produced (M_{H2}), during life time (t) as follows:

$$COH = \frac{C_{elec} + C_{wt}}{M_{H2} \times t} \quad (\$/kg) \tag{63.8}$$

63.5 Results and Discussions

Figure 63.3 shows the monthly average wind speed for the location of Hassi R'mel, it can be seen that the highest speed value is determined as 8.51 m/s in May, while the lowest average wind speed of 4.97 m/s occurred in July, with significant shift in wind speeds exceeding 3.5 m/s. The high wind speeds are recorded during the months of March, April, and May. The annual average wind speed is 6.37 m/s at 10 m.

The evolution of the annual average wind speed for hub heights ranging from 10 to 100 m is shown in Fig. 63.4. We can notice that the rate of increasing is significant between 100 and 10 m with a difference of 3.96 m/s.

Table 63.3 shows the capacity factor, the energy and hydrogen produced for both used wind turbines (Nordex N54, and Nordex N100). The analysis shows that, the energy produced by Nordex N54 at 60 m, and Nordex N100 at 100 m are 2771.5 MWh/yr, and 9128.1 MWh/yr respectively. Besides, it is found that the amounts of hydrogen produced by Nordex N54 at 60 m, and Nordex N100 at 100 m are 34.893 tH_2/yr, and 114.92 tH_2/yr respectively. In term of production, at 100 m, the Nordex N100 wind turbine produces more than three times the quantities of energy and hydrogen produced by the Nordex N54 at 60 m.

Table 63.4 depicted the cost of energy and hydrogen produced by the used wind turbines.

In term of cost, the low cost of energy and hydrogen are occurred by the N100 wind turbine at 100 m, with 0.0533 \$/kWh and 25.27 \$/kg respectively.

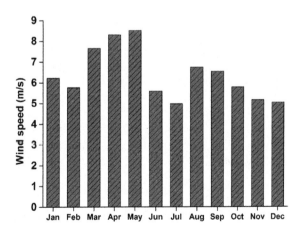

Fig. 63.3 Monthly wind speed at 10 m

Fig. 63.4 Wind speed evolution with heights

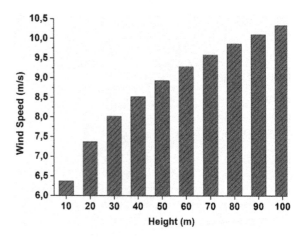

Table 63.3 Energy and hydrogen production

Heading level	Nordex N54	Nordex N100
Hub height (m)	60	100
Capacity factor CF	31.0	41.6
Energy production (MWh/yr)	2771.5	9128.1
Hydrogen production (tH_2/yr)	34.893	114.92

Table 63.4 Cost of energy and hydrogen

Heading level	Nordex N54	Nordex N100
Hub height (m)	60	100
Energy cost ($/kWh)	0.0702	0.0533
Hydrogen production ($/kg H_2)	33.21	25.27

63.6 Conclusion

This study relates to large-scale energy and hydrogen production using wind energy, it can be considered as preliminary results to estimate the amount of energy and hydrogen that can be generated from wind energy in the site of Hassi R'mel. The economic aspect in terms of unit cost of energy and hydrogen had prospected for both selected turbines. The yearly energy produced was 2771.5 MWh/yr, 9128.1 MWh/yr for Nordex N54 and Nordex N100 respectively. Besides, the amount of hydrogen produced by N54 and N100 wind turbines was 34.893 tons/yr and 114.92 tons/yr respectively. The study pointed out that the large-scale range wind turbine N100 is more suitable for energy and hydrogen production in the region of Hassi R'mel with low costs.

References

1. A.B. Stambouli, Z. Khiat, S. Flazi, Y. Kitamura, A review on the renewable energy development in Algeria: current perspective, energy scenario and sustainability issues. Renew. Sustain. Energy Rev. **16**(7), 4445–4460 (2012)
2. F. Chellali, A. Khellaf, A. Belouchrani, A. Recioui, A contribution in the actualization of wind map of Algeria. Renew. Sustain. Energy Rev. **15**(2), 993–1002 (2011)
3. A. Khalilnejad, G. Riahy, A hybrid wind-PV system performance investigation for the purpose of maximum hydrogen production and storage using advanced alkaline electrolyzer. Energy Convers. Manag. **80**, 398–406 (2014)
4. B. Olateju, A. Kumar, Hydrogen production from wind energy in Western Canada for upgrading bitumen from oil sands. Energy **36**(11), 6326–6339 (2011)
5. Ø. Ulleberg, T. Nakken, A. Ete, The wind/hydrogen demonstration system at Utsira in Norway: evaluation of system performance using operational data and updated hydrogen energy system modeling tools. Int. J. Hydrogen Energy **35**(5), 1841–1852 (2010)
6. D. Bechrakis, E. McKeogh, P. Gallangher, Simulation and operational assessment for a small autonomous wind–hydrogen energy system. Energy Convers. Manag. **47**(1), 46–59 (2006)
7. M. Gökçek, M.S. Genç, Evaluation of electricity generation and energy cost of wind energy conversion systems (WECSs) in Central Turkey. Appl. Energy **86**(12), 2731–2739 (2009)
8. G. Genc, M. Celik, M. Serdar Genc, Cost analysis of wind-electrolyzer-fuel cell system for energy demand in Pınarbası–Kayseri. Int. J. Hydrogen Energy **37**(17), 12158–12166 (2012)
9. R. Boudries, R. Dizene, Potentialities of hydrogen production in Algeria. Int. J. Hydrogen Energy **33**(17), 4476–4487 (2008)
10. S. Rahmouni, B. Negrou, N. Settou, J. Dominguez, A. Gouareh, Prospects of hydrogen production potential from renewable resources in Algeria. Int. J. Hydrogen Energy **42**(2), 1383–1395 (2017)
11. L. Aiche-Hamane, M. Belhamel, B. Benyoucef, M. Hamane, Feasibility study of hydrogen production from wind power in the region of Ghardaia. Int. J. Hydrogen Energy **34**(11), 4947–4952 (2009)
12. F. Meziane, A. Khellaf, F. Chellali, Study and dimensioning of a wind-electrolyzer-fuel cell system for the power supply of an isolated site, in *Revue des Energies Renouvelables SIENR'12 Ghardaia* (2012), pp. 381–391
13. M. Douak, N. Settou, Estimation of hydrogen production using wind energy in Algeria. Energy Procedia **74**, 981–990 (2015)

Chapter 64
Analysis and Design of PEM Fuel Cell/Photovoltaic System to Supply a Traffic Light Signals in Ouargla City Based on Field Experience

Abdelmoumen Gougui, Ahmed Djafour, Taha Hamidatou, S. Eddine Khennour, and Mohammed Bilal Danoune

Abstract This paper seeks to address an application using a small PEM fuel cell system (Heliocentris FC50), this one uses a solar hydrogen fuel in order to power a known electric load, hence, a series of experiments have been conducted in LAGE Laboratory at Ouargla University, that the effect of operating temperature on (I-V) fuel cell characteristics have been investigated, another tests were conducted to determine the relation between the hydrogen input flow and the available output power from the fuel cell, hence the photovoltaic-hydrogen system components are properly sized in order to fulfil the daily cycle energy needs, a computer program also has been developed to size system components in order to match the load of the site in the most effective way. Consequently, one scenario configuration is proposed, which covered the right balance and no power supply interruption to the proposed load.

Keywords Solar energy · Photovoltaic · Fuel cell · Solar hydrogen · Design · Scenario

64.1 Introduction

Environmental eco-system concerns dominate the world today, especially with global problems like ozone layer expansion. Therefore, experts are trying to solve these issues in different ways, among their contributions is directed toward curtailing the use of fossil fuels, and replacing them with alternative sources [1], the photovoltaic is being considered as a promising technology, on the one hand, solar energy is the

---

A. Gougui (✉) · A. Djafour · T. Hamidatou · S. E. Khennour · M. B. Danoune
Laboratoire LAGE, Faculté des Sciences Appliquées, University of Ouargla, Ouargla 30000, Algeria
e-mail: abdelmoumengougui@gmail.com; gougui.abdelmoumen@univ-ouargla.dz

M. B. Danoune
e-mail: danoune.mohammed.bilal@gmail.com; danoune.m_bilal@univ-ouargla.dz

© The Editor(s) (if applicable) and The Author(s), under exclusive license to Springer Nature Singapore Pte Ltd. 2021
A. Khellaf (ed.), *Advances in Renewable Hydrogen and Other Sustainable Energy Carriers*, Springer Proceedings in Energy,
https://doi.org/10.1007/978-981-15-6595-3_64

most non-conventional energy source gaining interest throughout the world cities, however, one of the PV system's disadvantage is the unexpected fluctuation of yield electric power owing to the unstable solar radiation and surface temperature [2] for this reason, it is worthwhile to asylum the hybridization of multiple sources in order to improve system efficiency and the accuracy of the supply [3].

The traffic road signals are a promising method for facilitating road comfort and human being security, unfortunately, in most cases, they are supplied by the utility grid, this one is unpredictable repeated power failure, which arises due to lack of fossil fuel at generation stations [4], in spite of, the traffic light system is considered to be one of the most critical applications that can solve the congestion problems of the streets. In this regard, the photovoltaic system is one of the optimum solutions for these systems, but there are still two principal hurdles to the use of PV systems: the high installation cost and the low energy conversion efficiency. Omran et al. [4], the objective of this paper is to promote the use of hybrid system, that the key components include a PV modules as a primary source, while the fuel cell, the electrolyzer, and hydrogen tanks are used as a backup and a long-term storage system, that most of these above-mentioned components are probably sized. Meanwhile, an experimental studies have been carried out to verify the electrical power output of PEM fuel cell according to V–I curve, in line with this another experimental results such as the (I, H_2) and (P, I) characteristics when the fuel cell supplies the traffic light signal panel.

64.2 System Description

The test bench is mainly constituted by a PEM fuel cell (FC50), which contains a 10 single cells lumped together, the produced hydrogen stored in a metal hydride canister with 225Nl at 10 bar type Ovonic^{TM}, a low-pressure regulator valve placed between the fuel cell and the storage bottle in order to maintain replenishes hydrogen adapted for fuel cell stack.

64.2.1 Conducted Tests

Two-fold tests have been conducted, firstly, the fuel cell was extensively tested at various working temperatures, that the current-voltage characteristics at each temperature were determined, the second fold of experiments is consisted of powering one traffic light signals module by the 50 W PEM fuel cell via a DC-DC converter, hence one lighting cycle has been identified, furthermore, the rated hydrogen flow adequate to feed the traffic light equipment have been determinate through (Figs. 64.1 and 64.2) which obtain during supplying a variable DC load.

Fig. 64.1 (Q-I) Graph of a fuel cell

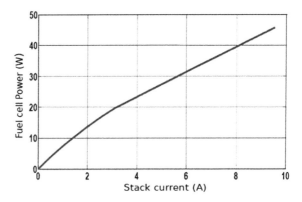

Fig. 64.2 (P-I) Characteristics of the fuel cell

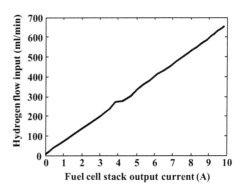

64.2.2 The Considered Powered System

In this section, we consider a crossroad of two two-way streets, there are four lanes on each corner of this intersection there is a traffic light module, for each module, there are three subsequent phases, green, amber, and red colour.

The average cycle load for one traffic light module was obtained from the study conducted at LAGE Laboratory, as shown in Fig. 64.3. The generation and load are assumed to keep constant in each cycle interval, further the switching time sequence and consumption power of each traffic colour illustrated below:

$T_{Amber} = 3$ s with $P = 2.3$ W. $T_{Red} = 18$ s with $P = 2.5$ W. $T_{Red-Amber} = 3$ s with $P = 6.5$ W. $T_{Green} = 18$ s with $P = 1.2$ W. Thus, the total rated cycle time about 42 s and the average power cycle is 2.22 W.

Fig. 64.3 Load profile during two cycles period

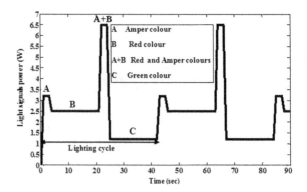

64.3 Experimental Results and Discussion

One of the experiment goals is to assess the influence of the fuel cell stack temperature on the polarization behavior (Fig. 64.4), from the illustrated curves we can see that, the trends show a similar behavior variations of (I-V) characteristics, along with the sharp break of the polarization curves are similar form in the whole temperature range investigated (39–26 °C). In particular, the polarization curve at 39 °C shows a high open-circuit voltage with 8.8 V, this result was expected, that the activation of the reactions is facilitated at higher temperatures, in turn, the sudden voltage drop is mainly due to a low ionic conductivity of the membrane [5].

The second experimental goal is to determinate the electrochemical relationship between the current consumption from the load and the input hydrogen flow to the fuel cell through Figs. 64.1 and 64.2, that the FC stack consumes hydrogen according to the power demand with linear relation as presented in Fig. 64.1 like mentioned by [6].

Fig. 64.4 Experimental polarization curves point for PEMFC stack at different temperatures

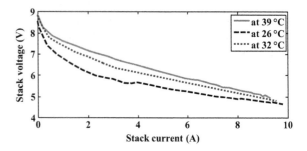

64.3.1 Sizing of Experimental System

An individual traffic light signal module is tested in real-life conditions at LAGE Laboratory, which the load cycle power based on field measurement, that the load average power is 2.22 W, which represents a daily energy consumption of 53.28 Wh, that high frequencies cycles above 2057 cycles/day.

64.4 Process Scenario Configuration of the Studied System

During the morning time, the solar panels feed-in priority the required load and the rest of the energy would be adapted to charge the batteries, when these batteries are not be fully charged hence the $SOC_b < SOC_{Max}$. In the other situation, when energy produced by PV source is smaller than the load demand, the batteries would support the principal power source by supplying the power lack to the load, the link between the PV panels and above-cited elements are ensured with a bi-directional converter type PR2020 based on PWM strategies, in this analysis, the electrolyser was turned on to begin producing hydrogen, which is delivered to the hydrogen storage tanks thereby, the storage tanks magnitude are calculated, the calculation was done for a longer period of hydrogen production, hence the electrolyser operating in the extra operation hours after sunset, obviously, in order to size the installation appropriately, the amount of hydrogen produced by the electrolyzer must be equal to or greater than the daily amount of hydrogen used by the fuel cell, whilst, in the night time, the FC50 fuel cell will operates automatically to supply the traffic light devices with necessary power, that it consumes the hydrogen produced by the PEM electrolyser, that the evaluation of hydrogen amount according to experimental results which are represented in Figs. 64.1 and 64.2. A self-designed DC-DC converter was interconnected between the PEM fuel cell system and the load, however, in the typical autonomy days, the batteries are supplied the traffic light signals and the PEM electrolyser, in turn in the night time, the fuel cell will assume a reliable electricity supply to feed the proposed load.

64.4.1 Sizing Process Calculation

The total amount of electrical energy consumption from the PEM electrolyser and the four traffic modules during a day, which is calculated by the following equation:

$$E_C = T(ele) * C(ele) + T(ts) * C(ts) \qquad (64.1)$$

T(ele) the operation hours of the electrolyser with 8.9 h.
C(ele) the PEM electrolyser rated power consumption.

C(ts)   the power consumption of traffic light signals, which is 8.7 W for each panel.
T(ts)   the operation hours of the light signals in 24 h.

The generator required power (P_G) in kW can express followed [8]:

$$p_G = \frac{Ec}{H_d * K} \quad (64.2)$$

With,

E_c is the energy consumed per day (kWh/day).
H_d is the average daily sunshine hours in our case equal to 8.9 h/day [7].
K   Correction coefficient for systems with batteries. The value used in our calculations is equal to 0.68 [8].

The required number of modules N_T is calculated as follow:

$$N_T = \frac{P_G}{Pu_{module}} \quad (64.3)$$

With, Pu_{module}: the unit power of the module, it can be expressed as:

$$Pu_{module} = \frac{Es * S * \eta}{D_{dAV}} \quad (64.4)$$

E_s the annual average value of solar energy, in our situation E_s = 5880 Wh/m² day.
D_{dAV} average day duration (hours). In our case, D_{dAV} = 8.9 h/day [7].
S   module surface (m²) and η module efficiency, we choose photovoltaic module A120 (mono-crystalline), the technical specification of this module mentioned in [9].

The storage capacity in C_{20} that is necessary is calculated by [10]:

$$C_{20} = \frac{E_{c/day} * D_{Aut}}{R_T * P_d * V} \quad (64.5)$$

With,

R_T temperature reduction coefficient (%).
P_d Maximum authorized depth of discharge (%).
V   Working voltage (V).

We choose Banner C20-963.51 battery of 12 V, 180 Ah in C_{20} with R_T = 0.9 and P_d = 0.8 [11].

In this case during the night, the fuel cell net power of 40 W and considering auxiliary systems consuming 17% of this power, which meet the need energy for

Table 64.1 System sizing results

Parameters	Resultants
The consumption EC/day [Wh/d]	1284
Generator power PG [W]	94
Number of module	1
Module surface [m^2]	1.46
Storage battery capacity in C20	342.6
Number of battery branches	2
Number of battery in series	1
Hydrogen generator number	1
Fuel cell 50 PEM number	1
Hydrogen tank [l]	64.52

the operation of the system. In view of the experimental data, the fuel cell generates a current of 1.8 A with 8.9 W to feed the fourth traffic signal panels according to Fig. 64.3 in this case, the fuel cell consumes 110 ml/min, thus 6.6 l/h according to the trends Fig. 64.3, from these values, the choice of the hydrogen tank, it was considered four bottles of 20Nl type MH-20 tank [12], the second load device is the electrolyser chosen, this one type 1000 PEM electrolyser, its lifetime assumed to be 20 years according to the manufacturing data [13].

64.4.2 System Sizing Results

The sizing results are calculated from the mathematical program developed in MATLAB environments are shown in Table 64.1.

64.5 Conclusion

In this work, we have presented a design of photovoltaic-hydrogen-powered four traffic light signal, this design followed by a number of experiments conducted on a PEM fuel cell FC50, a configuration mode is recommended to supply of whole proposed system, wherever satisfy the required power demand at all time throughout the day, in such a way the proposed sizing system based on both MATLAB simulation, and also laboratory field tests, the PV system represents the smallest surface of 1.46 m^2 for only one street intersection. As experimental results, the polarization curves of the fuel cell at various operating temperatures highlighted that the PEM fuel cell performance was improved with increasing operating temperature from 26 to 39 °C, another finding from these results is that the FC stack consumes hydrogen according to the power demand with a proportional relationship, in view of close to reality, the proposed configuration requires a switching algorithm control, which is the principal advantage to manage energy in hybrid power systems, for this a future work will focus on this sector.

References

1. H. Moghbelli, K. Ellithy, Z. Eslami, R. Vartanian, Investigation of solar energy applications with design and implementation of photovoltaic traffic light signal system for Qatar, in *Proceedings of ICREPQ. 9* (2009)
2. T.F. El-Shatter, M.N. Eskandar, M.T. El-Hagry, Hybrid PV/fuel cell system design and simulation. Renew. Energy **27**(3), 479–485 (2002)
3. O. Nadjemi, T. Nacer, A. Hamidat, H. Salhi, Optimal hybrid PV/wind energy system sizing: application of cuckoo search algorithm for Algerian dairy farms. Renew. Sustain. Energy Rev. **70**, 1352–1365 (2017)
4. A.E.-F.A. Omran, F.H. Fahmy, A.E.-S.A. Nafeh, H.K. Yousef, Sizing, modeling and control of photovoltaic traffic light system. IOSR J. Electr. Electron. Eng. **11**(6), 25–36 (2016)
5. M.G. Santarelli, M.F. Torchio, P. Cochis, Parameters estimation of a PEM fuel cell polarization curve and analysis of their behavior with temperature. J. Power Sources **159**(2), 824–835 (2006)
6. M.Y. El-Sharkh, A. Rahman, M.S. Alam, P.C. Byrne, A.A. Sakla, T. Thomas, A dynamic model for a stand-alone PEM fuel cell power plant for residential applications. J. Power Sources **138**(1–2), 199–204 (2004)
7. *National Office of Meteorological (NOM) of Ouargla region during (2004–2017)*
8. H. Abdi, N. Ait Messaoudene, M. Omari, Y. Bekhta, *Etude et dimensionnement d'une installation solaire. De production d'hydrogene* (WIH$_2$ Ghardaïa, Algérie, 2007)
9. https://www.solarboutik.com. Visited in May 2018
10. A. Labouret, M. Villoz, *Solar Photovoltaic Energy*, 4th edn. (Dunod, Paris, 2009)
11. B.O. Bilal, P.A. Ndiaye, C.M.F. Kebe, V. Sambou, M. Ndongo, Methodology to size an optimal standalone hybrid solar-wind-battery system using genetic algorithm. Int. J. Phys. Sci. **7**(18), 2647–2655 (2012)
12. https://www.pragma-industries.com/products/hydrogen-storage/#mh150-metal-hydride-tank. Visited in Oct 2018
13. http://www.hydrogen-generator-for-car.cn/product%20new%20prices.html. Visited in November 2018

Chapter 65
Assessment of Hydrogen Production from Geothermal Thermoelectric Generator

M. M. Hadjiat, S. Ouali, K. Salhi, A. Ait Ouali, and E. H. Kuzgunkaya

Abstract In this paper, a novel system for hydrogen production is presented. Usually, an Organic Rankine Cycle (ORC) machine driven by geothermal energy is used to produce electric energy and then hydrogen. This method is efficient for high geothermal enthalpy only. In our system, we use thermoelectric generators TEG instead of ORC machine. For low geothermal temperature resources, (temperature below 100 °C), the thermoelectric generator TEG can provide electricity with a low heat source temperature, it has no moving parts, and is compact, quiet, highly reliable and environmentally friendly as it has no greenhouse gas emissions. Therefore, the system can operate over an extended period with minimal maintenance. This new model of electric generator allows us to configure the voltage and the current according to the type of electrolyze by using some TEG connected in parallel and in series. For this study, we used the thermoelectric module TEC1-12706. The simulation of the system show that the production of hydrogen depend of the temperature difference between hot side and cold side of the thermoelectric module. As example, the result show that for $\Delta T = 40$ °C, produced hydrogen is 0.02 m^3/h.

Keywords Thermoelectric generator · Geothermal · Seebeck effect · Energy efficiency

65.1 Introduction

Nowadays, global warming and the emission of greenhouse gases caused by the excessive consumption of fossil fuels represent a danger to the planet and the future of humanity. It is therefore urgent to find alternative energies. Several solutions exist and each solution has its advantages and disadvantages. Hydrogen represents a

---

M. M. Hadjiat (✉) · S. Ouali · K. Salhi · A. Ait Ouali
Centre de Développement des Energies Renouvelables, CDER, Bouzaréah 16340, Algiers, Algeria
e-mail: m.hadjiat@cder.dz

E. H. Kuzgunkaya
Geothermal Energy Research and Application Center, Izmir Institute of Technology, Urla, İzmir, Turkey

promising energetic vector if it is produced by renewable energy. In fact, the production of hydrogen by electrolysis of water using geothermal energy is a non-polluting and environmentally friendly process, this method has many advantages. Indeed, geothermal energy is provided in the form of heat constantly and durably. This energy is converted by ORC machine into electricity and then hydrogen. The hydrogen can be stored and transported easily, and this is a huge advantage of this process. The production of hydrogen by geothermal energy has been widely discussed among researcher in the word. Ouali et al. [1] propose to produce hydrogen from hydrogen sulfide in geothermal area. It is a cost-effective process because it need low energy than production hydrogen by electrolysis. Balta et al. [2] analyzed high-temperature electrolysis process where geothermal water is used as the heat source. The same group [3] investigated various options for geothermal-based hydrogen production systems and their technical, operational and efficiency aspects. Kanoglu et al. [4] developed four models for hydrogen production by geothermal energy and analyzed these models thermodynamically. Yilmaz et al. [5] considered seven models for hydrogen production and liquefaction by geothermal energy, and their thermodynamic and simple economic analyses were performed. It is estimated that the cost of hydrogen production and liquefaction ranges between 0.979 and 2.615 $/kg H_2 depending on the model. The results show that the cost of hydrogen production and liquefaction decreases as the geothermal water temperature increases. Therefore, this system is efficient for high enthalpy geothermic. In Algeria, the geothermal energy can be considered as low enthalpy, the temperature of the most of hot spring is situated in the range of 30 and 70 °C. Therefore, all system based on ORC machine is not cost effective. Holdmann [6] reported ORC in Chena Hot Spring, Alaska, geothermal heat sources with temperatures 73 °C, can produce a capacity of 210 kW and an efficiency of 8.2%, however, water cooling fluid temperature is 4.4 °C during all the year and this, water cooling temperature at 4 °C in Algeria is not available. In this paper, we propose a new model based on Thermoelectric Generator TEG to produce hydrogen from geothermal energy at low enthalpy. Thermoelectric devices possess various advantages compared to other power-generation systems. TEGs are branded attractive power-generation systems, because they are silent solid-state devices with no moving parts, environmental friendly, scalable from small to giant heat sources, and highly reliable. They also have extended lifetime and ability to utilize low-grade thermal energy to generate electrical energy [7].

65.2 Description of the System

In this study, a thermoelectric module is considered as power generator for hydrogen production by electrolysis. Figure 65.1 show the schema of the system. The geothermal water is used as the heat source of the TEG hot side, the cold side is cooled by cold water. Another part of geothermal water go to heat exchanger to heat electrolysis water. The generated electric power is used to produce hydrogen with PEM water electrolysis. The used thermoelectric module is TEC1-12706.

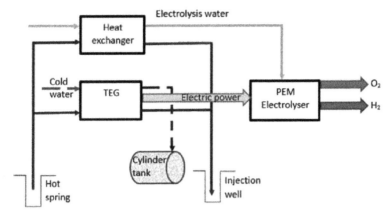

Fig. 65.1 Schema of the hydrogen production system

65.3 Description of Thermoelectric Modules

Typical TE modules (also named Seebeck cells) are composed by a set of semiconductor components formed from two different materials. As shown in Fig. 65.2, these components are connected thermally in parallel and electrically in series. Two ceramic plates are stuck on each side for electrical insulation. The voltage is created in the TE module when a difference of temperature appear between cold side and hot side of the TE. The thermoelectric modules used in this work is TEC1-12706. Table 65.1 shows the specifications of the thermoelectric module used in this study.

The behavior of a thermoelectric couple is determined by three essential parameters: K the thermal resistance, R the electrical resistance and α the Seebeck coefficient.

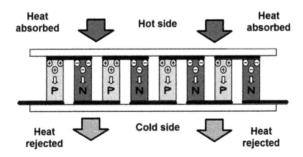

Fig. 65.2 Cross section of TE module

Table 65.1 Specifications of TEC1-12706

Type	N	I_{max} (A)	V_{max} (V)	Q_{max} (W)	ΔT (K)
TEC1-12706	127	6	15.2	56.5	68

Knowing that in a module there is N pair and assuming that all couples are the same, the parameters of a Peltier module are [8]:

$$\alpha_m = \alpha N \tag{65.1}$$

$$R_m = RN \tag{65.2}$$

$$K_m = KN \tag{65.3}$$

However, these coefficients can be calculated using parameters V_{max}, I_{max} and ΔT_{max}, which are given by the manufacturer in Table 65.1.

$$\alpha_m = \frac{V_{max}}{T_h} \tag{65.4}$$

$$R_m = \frac{V_{max}(T_h - \Delta T_{max})}{I_{max} T_h} \tag{65.5}$$

$$\frac{1}{K_m} = \frac{\Delta T_{max}}{I_{max} V_{max}} \frac{2T_h}{(T_h - \Delta T_{max})} \tag{65.6}$$

TEG is characterized by numerous performance expressions, including heat absorbed on the hot side, heat rejected on the cold side, power output, voltage induced and current flowing in electrical circuit with load resistor. Expressions of heat flow through the hot and cold junctions for N semiconductor thermocouples can therefore be expressed as follows:

$$Q_c = \alpha_m I T_c + \frac{1}{2} R_m I^2 + K_m (T_h - T_c) \tag{65.7}$$

$$Q_h = \alpha_m I T_h - \frac{1}{2} R_m I^2 + K_m (T_h - T_c) \tag{65.8}$$

For our research, we evaluate the current I (A) and voltage V (V) of the used TE module experimentally, to supply heat energy to the hot side of TE module, electric heater of 150 W is used. An aluminum heat sink fins was mounted on the cold side of the modules. Temperature measurements were made with type K thermocouple. The current and tension were recorded by using an EXTECH EX542 data logger. Whereas the temperature were recoded with an EXTECH SDL200 data logger. The power generated can be expressed as follows:

$$P = VI \tag{65.9}$$

65.4 Result and Discussion

Figures 65.3 and 65.4 show the variation of the voltage and the current characteristics of the used TE module respectively as a function of the temperature difference (ΔT). Thus, we can verify that the voltage and current increase when the difference between the hot and cold side of the TE module increase [9]. To produce hydrogen by electrolysis, its requires a voltage greater than or equal to 1.8 V and a current greater than 6 A, this is possible if we use several modules. If one TE module at ΔT generate a tension V and current I, then for a system of N modules connected in parallel, the tension remain equal to V and current is equal to N × I. and for N modules connected in series, the tension is equal to N × V and the current remain equal to I.

Figure 65.5 show 2N module connected in series and parallel. As shown in Figs. 65.3 and 65.4 one module TEC1-12706 at ΔT equal to 30 °C can produce

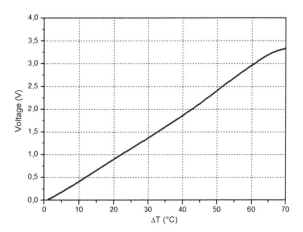

Fig. 65.3 Voltage of TE module

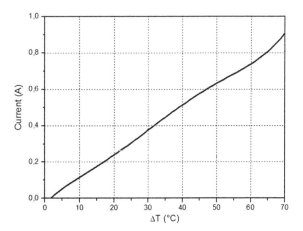

Fig. 65.4 Current of TE module

Fig. 65.5 Schema of 2 × N modules connected in series and parallel

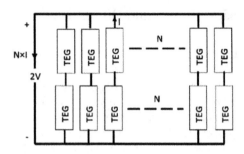

tension V = 1.4 V and current I = 0.4 A. Then, according to Fig. 65.5, for a system of 60 (N = 30) modules the tension produced is 2.8 V and current is 12 A as shown in Fig. 65.6a, b.

Yilmaz and Kanoglu [10] reported that under realistic operating conditions, 3810 kW power could be produced in a binary geothermal power plant. The produced power used for the electrolysis process can produced hydrogen at a rate of 0.0340 kg/s. based on this result and according to the power generated by our system as shown in Fig. 65.6c, we had determine the produced hydrogen as function of ΔT and presented in Fig. 65.6d.

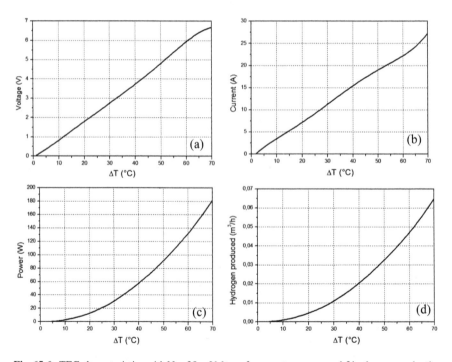

Fig. 65.6 TEG characteristics with N = 30. **a** Voltage, **b** current, **c** power and **d** hydrogen production

65.5 Conclusion

A hydrogen production system has been developed for the production of hydrogen by electrolysis of water powered by geothermal energy. For a simple and low cost application, commercially available modules have been tested. The performance of energy conversion of Bismuth Telluride TE modules has been evaluated experimentally. The results show that when the temperature difference between hot side and cold side of the TE module increase, the production of hydrogen increase also. The system provides a viable option for renewable energy powered hydrogen production.

References

1. S. Ouali, S. Chader, M. Belhamel, M. Benziada, The exploitation of hydrogen sulfide for hydrogen production in geothermal areas. Int. J. Hydrogen Energy **36**, 4103–4109 (2011)
2. M.T. Balta, I. Dincer, A. Hepbasli, Thermodynamic assessment of geothermal energy use in hydrogen production. Int. J. Hydrogen Energy **34**(7), 2925–2939 (2009)
3. M.T. Balta, I. Dincer, A. Hepbasli, Potential methods for geothermal-based hydrogen production. Int. J. Hydrogen Energy **35**(10), 4949–4961 (2010)
4. M. Kanoglu, A. Bolatturk, C. Yilmaz, Thermodynamic analysis of models used in hydrogen production by geothermal energy. Int. J. Hydrogen Energy **35**(16), 8783–8791 (2010)
5. C. Yilmaz, M. Kanoglu, A. Bolatturk, M. Gadalla, Economics of hydrogen production and liquefaction by geothermal energy. Int. J. Hydrogen Energy **37**(2), 2058–2069 (2012)
6. G. Holdmann, *400 kW geothermal power plant at Chena hot springs* [Final Report] (Alaska, Chena Power, 2007), https://geothermalcommunities.eu/assets/elearning/7.13.FinalProjectReport_ChenaPowerGeothermalPlant.pdf
7. E. Kanimba, Z. Tian, Modeling of a thermoelectric generator device, in *Thermoelectrics for Power Generation—A Look at Trends in the Technology* (IntechOpen, 2016), pp. 461–479
8. S. Lineykin, S. Ben-Yaakov, User-friendly and intuitive graphical approach to the design of thermoelectric cooling systems. Int. J. Refrig. **30**, 798–804 (2007)
9. D. Champier, J.P. Bedecarrats, M. Rivaletto, Thermoelectric power generation from biomass cook stoves. Energy **35**(2), 935–942 (2010)
10. C. Yilmaz, M. Kanoglu, Thermodynamic evaluation of geothermal energy powered hydrogen production by PEM water electrolysis. Energy **69**, 592–602 (2014)

Author Index

A
Abada, B., 161, 171
Abada, S., 177
Abadlia, Issam, 85
Abboudi, Saïd, 379
Abdat, M. Y., 177
Abdelsadek, Zoulikha, 299
Abdi, H., 247
Adjabi, Mohamed, 85
Aider, Nadia, 299, 443
Ait Messaoudene, N., 247
Ait Ouali, A., 507
Akram, Hanane, 47
Alliche, Mounir, 231
Alloune, R., 177
Amine, Chetouane Mohammed, 405
Amrouche, Fethia, 185, 315
Aouachria, Z., 55, 77
Aourag, Hafid, 93
Attia, S., 105
Aziza, M. A., 177
Azri, Y. M., 285

B
Baadj, S., 307
Babbou, Allal, 379
Badji, A., 291
Bagtache, R., 25, 331
Bali, Ferroudja, 299, 443
Bammoune, Abdelaziz, 421
Bebon, Ludovic, 1
Bekhechi, Smain, 93
Belasri, A., 307
Belhadi, Akila, 339, 429

Belkadi, Mustapha, 413
Bellal, Abdelmalek, 451
Benamira, M., 11, 19, 129
Benamrouche, N., 239, 291
Benayyad, Amina, 279
Benbouzid, M., 271
Bendjebbar, M., 459
Bendjedia, B., 255
Bengourina, M. R., 459
Benhabylès, L., 285
Ben-Moussa, Hocine, 195
Benrabah, A., 271
Benyahia, N., 151, 239, 291
Benyelloul, Kamel, 93, 113
Benzaoui, Ahmed, 185
Bessah, R., 177
Bessekhouad, Y., 105
Boucheffa, Youcef, 41
Boudjedien, K., 25
Boudjellal, L., 339
Boudries, Rafika, 481
Bouferrouk, Abdessalem, 371
Bouhadda, Youcef, 93, 113, 379
Bouhala, A., 129
Boukhnifer, M., 255
Boulahaouache, A., 331
Boumaza, S., 339
Bououdina, M., 113
Bouremmad, F., 387
Bousbaine, A., 151, 239, 291
Boussetla, Selsabil, 55, 61, 69
Bouzeria, Hamza, 85
Brahimi, R., 339

Author Index

C
Caldes, M. T., 19
Chapelle, David, 1
Chaurasia, Amit Kumar, 263
Chellali, F., 491
Cherif, Ali, 121, 137, 145
Chibane, Lemnouer, 451

D
Dadda, Bachir, 379, 421
Danane, F., 177
Danoune, Mohammed Bilal, 203, 499
Denoun, H., 291
Derbal-Mokrane, Halima, 185
Djafour, Ahmed, 203, 499
Djebarri, Baya, 299, 443
Djellouli, Abdelkader, 93
Djermouni, Mohamed, 211, 221
Doufar, N., 11

E
Erickson, P. A., 315

F
Fenineche, N., 113

G
Gal La Salle Le, A., 19
Gougui, Abdelmoumen, 203, 499
Gulgun, M. A., 387

H
Haddad, Meriem, 429
Hadef, Amar, 55, 61, 69, 77
Hadjiat, M. M., 507
Hadji, F., 387
Halliche, Djamila, 299, 443
Hamidatou, Taha, 499
Hamouche, Houria, 435
Harrag, Abdelghani, 355, 363
Hassaine, L., 459
Hazi, Mourad, 395
Helaïli, N., 105
Holgado, Juan Paul, 443

I
Innocent, Christophe, 279
Insaf, Tou, 99

J
Joubert, O., 19

K
Kabouche, Nourdine, 473, 481, 491
Kameche, Mostefa, 279
Kebaili, Hakima, 279
Khelladi, Fatma Zohra, 231
Khennour, S. Eddine, 499
Kherrat, M., 161, 171
Khodja, Khadidja, 93
Khoucha, F., 271
Kouachi, Kahina, 47
Kuzgunkaya, E. H., 507

L
Lahmar, H., 11, 129
Laouedj, Samir, 421
Lassouane, F., 323

M
Mameri, Abdelbaki, 55, 61, 69, 77
Maynadier, Anne, 1
Melouki, Redouane, 41
Menia, Sabah, 467, 473
Meriem, Sadi, 99
Messaadia, L., 129
Messaoudi, Djilali, 371, 405
Meziane, Fares, 473, 481, 491
Meziani, D., 25
Mohammedi, Abdallah, 195
Mohammedi, K., 491
Mokhtara, Charafeddine, 371, 405
Mondal, Prasenjit, 263
Moualek, R., 239, 291
Mounia, Azri Yamina, 99
Mraoui, A., 161, 171

N
Naceur, M. W., 247
Nasseri, Lyes, 121, 137, 145
Nebbali, Rachid, 121, 137, 145
Negrou, Belkhir, 371, 405
Nouicer, Ilyes, 467, 473, 481, 491

O
Omari, Mahmoud, 33
Omari, Mohamed Nazim, 185
Ouadha, Ahmed, 211, 221

Author Index

Ouali, S., 507
Ouarzki, Imane, 395
Ould Dris, Aissa, 395

P
Park, J. W., 315

R
Rabehi, Ilhem Nadia, 347
Rebhi, Redha, 231
Rizoug, N., 255
Rodríguez-Ramos, Inmaculada, 47

S
Saad, A., 177
Sadi, M., 285
Sahli, Youcef, 195
Saidia, L., 307
Salhi, K., 507
Salhi, N., 331
Sebai, I., 25, 331
Selmane, Fouad, 221
Settou, Belkhir, 405
Settou, Noureddine, 371, 405
Shabat, Mohammed M., 435
Shawuti, S., 387

Smadi, A. A., 271
Smaili, Arezki, 413

T
Tamalouzt, S., 151, 291
Tamerabet, Monsaf, 195
Tebibel, Hammou, 467
Telmani, K., 129
Thiébaud, Frédéric, 1
Tibermacine, Warda, 33
Touahra, Fouzia, 299, 443
Tou, I., 285
Trari, Mohamed, 25, 105, 129, 331, 339, 429

V
Varnhagen, S., 315

Y
Yahi, Nora, 47
Yahmi, Ismael, 185
Yaiche, M. R., 473
Yao, Yufeng, 371

Z
Zarrit, Rida, 379

CPSIA information can be obtained
at www.ICGtesting.com
Printed in the USA
LVHW011144290821
696386LV00001B/23